TOPICS IN ADAPTIVE OPTICS

Edited by **Robert K. Tyson**

Topics in Adaptive Optics
Edited by Robert K. Tyson

Published by InTech

Janeza Trdine 9, 51000 Rijeka, Croatia

Publishing Process Manager Dejan Grgur

Technical Editor Teodora Smiljanic

Cover Designer InTech Design Team

Additional hard copies can be obtained from orders@intechopen.com

Topics in Adaptive Optics, Edited by Robert K. Tyson
p. cm.
ISBN 978-953-307-949-3

For EU product safety concerns:
IN TECH d.o.o., Prolaz Marije Krucifikse Kozulić 3, 51000 Rijeka, Croatia,
info@intechopen.com or visit our website at intechopen.com.

Meet the editor

Dr Robert K. Tyson is an Associate Professor of Physics and Optical Science at the University of North Carolina at Charlotte, Charlotte, North Carolina, USA. He has a BS in physics from Penn State University, and MS and PhD degrees from West Virginia University. His research interests include atmospheric turbulence compensation, novel wavefront sensing techniques, diffraction theory, and Fourier optics. Prior to his faculty appointment, Dr Tyson worked in the aerospace industry on various strategic and tactical laser weapons and imaging system designs for 20 years. He authored seven books on adaptive optics, including Principles of Adaptive Optics, 3rd edition, CRC Press.

Contents

Preface

Advances in adaptive optics technology and applications move forward at a rapid pace. The basic idea of wavefront compensation in real-time has been around since the mid 1970s. The first widely used application of adaptive optics was for compensating atmospheric turbulence effects in astronomical imaging and laser beam propagation. While some topics have been researched and reported for years, even decades, new applications and advances in the supporting technologies occur almost daily. This book brings together 11 original chapters on topics related to adaptive optics, written by an international group of invited authors.

In Section 1, Dr Avila and Dr Rasouli support this volume with 2 chapters on optical turbulence aberration profiles, measurement, and characterization. Dr Avila's chapter concentrates on a description of the turbulent atmosphere and parameters necessary to characterize it, such as index of refraction fluctuations, scintillation, and wind profiles. Dr Rasouli's chapter describes turbulence characterization and wavefront sensing by exploiting the Talbot effect and moiré deflectometry.

In Section 2, specific applications and solutions are addressed. Dr Schmidt and Dr Neuhäuser show how adaptive optics is used on large telescopes, such as the array of four 8.2 meter telescopes known as the Very Large Telescope (VLT), to directly image faint objects such as brown dwarfs and extrasolar planets orbiting nearby stars. Professor Rao and colleagues present a chapter describing high resolution image post-processing needed to compensate residual aberrations not corrected by the real-time adaptive optics. This section concludes with a chapter by Professors Zou and Wattellier, who study the adaptive optics ability to compensate wavefronts and spot shapes caused by thermal effects in a high peak power laser.

Section 3 is devoted to adaptive optics and the human eye. Dr Doble provides a chapter overview of the structure and optical aberrations of the eye with parts devoted to adaptive optics retinal imaging and vision testing. Professor Lu and colleagues present a chapter showing the operation of an adaptive optics scanning laser ophthalmoscope, with results for retinal imaging including super-resolution and real-time tracking.

The final section, Section 4, concentrates on the subsystems of adaptive optics, namely the wavefront sensors and deformable mirrors. The first chapter, by Mr. Akondi, provides various solutions for improving a Shack-Hartman wavefront sensor. These include such topics as improved centroiding algorithms, spot shape recognition, and applying a small dither to the optics. A chapter by Dr Li and colleagues is an extensive investigation of the measurement error associated with a Shack-Hartmann sensor. A chapter by Dr Magdaleno and Dr Rodriguez provides a detailed design of a wavefront reconstructor using field-programmable gate array technology that can bridge the gap to high-speed operation for the calculational burden of advanced large telescopes. The final chapter, by Dr Bonora and colleagues, describes a number of novel designs of micro-electro-mechanical systems (MEMS) deformable mirrors. The novel designs include resistive actuators, photo-controlled deformations, and a solution to the "push-pull" problem of conventional electrostatic MEMS deformable mirrors.

Robert K. Tyson, PhD
Associate Professor
Department of Physics and Optical Science
University of North Carolina at Charlotte
Charlotte, North Carolina
USA

Part 1

Atmospheric Turbulence Measurement

1

Optical Turbulence Profiles in the Atmosphere

Remy Avila
Centro de Física Aplicada y Tecnología Avanzada,
Universidad Nacional Autónoma de México
Centro de Radioastronomía y Astrofísica,
Universidad Nacional Autónoma de México
México

1. Introduction

Turbulence induces phase fluctuations on light waves traveling through the atmosphere. The main effect of those perturbations on imaging systems is to diminish the attainable angular resolution, whereas on free-space laser communications the turbulence drastically affects system performances. Adaptive Optical (AO) methods are aimed at reducing those cumbersome effects by correcting the phase disturbances introduced by atmospheric turbulence. The development of such methods would not have seen the light without research of the turbulent fluctuations of the refractive index of air, the so called Optical Turbulence (OT). It is necessary to study the statistical properties of the perturbed wavefront to design specific AO systems and to optimize their performances. Some of the useful parameters for this characterization and their impact on AO are the following: The Fried parameter r_0 (Fried, 1966), which is inversely proportional to the width of the image of point-like source (called "seeing"), leads to the determination of the spatial sampling of the wavefront for a given degree of correction (Rousset, 1994). r_0 is the dominant parameter in the calculation of the phase fluctuation variance. The coherence time τ_0 (Roddier, 1999), during which the wavefront remains practically unchanged, is needed to determine the temporal bandwidth of an AO system and the required brightness of the reference sources used to measure the wavefront. The isoplanatic angle θ_0 (Fried, 1982), corresponding to the field of view over which the wavefront perturbations are correlated, determines the angular distance between the corrected and the reference objects, for a given degree of correction.

The parameters described in the last paragraph depend on the turbulence conditions encountered by light waves along its travel through the atmosphere. The principal physical quantities involved are the vertical profile of the refractive index structure constant $C_N^2(h)$, which indicates the optical turbulence intensity, and the vertical profile of the wind velocity $\mathbf{V}(h)$, where h is the altitude above the ground.

The profiles $C_N^2(h)$ and $\mathbf{V}(h)$ can be measured with balloons equipped with thermal micro sensors and a GPS receiver. This method enables detailed studies of optical turbulence and its physical causes but it is not well suited to follow the temporal evolution of the measured parameters along the night nor to gather large enough data series to perform statistical studies. To do so, it is convenient to use remote sensing techniques like Scintillation Detection and Ranging (SCIDAR) (Rocca et al., 1974) and its modern derivatives like Generalized

SCIDAR (Avila et al., 1997; Fuchs et al., 1998) and Low Layer SCIDAR (Avila et al., 2008). Those techniques make use of statistical analysis of double star scintillation images recorded either on the telescope pupil plane (for the classical SCIDAR) or on a virtual plane located a few kilometers below the pupil. Because of this difference, the classical SCIDAR is insensitive to turbulence within the first kilometer above the ground. The Generalized SCIDAR can measure optical turbulence along the whole path in the atmosphere but with an altitude resolution limited to 500 m on 1 to 2-m-class telescopes and the Low Layer SCIDAR can achieve altitude sampling as thin as 8 m but only within the first 500 m using a portable 40-cm telescope. Another successful remote optical-turbulence profiler that has been developed and largely deployed in the last decade is the Slope Detection and Ranging (SLODAR) which uses statistical analysis of wavefront-slope maps measured on double stars (Butterley et al., 2006; Wilson et al., 2008; Wilson et al., 2004). In this book chapter, only SCIDAR related techniques and results are presented.

In § 2 I introduce the main concepts of atmospheric optical turbulence, including some effects on the propagation of optical waves and image formation.The Generalized SCIDAR and Low Layer SCIDAR techniques are explained in § 3. § 4 is devoted to showing some examples of results obtained by monitoring optical turbulence profiles with the afore-mentioned techniques. Finally a summary of the chapter is put forward in § 5.

2. Atmospheric optical turbulence

2.1 Kolmogorov turbulence

The turbulent flow of a fluid is a phenomenon widely spread in nature. In his book "La turbulence", Lesieur (1994) gives a large number of examples where turbulence is found. The air circulation in the lungs as well as gas movement in the interstellar medium are turbulent flows. A spectacular example of turbulence is shown in Fig. 1 where an image of a zone of Jupiter's atmosphere is represented.

Since Navier's work in the early 19th century, the laws governing the movement of a fluid are known. They are expressed in the form of the Navier-Stokes equations. For the case of a turbulent flow, those equations are still valid and contain perhaps all the information about turbulence. However, the stronger the turbulence, the more limited in time and space are the solutions of those differential equations. This non-deterministic character of the solutions is the reason for which a statistical approach was needed for a theory of turbulence to see the day. We owe this theory to Andrei Nikolaevich Kolmogorov. He published this work in 1941 in three papers (Kolmogorov, 1941a;b;c), the first being the most famous one. A rigorous treatment of Kolmogorov theory is given by Frisch (1995).

Since 1922, Richardson described turbulence by his poem:

Big whorls have little whorls,
Which feed on their velocity;
And little whorls have lesser whorls,
And so on to viscosity
(in the molecular sense).

Kinetic energy is injected through the bigger whorls, whose size L is set by the outer scale of the turbulence. The scale l at which the kinetic energy is dissipated by viscosity is called the

Fig. 1. Turbulence in Jupiter's atmosphere. Each of the two large eddies in the center have a dimension of 3500 km along the nort-south (up-down) direction. Image was obtained by the Galileo mission on may 7 1997. (source: `http://photojournal.jpl.nasa.gov/animation/PIA01230`

inner scale and corresponds to the smallest turbulent elements. Kolmogorov proposed that the kinetic energy is transferred from larger to smaller eddies at a rate ϵ that is independent from the eddy spatial scale ρ. This is the so called "Kolmogorov cascade". Under this hypothesis, in the case of a completely developed turbulence and considering homogenous and isotropic three-dimensional velocity fluctuations, Kolmogorov showed that the second order structure function[1] is written as:

$$D_v(\rho) \propto \epsilon^{2/3}\rho^{2/3}\,, \tag{1}$$

[1] the second order structure function is commonly called just structure function

for scales ρ within the inertial scale defined as $l \ll \rho \ll L$. In the terrestrial atmosphere, l is of the order of a few millimeters. The outer scale L is of the order of hundreds of meters in the free atmosphere and close to the ground it is given approximately by the altitude above the ground.

2.2 Refractive index fluctuations

The perturbations of the phase of electromagnetic waves traveling through a turbulent medium like the atmosphere are due to fluctuations of the refractive index N, also called optical turbulence. In the domain of visible wavelengths, the optical turbulence is principally provoked by temperature fluctuations. In the mid infrared and radio ranges, the water vapor content is the dominant factor.

I underline that it is not the turbulent wind velocity field (dynamical turbulence) which is directly responsible of the refractive index fluctuations. Coulman et al. (1995) propose the following phenomenological description of the appearance of optical turbulence. First, dynamical turbulence needs to be triggered. For that to happen, vertical movements of air parcels have to be strong enough to break the stability imposed by the stratification in the atmosphere. In the free atmosphere this occurs when the power associated to wind shear (wind velocity gradient) exceeds that of the stratification. The quotient of those energies is represented by the mean Richardson number:

$$\mathrm{Ri} = \frac{g}{\theta} \frac{\partial \bar{\theta} / \partial z}{|\partial \bar{\mathbf{u}} / \partial z|^2} \tag{2}$$

where g is the acceleration of gravity, $\bar{\theta}$ is the mean potential temperature, $\bar{\mathbf{u}}$ is the mean wind velocity and z is the vertical coordinate. If Ri is higher than $1/4$, then the air flow is laminar. If Ri is lower than $1/4$ but positive, then the flow is turbulent and if Ri is negative then the flow is convective. When turbulence develops in a zone of the atmosphere, one expects that air at slightly different temperatures mix together, which generates optical turbulence. After some time, the temperature within that layer tends to an equilibrium and although turbulent motions of air may prevail, no optical turbulence is present. Only at the boundaries of that turbulent layer air at different temperatures may be mixing, giving birth to thin otical turbulence layers. This phenomenology would explain the relative thinness found in the optical turbulence layers (tens of meters) measured with instrumented balloons and the fact the layers tend to appear in pairs (one for each boundary of the correspoding dynamical turbulent layer). If the fluctuations of N are not substantially anisotropic within a layer, then the outer scale L_0 of the optical turbulence in that layer cannot exceed the layer thickness. Distinction must be made between the outer scale of the dynamical turbulence - which is of the order of hundreds of meters - and that of the optical turbulence L_0 which has been shown to have a median value of about 25 m at some sites (Martin et al., 1998) or even 6 m at Dome C (Ziad et al., 2008). Those measurements were carried on with the dedicated instrument called Generalized Seeing Monitor (former Grating Scale Monitor). The optical turbulence inner scale l_0 keeps the same size as l.

Based upon the theory of the temperature field micro-sctructure in a turbulent flow (Obukhov, 1949; Yaglom, 1949), Tatarski studied the turbulent fluctuations of N in his research of the propagation of waves in turbulent media, published in russian in 1959 and then translated in english (Tatarski, 1961). He showed that the refractive index structure function has a similar expression as Eq. 1:

$$D_N(\rho) = C_N^2\,\rho^{2/3}\,,\text{for } l_0 \ll \rho \ll L_0\,. \tag{3}$$

The refractive index structure constant, C_N^2 determines the intensity of optical turbulence.

When an electromagnetic wave, coming from an astronomical object, travels across an optical-turbulence layer, it suffers phase fluctuations due to the fluctuations of N within the layer. At the exit of that layer, one can consider that the wave amplitude is not affected because the diffraction effects are negligible along a distance equal to the layer thickness. This is the approximation known as *this screen*. However, the wave reaching the ground, having gone through multiple *this screens* along the lowest 20 km of the atmosphere approximately, carries amplitude and phase perturbations. In the weak perturbation hypothesis, which is generally valid at astronomical observatories when the zenith angle does not exceed 60°, the power spectrum of the fluctuations of the complex amplitude $\Psi(\mathbf{r})$ ($\mathbf{r}$ indicating a position on the wavefront plane) can be written as

$$W_\Psi(\mathbf{f}) = W_\varphi(\mathbf{f}) + W_\chi(\mathbf{f})\,, \tag{4}$$

where $W_\varphi(\mathbf{f})$ and $W_\chi(\mathbf{f})$ stand for the power spectra of the phase and the amplitude logarithm fluctuations, respectively, and $\mathbf{f}$ represents the spatial frequency on the wavefront plane. For a wavelength λ, the expressions of those power spectra are Roddier (1981):

$$W_\Psi(\mathbf{f}) = 0.38\lambda^{-2}f^{-11/3}\int \mathrm{d}h\; C_N^2(h) \tag{5}$$

$$W_\varphi(\mathbf{f}) = 0.38\lambda^{-2}f^{-11/3}\int \mathrm{d}h\; C_N^2(h)\cos^2\left(\pi\lambda h f^2\right) \tag{6}$$

$$W_\chi(\mathbf{f}) = 0.38\lambda^{-2}f^{-11/3}\int \mathrm{d}h\; C_N^2(h)\sin^2\left(\pi\lambda h f^2\right)\,, \tag{7}$$

which are valid for spatial frequencies within the inertial zone. All the equations given in this Chapter refer to observations made at the zenith. When observations are carried on with zenith angle z, the altitude variable h is to be replaced by $h/\cos(z)$.

Fried (1966) gave a relation analogous to Eq. 4 for the structure functions:

$$D_\Psi(\mathbf{r}) = D_\varphi(\mathbf{r}) + D_\chi(\mathbf{r})\,, \tag{8}$$

and proposed the simple expression

$$D_\Psi(\mathbf{r}) = 6.88\left(\frac{r}{r_0}\right)^{5/3}\,, \tag{9}$$

where r_0 is the well known Fried's parameter, given by

$$r_0 = \left[0.423\left(\frac{2\pi}{\lambda}\right)^2\int \mathrm{d}h\; C_N^2\,(h)\right]^{-3/5}\,. \tag{10}$$

This parameter defines the quality of point-source long-exposure-time images, of which the angular size is proportional to λ/r_0, for a telescope larger than r_0. Fried's parameter can be interpreted as the size of a telescope which in a turbulence-free medium would provide the same angular resolution as that given by an infinitely large telescope with turbulence.

2.3 Stellar scintillation

The fluctuations of the wave amplitude at ground level translate into intensity fluctuations. This is responsible of stellar scintillation which is visible with the naked eye. The power spectrum $W_I(\mathbf{f})$ of the spatial fluctuations of the intensity $I(\mathbf{r})$ is written in terms of that of the amplitude logarithm as

$$W_I(\mathbf{f}) = 4W_\chi(\mathbf{f})\,. \tag{11}$$

The spatial autocovariance of scintillation produced by a turbulent layer at an altitude h, strength C_N^2 and thickness δh is given by

$$C(r) = C_\mathrm{N}^2(h)\delta h\; K(r,h), \tag{12}$$

where r stands for the modulus of the position vector $\mathbf{r}$ and $K(r,h)$ is given by the Fourier transform of the power spectrum W_I of the irradiance fluctuations (Eqs. 11 and 7). The expression for $K(r,h)$ in the case of Kolmogorov turbulence and weak perturbation approximation is (Prieur et al., 2001) :

$$K(r,h) = 0.243k^2 \int_0^\infty \mathrm{d}f\; f^{-8/3} \sin^2\left(\pi\lambda h f^2\right) J_0\left(2\pi f r\right), \tag{13}$$

where $k = 2\pi/\lambda$. Note that the scintillation variance, $\sigma_I^2 = C(0)$, is proportional to $h^{5/6}$ (as can be easily deduced from Eqs. 12 and 13 by changing the integration variable to $\xi = h^{1/2}f$). Therefore as the turbulence altitude is lower, σ_I^2 decreases. In the limit, a single layer at ground level ($h = 0$) produces no scintillation. In § 3 I present a method for ground turbulence to produce detectable scintillation. It is the Generalized SCIDAR principle.

The scintillation index is defined as $\sigma_I^2/\langle I\rangle^2$, $\langle I\rangle$ being the mean intensity. Typical values for stellar scintillation index are of the order of 10% in astronomical sites at the zenith. A thorougful treatment of stellar scintillation is presented by Dravins et al. (1997a;b; 1998).

3. Generalized SCIDAR based techniques

3.1 Generalized SCIDAR

The Scintillation Detection and Ranging (SCIDAR) technique, proposed by Vernin & Roddier (1973), is aimed at the measurement of the optical-turbulence profile. The method and the physics involved have thoroughly been treated by a number of authors (Klückers et al., 1998; Prieur et al., 2001; Rocca et al., 1974; Vernin & Azouit, 1983a;b). Here I only recall the guidelines of the principle.

The SCIDAR method consists of the following: Light coming from two stars separated by an angle ρ and crossing a turbulent layer at an altitude h casts on the ground two identical scintillation patterns shifted from one another by a distance ρh. The spatial autocovariance of the compound scintillation exhibits peaks at positions $\mathbf{r} = \pm\rho h$ with an amplitude proportional to the C_N^2 value associated to that layer. The determination of the position and amplitude of those peaks leads to $C_\mathrm{N}^2(h)$. This is the principle of the so-called Classical

SCIDAR (CS), in which the scintillation is recorded at ground level by taking images of the telescope pupil while pointing a double star. As the scintillation variance produced by a turbulent layer at an altitude h is proportional to $h^{5/6}$, the CS is blind to turbulence close to the ground, which constitutes a major disadvantage because the most intense turbulence is often located at ground level (Avila et al., 2004; Chun et al., 2009).

To circumvent this limitation, Fuchs et al. (1994) proposed to optically shift the measurement plane a distance h_{gs} below the pupil. For the scintillation variance to be significant, h_{gs} must be of the order of 1 km or larger. This is the principle of the Generalized SCIDAR (GS) which was first implemented by Avila et al. (1997). In the GS, a turbulent layer at an altitude h produces autocovariance peaks at positions $\mathbf{r} = \pm\rho(h + h_{gs})$, with an amplitude proportional to $(h + h_{gs})^{5/6}$. The cut of the peak centered at $\mathbf{r} = \rho(h + h_{gs})$, along the direction of the double-star separation is given by

$$\mathcal{C}\left(r - \rho\left(h + h_{gs}\right)\right) = C_N^2(h)\delta h\, K\left(r - \rho\left(h + h_{gs}\right), h + h_{gs}\right). \tag{14}$$

In the realistic case of multiple layers, the autocovariance corresponding to each layer adds up because of the statistical independence of the scintillation produced in each layer. Hence, Eq. 14 becomes:

$$\mathcal{C}_{\text{multi}}\left(r - \rho\left(h + h_{gs}\right)\right) = \int_{-h_{gs}}^{+\infty} \mathrm{d}h\, C_N^2(h)\delta h\, K\left(r - \rho\left(h + h_{gs}\right), h + h_{gs}\right). \tag{15}$$

For h between $-h_{gs}$ and 0, $C_N^2(h) = 0$ because that space is virtual. To invert Eq. 15 and determine $C_N^2(h)$, a number of methods have been used like Maximum Entropy (Avila et al., 1997), Maximum likelihood (Johnston et al., 2002; Klückers et al., 1998) or CLEAN (Avila et al., 2008; Prieur et al., 2001).

The altitude resolution or sampling interval of the turbulence profile is equal to $\Delta d/\rho$, where Δd is the minimal measurable difference of the position of two autocorrelation peaks. The natural value of Δd is the full width at half maximum L of the aucorrelation peaks: $L(h) = 0.78\sqrt{\lambda(h - h_{gs})}$ (Prieur et al., 2001), where λ is the wavelength. However, Δd can be shorter than L if the inversion of Eq. 15 is performed using a method that can achieve super-resolution like Maximum Entropy or CLEAN. Both methods have been used in GS measurements (Prieur et al., 2001). Fried (1995) analized the CLEAN algorithm and its implications for super-resolution. Applying his results for GS leads to an altitude resolution of

$$\Delta h = \frac{2}{3}\frac{L}{\rho} = 0.52\frac{\sqrt{\lambda(h + h_{gs})}}{\rho}. \tag{16}$$

The maximum altitude, h_{max} for which the C_N^2 value can be retrieved is set by the altitude at which the projections of the pupil along the direction of each star cease to be overlapped, as no correlated speckles would lie on the scintillation images coming from each star. Figure 2 illustrates the basic geometrical consideration involved in the determination of h_{max} . Note that h_{max} does not depend on h_{gs}. The maximum altitude is thus given by

$$h_{max} = \frac{D}{\rho}, \tag{17}$$

where D is the pupil diameter.

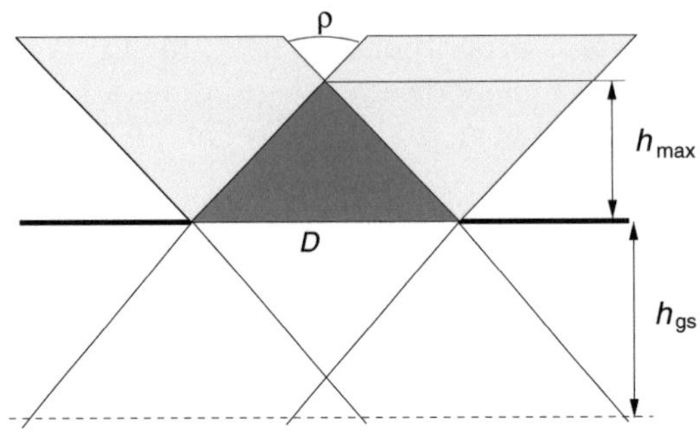

Fig. 2. Schematic for the determination of the maximum altitude h_{max} for which the C_N^2 value can be retrieved. The altitude of the analysis plane h_{gs} is represented only to make clear that this value is not involved in the calculation of h_{max}.

The procedure to estimate the scintillation autocovariance $\mathcal{C}$ is to compute the mean autocorrelation of double-star scintillation images, normalized by the autocorrelation of the mean image. In the classical SCIDAR - where images are taken at the telescope pupil - this computation leads analytically to $\mathcal{C}$ (Rocca et al., 1974). However for the GS, Johnston et al. (2002) pointed out that the result of this procedure is not equal to $\mathcal{C}$. The discrepancy is due to the shift of the out-of-focus pupil images produced by each star on the detector. Those authors analyzed this effect only for turbulence at ground level ($h = 0$) and Avila & Cuevas (2009) generalized the analysis to turbulence at any height. The effect of this misnormalization is to overestimate the C_N^2 values. The relative error is a growing function of the turbulence altitude h, the star separation ρ, the conjugation distance h_{gs} and a decreasing function of the telescope diameter D. Some configurations lead to minor modification of $C_N^2(h)$ like in Avila et al. (2011) but others provoke large discrepancies like in García-Lorenzo & Fuensalida (2011).

3.2 Low layer SCIDAR

The Low Layer SCIDAR (LOLAS) is aimed at the measurement of turbulence profiles with very high altitude-resolution but only within the first kilometer above the ground at most. The interest of such measurements resides in the need of them for constraining the design and performance estimations of adaptive optics systems dedicated to the correction of wavefront deformations induced near the ground - the so-called Ground Layer Adaptive Optics (GLAO).

LOLAS concept consists of the implementation of the GS technique on a small dedicated telescope, using a very widely separated double star. For example, for $h_{\text{gs}} = 2$ km, $h = 0$, $\lambda = 0.5\ \mu$m, $D = 40$ cm and star separations of $180''$ and $70''$, the altitude resolution Δh equals 19 and 48 m, while the maximum sensed altitude h_{max} equals 458 and 1179 m, respectively. GS uses a larger telescope (at least 1-m diameter) and closer double stars, so that the entire altitude-range with non-negligible C_N^2 values is covered ($h_{\text{max}} \gtrsim 30$ km).

The altitude of the analysis plane, h_{gs} is set 2 km below the ground, as a result of a compromise between the increase of scintillation variance, which is proportional to $|h + h_{\text{gs}}|^{5/6}$, and the reduction of pupil diffraction effects. Indeed, pupil diffraction caused by the virtual distance between the pupil and the analysis planes provokes that Eq. 15 is only an approximation. The larger h_{gs} or the smaller the pupil diameter, the greater the error in applying Eq. 15. Numerical simulations to estimate such effect have been performed and the pertinent corrections are applied in the inversion of Eq. 15.

The pixel size projected on the pupil, $d_{\rm p}$, is set by the condition that the smallest scintillation speckles be sampled at the Nyquist spatial frequency or better. The typical size of those speckles is equal to $L(0)$. Taking the same values as above for $h_{\rm gs}$ and λ, yields $L(0) = 2.45$ cm. I chose $d_{\rm p} = 1$ cm, which indeed satisfies the Nyquist criterion $d_{\rm p} \leq L(0)/2$. The altitude sampling of the turbulence profiles is $\delta_h = d_p/\rho$. Note, from the two last expressions and Eq. 16, that the altitude resolution Δh and the altitude sampling δ_h are related by $\delta_h \leq (3/4)\Delta h$ for $h = 0$.

3.3 Measurement of velocity profiles

Wind-velocity profiles $\mathbf{V}(h)$ can be computed from the mean cross-correlation of images taken at times separated by a constant delay Δt. Note that the mean autocorrelation and mean cross-correlation need to be normalized by the autocorrelation of the mean image. Hereafter, I will refer to this mean-normalized cross-correlation simply as *cross-correlation*. The method is based on the following principle:

Let us assume that the turbulent structures are carried by the mean wind without deformation. This assumption is known as *Taylor hypothesis*, and is valid for short enough time intervals. In this case, the scintillation pattern produced by a layer at altitude h, where the mean horizontal wind velocity is $\mathbf{V}(h)$, moves on the analysis plane a distance $\mathbf{V}(h)\,\Delta t$ in a time Δt. If the source was a single star, the cross-correlation of images separated by a lapse Δt, would produce a correlation peak located at the point $\mathbf{r} = \mathbf{V}(h)\,\Delta t$, on the correlation map. By determining this position, one can deduce $\mathbf{V}(h)$ for that layer. As a double star is used, the contribution of the layer at altitude h in the cross-correlation consists of three correlation peaks, which we call a triplet: a central peak located at $\mathbf{r} = \mathbf{V}(h)\,\Delta t$ and two lateral peaks separated from the central one by $\pm\boldsymbol{\rho}H$, where H is the distance from the analysis plane to the given layer $H = h + h_{\rm gs}$ and $\boldsymbol{\rho}$ is the angular separation of the double star. The cross-correlation can be written as:

$$C_{\rm c}^{**}(\mathbf{r},\Delta t) = \int_0^\infty {\rm d}h\, C_{\rm N}^2(h)\ \{\ a\,\mathcal{C}_{\rm c}(\mathbf{r}-\mathbf{V}(h)\Delta t, H) + b\,[\,\mathcal{C}_{\rm c}(\mathbf{r}-\mathbf{V}(h)\Delta t - \boldsymbol{\rho}H, H) + \mathcal{C}_{\rm c}(\mathbf{r}-\mathbf{V}(h)\Delta t + \boldsymbol{\rho}H, H)\,]\,\}. \tag{18}$$

$\mathcal{C}_{\rm c}$ is the theoretical cross-correlation of the scintillation produced by a layer at an altitude h and unit $C_{\rm N}^2$. It differs from the theoretical autocorrelation $\mathcal{C}$ only by an eventual temporal decorrelation of the scintillation (partial failure of Taylor hypothesis) and an eventual fluctuation of $\mathbf{V}(h)$ during the integration time. The decorrelation would make $\mathcal{C}_{\rm c}$ smaller than $\mathcal{C}$, and the fluctuation of $\mathbf{V}(h)$ would make $\mathcal{C}_{\rm c}$ smaller and wider than $\mathcal{C}$ (Caccia et al., 1987). Those effects do not affect the determination of $\mathbf{V}(h)$, as the only information needed here is the position of each correlation peak. Note that $\mathcal{C}$ takes into account the spatial filtering introduced by the detector sampling. The factors a and b are given by the magnitude difference of the two stars Δm through:

$$a = \frac{1+\alpha^2}{(1+\alpha)^2} \quad \text{and} \quad b = \frac{\alpha}{(1+\alpha)^2}, \quad \text{with} \quad \alpha = 10^{-0.4\Delta m}. \tag{19}$$

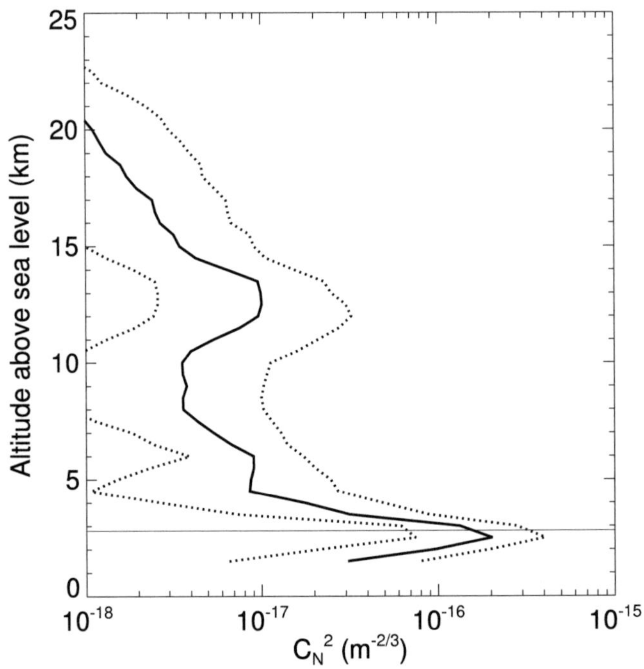

Fig. 3. Median (full line), 1st and 3rd quartiles (dashed lines) of the $C_N^2(h)$ values obtained with the GS at both telescopes, during 1997 and 2000 campaigns. The horizontal axis represents C_N^2 values, in logarithmic scale, and the vertical axis represents the altitude above sea level. The horizontal lines indicate the observatory altitude. Dome seeing has been removed.

4. Examples of turbulence and wind profiles results

4.1 $C_N^2(h)$ profiles with generalized SCIDAR

Turbulence profiles have been measured with the GS technique at many astronomical sites by a number of authors (Avila et al., 2008; Avila et al., 2004; 1998; 1997; 2001; Egner & Masciadri, 2007; Egner et al., 2007; Fuchs et al., 1998; Fuensalida et al., 2008; García-Lorenzo & Fuensalida, 2006; Kern et al., 2000; Klückers et al., 1998; Prieur et al., 2001; Tokovinin et al., 2005; Vernin et al., 2007; Wilson et al., 2003). Here I will summarize the results presented by Avila et al. (2004) which have been corrected for the normalization error by Avila et al. (2011).

Two GS observation campaigns have been carried out at the Observatorio Astronómico Nacional de San Pedro Mártir (OAN-SPM) in 1997 and 2000, respectively. The OAN-SPM , held by the Instituto de Astronomía of the Universidad Nacional Autónoma de México, is situated on the Baja California peninsula at 31° 02′ N latitude, 115° 29′ W longitude and at an altitude of 2800 m above sea level. It lies within the North-East part of the San Pedro Mártir (SPM) National Park, at the summit of the SPM sierra. Cruz-González et al. (2003) edited in a single volume all the site characteristics studied so far. In 1997, the GS was installed at the 1.5-m and 2.1-m telescopes (SPM1.5 and SPM2.1) for 8 and 3 nights (1997 March 23–30 and April 20–22 UT), whereas in 2000, the instrument was installed for 9 and 7 nights (May 7–15 and 16–22 UT) at SPM1.5 and SPM2.1. The number of $C_N^2(h)$ samples obtained in 1997 and 2000 are 3398 and 3016, respectively, making a total of 6414. The altitude scale of the profiles refers to altitude above sea level (2800 m at OAN-SPM). In GS data, part of the turbulence measured at the observatory altitude is produced inside the telescope dome. For site characterization, this contribution must be subtracted. In all the analysis presented here, dome turbulence has been removed using the procedure explained by Avila et al. (2004).

The median $C_N^2(h)$profile together with the first and third quartiles profiles are shown in Fig. 3. Almost all the time the most intense turbulence is located at the observatory altitude. There are marked layers centered at 6 and 12 km approximately above sea level. Although those layers appear clearly in the median profile, they are not present every night.

From a visual examination of the individual profiles, one can determine five altitude slabs that contain the predominant turbulent layers. These are]2,4],]4,9],]9,16],]16,21] and]21,25] km above sea level. In each altitude interval of the form $]h_l,h_u]$ (where the subscripts l and u stand for "lower" and "upper" limits) and for each profile, I calculate the turbulence factor

$$J_{h_l,h_u} = \int_{h_l}^{h_u} \mathrm{d}h \, C_N^2(h), \tag{20}$$

and the correspondent seeing in arc seconds:

$$\epsilon_{h_l,h_u} = 1.08 \times 10^6 \lambda^{-1/5} J_{h_l,h_u}^{3/5}. \tag{21}$$

For the turbulence factor corresponding to the ground layer, $J_{2,4}$, the integral begins at 2 km in order to include the complete C_N^2 peak that is due to turbulence at ground level (2.8 km). Moreover, $J_{2,4}$ does not include dome turbulence. The seeing values have been calculated for $\lambda = 0.5\ \mu$m. In Fig. 4a the cumulative distribution functions of $\epsilon_{2,4}$ obtained at the SPM1.5 and the SPM2.1, calculated using the complete data set, are shown. The turbulence at ground level at the SPM1.5 is stronger than that at the SPM2.1. It is believed that this is principally due to the fact that the SPM1.5 is located at ground level, while the SPM2.1 is installed on top of a 20–m building. Moreover, the SPM2.1 building is situated at the observatory summit whereas the SPM1.5 is located at a lower altitude. The cumulative distributions of the seeing originated in the four slabs of the free atmosphere (from 4 to 25 km) are represented in Fig. 4b. The largest median seeing in the free atmosphere is encountered from 9 to 16 km, where the tropopause layer is located. Also in that slab the dynamical range of the seeing values is the largest, as can be noticed from the 1st and 3rd quartiles for example (0.″11 and 0.″39). Particularly noticeable is the fact that the seeing in the tropopause can be very small as indicated by the left-hand-side tail of the cumulative distribution function of $\epsilon_{9,16}$. The turbulence at altitudes higher than 16 km is fairly weak. Finally, Figs. 4c and 4d show the cumulative distribution of the seeing produced in the free atmosphere, $\epsilon_{4,25}$, and in the whole atmosphere, $\epsilon_{2,25}$, respectively.

From each C_N^2 profile of both campaigns one value of the isoplanatic angle θ_0 (Fried, 1982) has been computed, using the following expression:

$$\theta_0 = 0.31 \frac{r_0}{h_0}, \tag{22}$$

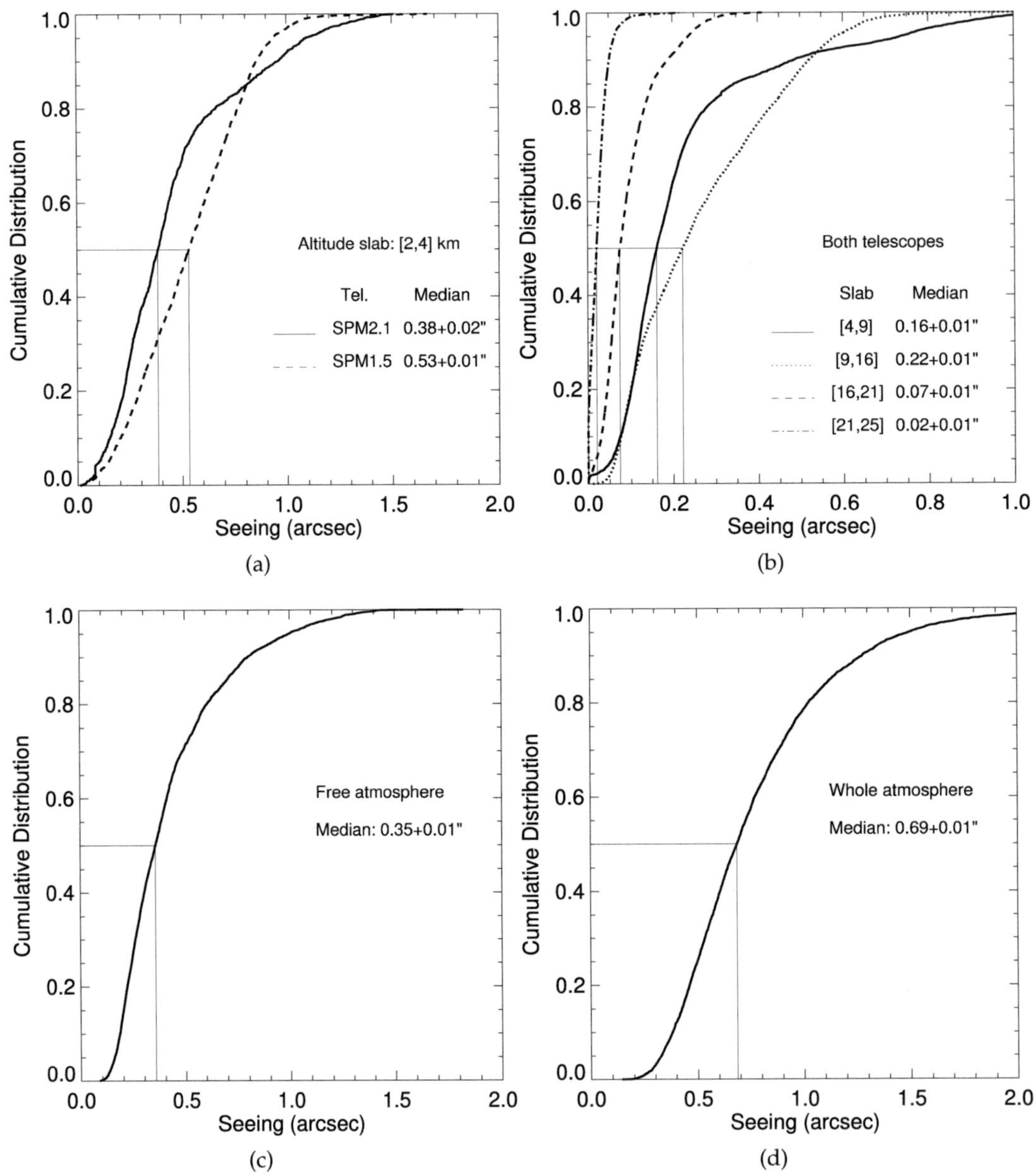

Fig. 4. Cumulative distributions of the seeing generated in different atmosphere slabs: (a)]2,4] km for the 2.1mT (full line) and the 1.5mT (dashed line) without dome seeing; (b)]4,9] km (full line),]9,16] km (dotted line),]16,21] km (dashed line),]21,25] km (dash–dotted line); (c) the free atmosphere (altitude higher than 4 km) and (d) the whole atmosphere, without dome seeing. The horizontal and vertical lines indicate the median values.

where r_0 is Fried's parameter defined in Eq. 10 and

$$h_0 = \left(\frac{\int \mathrm{d}h\, h^{5/3} C_N^2(h)}{\int \mathrm{d}h\, C_N^2(h)} \right)^{3/5} . \tag{23}$$

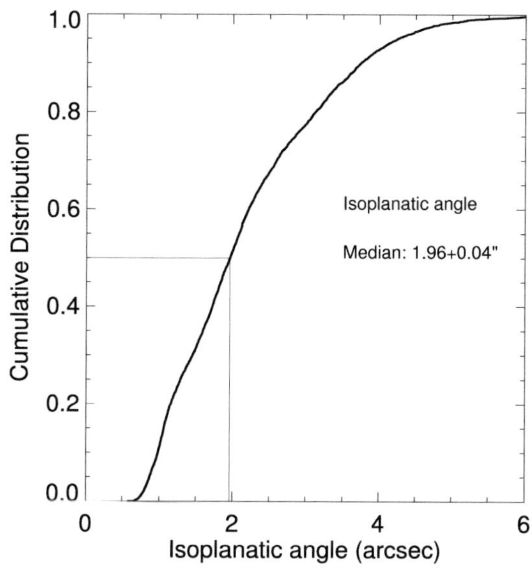

Fig. 5. Cumulative distribution of the isoplanatic angle θ_0 computed from each turbulence profile of both campaigns and both telescopes, using Eqs. 22, 10 and 23. The median value of θ_0 is $1\rlap{.}''96$ and the 1st and 3rd quartiles are $1\rlap{.}''28$ and $2\rlap{.}''85$.

The cumulative distribution function of θ_0 is shown in Fig. 5. The 1st, 2nd and 3rd quartiles values are equal to $1\rlap{.}''28$, $1\rlap{.}''96$ and $2\rlap{.}''85$.

4.2 $C_N^2(h)$ Profiles with low layer SCIDAR

The first results of LOLAS were obtained in September 2007 at Mauna Kea Observatory, as part of a collaboration between the Universidad Nacional Autónoma de México (UNAM), the University of Durham (UD) and the University of Hawaii (UH), under a contract with Gemini Observatory. The instrument was installed on the Coudé roof of the UH 2.2-m telescope. SLODAR and LOLAS instrument were implemented sharing the same telescope and camera. A detailed description of the campaign and results are reported by Chun et al. (2009).

To illustrate the highest altitude resolution that has so far been reached with LOLAS, Fig. 6 shows a C_N^2 profile obtained using as target a $199\rlap{.}''7$-separation double-star. The altitude resolution in vertical direction is 11.7 m and $\Delta C_N^2 = 1.6 \times 10^{-16}\text{m}^{-2/3}$. Note the ability for discerning a layer centered at 16 m from that at ground layer. Turbulence inside the telescope tube has been removed.

The quartile and 90 percentile profiles of all the C_N^2 measurements obtained with LOLAS at Mauna Kea are shown in Fg. 7. Given the consistent and very simple distribution of turbulence, the profiles were fit with an exponential form:

$$C_N^2(h)\text{d} = A\exp(-h/h_{\text{scale}}) + B, \tag{24}$$

where A, B and h_{scale} are constants, using a non-linear least-squares fit algorithm. The scaleheight h_{scale} of the turbulence within the ground layer increases as the integrated turbulence within the ground layer increases. The median scaleheight is 27.8 m.

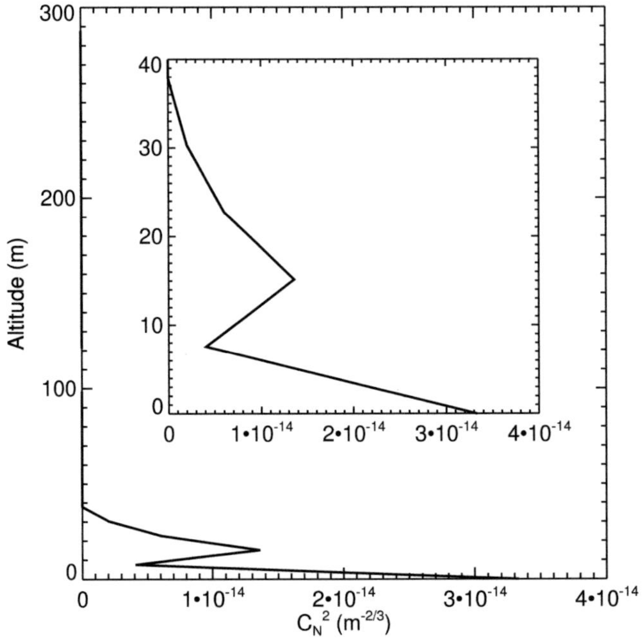

Fig. 6. Example of a turbulence profile with the highest altitude-resolution so far obtained with LOLAS. The data was taken in 2007 November 17 at 12:09 UT. The central frame shows an amplification of the profile in the low-altitude zone. The vertical axis represents altitude above the ground.

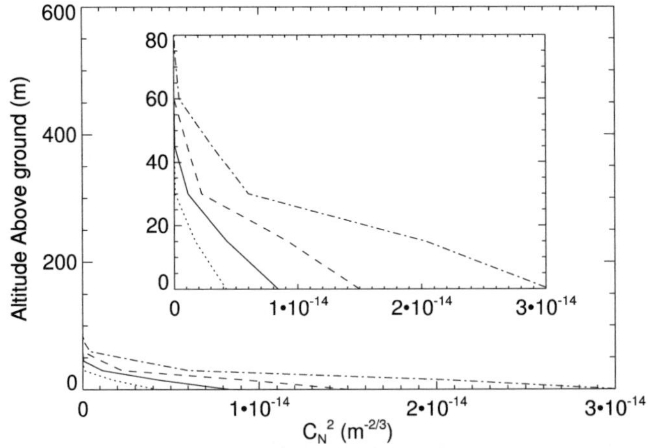

Fig. 7. Percentile profiles obtained with LOLAS at Mauna Kea. Dotter, solid, dashed and dot-dat-dashed lines represent the 25, 50,75 and 90 percentile values of C_N^2 as a function of altitude above the ground.

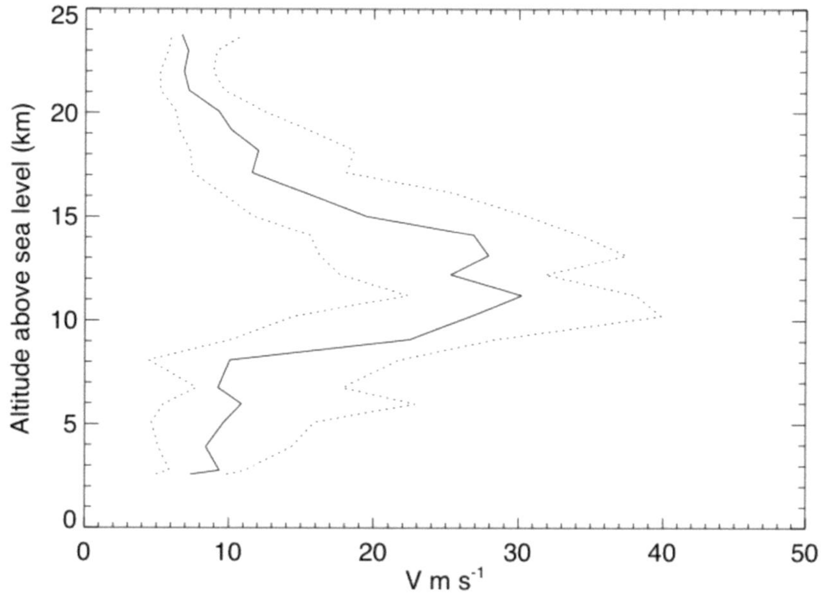

Fig. 8. Profiles of the median (solid line) and first and third quartile (dotted lines) of the wind speed.

4.3 $\mathbf{V}(h)$ profiles with generalized SCIDAR

Wind profiles were obtained at the OAN-SPM using the same data as that described in § 4.1. The median and first and third quartile values of the layer speed V as a function of height are shown in Fig. 8. It can be seen that the wind speed has similar low values within the first 4 km and above 16 km. In the jet stream zone (between 10 and 15 km), the wind speed sharply increases.

From the C_N^2 and V values of each detected layer, the coherence time τ of the wavefront deformations produced by that layer can be calculated using an expression analogous to that given by Roddier (1999):

$$\tau = 0.31 \frac{r_{0_{\rm ind}}}{V}, \tag{25}$$

where $r_{0_{\rm ind}}$ corresponds to the Fried's parameter that would occur if only the given layer was present:

$$r_{0_{\rm ind}} = \left[0.423 \left(\frac{2\pi}{\lambda} \right)^2 C_N^2 \Delta h \right]^{-3/5}, \tag{26}$$

where the wavelength $\lambda = 0.5\ \mu$m. The value of τ for a given layer sets the acceptable time delay of a deformable mirror for the mean square residual phase error of that layer, due solely to time delay, to be less than one radian. I have calculated $\tau(h)$ from each C_N^2 and $\mathbf{V}$ profiles, taking $\Delta h = 500$ m, i.e. equal to our C_N^2 profiles sampling. The median, first and third quartiles of $\tau(h)$ are shown in Fig. 9.

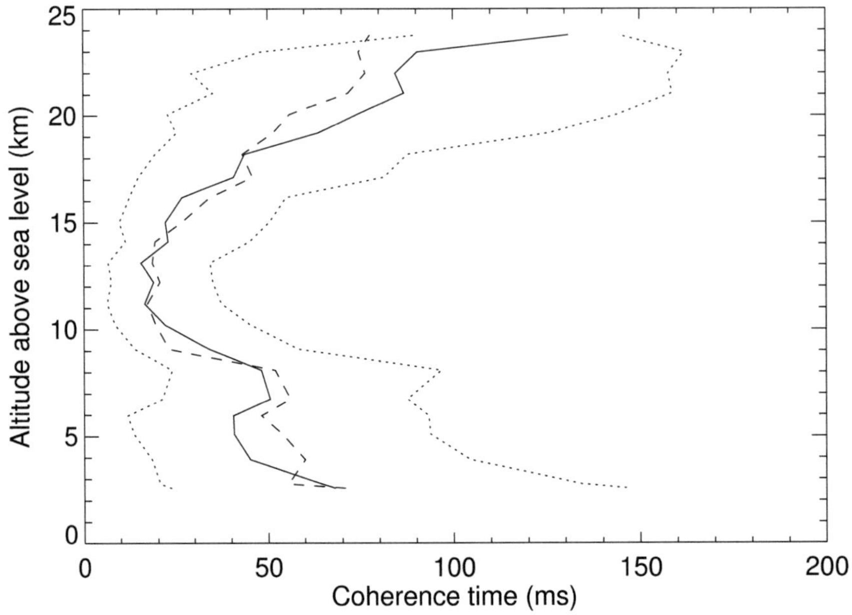

Fig. 9. Median (solid line), first and third quartiles (dotted lines) profiles of the coherence time for adaptive optics, as explained in the text (Eqs. 25 and 26). The dashed line represents the coherence time profile computed using Eq. 27.

It is interesting to note that the variation of τ with altitude, seems to be mainly governed by the variation of V. This is shown by the reasonably good agreement between the median of $\tau(h)$, and the median of the function

$$\tau_*(h) = 0.31 \frac{r_{0_{\mathrm{med}}}}{V(h)}, \tag{27}$$

where $r_{0_{\mathrm{med}}} = 1.8$ m, is the median value of $r_{0_{\mathrm{ind}}}$ for all altitudes and all turbulence profiles (we remind that $r_{0_{\mathrm{ind}}}$ is computed for 500-m slabs). The median profiles of $\tau(h)$ and $\tau_*(h)$ are shown in solid and dashed lines, respectively, in Fig. 9.

5. Summary

This chapter starts with a brief phenomenological description of the so-called optical turbulence in the atmosphere. Air flows can become turbulent when the stratification is broken due to wind shear or convection. In that case, air at different temperatures can be mixed, giving rise to a turbulent temperature field which translates into a turbulent field of the refractive index at optical wavelengths. This turbulent refractive index field is commonly known as optical turbulence. The strength of the optical turbulence is determined by the refractive index structure constant C_N^2.

After having introduced the main concepts of atmospheric optical turbulence, its effects on the propagation of optical waves and image formation are briefly presented. The spatial spectra of phase and intensity fluctuations are given for the weak perturbation approximation.

The Generalized SCIDAR (GS) technique for the measurement of the afore-mentioned vertical distributions is then presented. A recently developed application of this technique leads to the Low Layer SCIDAR (LOLAS), which is devoted to the measurement of optical turbulence profiles close to the ground with very high altitude-resolution. Those measurements are necessary for determining the expected performance and design of ground layer adaptive optics systems.

Results of GS and LOLAS measurements performed at San Pedro Mártir Observatory, Mexico and Mauna Kea Observatory, Hawaii, are finally shown, providing illustrative examples of the vertical distribution of optical turbulence in the atmosphere.

6. Acknowledgments

I am grateful to the OAN-SPM staff for their valuable support. Measurements at the OAN-SPM were carried out in the framework of a collaboration between the Instituto de Astronomía of the Universidad Nacional Autónoma de México (IA-UNAM) and the UMR 6525 Astrophysique, Université de Nice-Sophia Antipolis (France), supported by ECOS-ANUIES grant M97U01. Additional support was provided by grants J32412E and 58291 from CONACyT, IN118199, IN111403 and IN112606-2 from DGAPA-UNAM, and the TIM project (IA- UNAM). Funds for the LOLAS instrument construction and observations were provided by Gemini Observatory through con- tract number 0084699-GEM00445 entitled "Contract for Ground Layer Turbulence Monitoring Campaign on Mauna Kea".

7. References

Avila, R., Avilés, J. L., Wilson, R. W., Chun, M., Butterley, T. & Carrasco, E. (2008). LOLAS: an optical turbulence profiler in the atmospheric boundary layer with extreme altitude resolution, *Mon. Not. R. Astron. Soc.* 387: 1511–1516.

Avila, R. & Cuevas, S. (2009). On the normalization of scintillation autocovariance for generalized SCIDAR, *Optics Express* 17: 10926–+.

Avila, R., Masciadri, E., Vernin, J. & Sánchez, L. (2004). Generalized SCIDAR measurements at san pedro mártir: I. turbulence profile statistics, *Publ. Astron. Soc. Pac.* 116: 682–692.

Avila, R., Sánchez, L., Cruz-González, I., Castaño, V. M. & Carrasco, E. (2011). Recalibrated turbulence profiles at San Pedro Mártir, *Rev. Mex. Astron. Astrophys.* 47: 75–82.

Avila, R., Vernin, J. & Cuevas, S. (1998). Turbulence profiles with generalized scidar at San Pedro Mártir Observatory and isoplanatism studies, *Publ. Astron. Soc. Pac.* 110: 1106–1116.

Avila, R., Vernin, J. & Masciadri, E. (1997). Whole atmospheric-turbulence profiling with generalized scidar, *Appl. Opt.* 36(30): 7898–7905.

Avila, R., Vernin, J. & Sánchez, L. J. (2001). Atmospheric turbulence and wind profiles monitoring with generalized scidar, *Astron. Astrophys.* 369: 364.

Butterley, T., Wilson, R. W. & Sarazin, M. (2006). Determination of the profile of atmospheric optical turbulence strength from slodar data, *Mon. Not. R. Astron. Soc.* 369: 835–845.

Caccia, J. L., Azouit, M. & Vernin, J. (1987). Wind and C_n^2 profiling by single-star scintillation analysis, *Appl. Opt.* 26(7): 1288–1294.

Chun, M., Wilson, R., Avila, R., Butterley, T., Aviles, J.-L., Wier, D. & Benigni, S. (2009). Mauna Kea ground-layer characterization campaign, *Mon. Not. R. Astron. Soc.* 394: 1121–1130.

Coulman, C. E., Vernin, J. & Fuchs, A. (1995). Optical seeing-mechanism of formation of thin turbulent laminae in the atmosphere, *Appl. Opt.* 34: 5461–5474.

Cruz-González, I., Avila, R. & Tapia, M. (2003). *San Pedro Mártir: Astronomical Site Evaluation*, Vol. 19, Revista Mexicana de Astronomía y Astrofísica (serie de conferencias), Instituto de Astronomía, UNAM, México.

Dravins, D., Lindegren, L., Mezey, E. & Young, A. T. (1997a). Atmospheric Intensity Scintillation of Stars, I. Statistical Distributions and Temporal Properties, *Publ. Astron. Soc. Pac.* 109: 173–207.

Dravins, D., Lindegren, L., Mezey, E. & Young, A. T. (1997b). Atmospheric Intensity Scintillation of Stars. II. Dependence on Optical Wavelength, *Publ. Astron. Soc. Pac.* 109: 725–737.

Dravins, D., Lindegren, L., Mezey, E. & Young, A. T. (1998). Atmospheric Intensity Scintillation of Stars. III. Effects for Different Telescope Apertures, *Publ. Astron. Soc. Pac.* 110: 610–633.

Egner, S. E. & Masciadri, E. (2007). A g-scidar for ground-layer turbulence measurements at high vertical resolution, *Publ. Astron. Soc. Pac.* 119: 1441–1448.

Egner, S. E., Masciadri, E. & McKenna, D. (2007). Generalized SCIDAR Measurements at Mount Graham, *Publ. Astron. Soc. Pac.* 119: 669–686.

Fried, D. (1982). Anisoplanetism in adaptive optics, *J. Opt. Soc. Am. A* 72(1): 52–61.

Fried, D. L. (1966). Optical resolution through a randomly inhomogeneous medium for very long and very short exposures, *J. Opt. Soc. Am. A* 56(10): 1372–1379.

Fried, D. L. (1995). Analysis of the clean algorithm and implications for superresolution, *J. opt. Soc. Am. A* 12(5): 853–860.

Frisch, U. (1995). *The legacy of A. N. Kolmogorov*, Cambridge University Press, UK.

Fuchs, A., Tallon, M. & Vernin, J. (1994). Folding of the vertical atmospheric turbulence profile using an optical technique of movable observing plane, *in* W. A. Flood & W. B. Miller (eds), *Atmospheric Propagation and Remote Sensing III*, Vol. 2222, pp. 682–692.

Fuchs, A., Tallon, M. & Vernin, J. (1998). Focussiong on a turbulent layer: Principle of the generalized SCIDAR, *Publ. Astron. Soc. Pac.* 110: 86–91.

Fuensalida, J. J., García-Lorenzo, B. & Hoegemann, C. (2008). Correction of the dome seeing contribution from generalized-SCIDAR data using evenness properties with Fourier analysis, *Mon. Not. R. Astron. Soc.* 389: 731–740.

García-Lorenzo, B. & Fuensalida, J. J. (2006). Processing of turbulent-layer wind speed with Generalized SCIDAR through wavelet analysis, *Mon. Not. R. Astron. Soc.* 372: 1483–1495.

García-Lorenzo, B. & Fuensalida, J. J. (2011). Atmospheric optical turbulence at the Roque de los Muchachos Observatory: data base and recalibration of the generalized SCIDAR data, *Mon. Not. R. Astron. Soc.* pp. 1124–+.

Johnston, R. A., Dainty, C., Wooder, N. J. & Lane, R. G. (2002). Generalized scintillation detection and ranging results obtained by use of a modified inversion technique, *Appl. Opt.* 41(32): 6768–6772.

Kern, B., Laurence, T. A., Martin, C. & Dimotakis, P. E. (2000). Temporal coherence of individual turbulent patterns in atmospheric seeing, *Appl. Opt.* 39: 4879–4885.

Klückers, V. A., Wooder, N. J., Nicholls, T. W., Adcock, M. J., Munro, I. & Dainty, J. C. (1998). Profiling of atmospheric turbulence strength and velocity using a generalised SCIDAR technique, *Astron. Astrophys. Suppl. Ser.* 130: 141–155.

Kolmogorov, A. N. (1941a). Dissipation of energy in local isotropic turbulence, *Dokl. Akad. Nauk. SSSR* 32: 16.

Kolmogorov, A. N. (1941b). The local structure of turbulence in incompressible fluids with very high reynolds numbers, *Dokl. Akad. Nauk. SSSR* 30: 9.

Kolmogorov, A. N. (1941c). On degeneration (decay) of isotropic turbulence in an incompressible viscous liquid, *Dokl. Akad. Nauk. SSSR* 31: 538.

Lesieur, M. (1994). *La Turbulence*, Presses Universitaires de Grenoble, Grenoble.

Martin, F., Tokovinin, A., Ziad, A., Conan, R., Borgnino, J., Avila, R., Agabi, A. & Sarazin, M. (1998). First statistical data on wavefront outer scale at La Silla Observatory from the GSM instrument, *Astron. Astrophys.* 336: L49–L52.

Obukhov, A. M. (1949). Structure of the temperature field in a turbulent flow, *Izv. Akad. Nauk. SSSR, Ser Geograf. Geofiz.* 13: 58–69.

Prieur, J.-L., Daigne, G. & Avila, R. (2001). Scidar measurements at pic du midi, *Astron. Astrophys.* 371: 366–377.

Rocca, A., Roddier, F. & Vernin, J. (1974). Detection of atmospheric turbulent layers by spatiotemporal and spatioangular correlation measurements of stellar-light scintillation., *Journal of the Optical Society of America (1917-1983)* 64: 1000–1004.

Roddier, F. (1981). The effect of atmospheric turbulence in optical astronomy, *Progress in Optics* XIX: 281–376.

Roddier, F. (1999). *Adaptive optics in astronomy*, Cambridge University Press, United Kingdom.

Rousset, G. (1994). Wavefront Sensing, *in* D. M. Alloin & J. M. Mariotti (ed.), *NATO ASIC Proc. 423: Adaptive Optics for Astronomy*, pp. 115–+.

Tatarski, V. I. (1961). *Wave Propagation in a Turbulent Medium*, Dover Publications, Inc., New York.

Tokovinin, A., Vernin, J., Ziad, A. & Chun, M. (2005). Optical Turbulence Profiles at Mauna Kea Measured by MASS and SCIDAR, *Publ. Astron. Soc. Pac.* 117: 395–400.

Vernin, J. & Azouit, M. (1983a). Traitement d'image adapté au speckle atmosphérique. I-formation du speckle en atmosphère turbulente. propriétés statistiques, *Journal of Optics (Paris)* 14: 5–9.

Vernin, J. & Azouit, M. (1983b). Traitement d'image adapté au speckle atmosphérique. II-analyse multidimensionnelle appliquée au diagnostic à distance de la turbulence, *Journal of Optics (Paris)* 14: 131–142.

Vernin, J. & Roddier, F. (1973). Experimental determination of two-dimensional spatiotemporal power spectra of stellar light scintillation. evidence for a multilayer structure of the air turbulence in the upper troposphere, *J. opt. Soc. Am. A* 63: 270–273.

Vernin, J., Trinquet, H., Jumper, G., Murphy, E. & Ratkowski, A. (2007). OHP02 gravity wave campaign in relation to optical turbulence, *Environmental Fluid Mechanics* 7: 371–+.

Wilson, R., Butterley, T., Sarazin, M., Lombardi, G., Chun, M., Benigni, S., Weir, D., Avila, R. & Aviles, J.-L. (2008). SLODAR turbulence monitors for real-time support of astronomical adaptive optics, *Society of Photo-Optical Instrumentation Engineers (SPIE) Conference Series*, Vol. 7015 of *Society of Photo-Optical Instrumentation Engineers (SPIE) Conference Series*.

Wilson, R. W., Bate, J., Guerra, J. C., Sarazin, M. & Saunter, C. (2004). Development of a portable slodar turbulence profiler, *in* D. Bonnaccini, B. Ellerbroek & R. Raggazzoni (eds), *Advancements in Adaptive Optics*, Vol. 5490, SPIE, pp. 758–765.

Wilson, R. W., Wooder, N. J., Rigal, F. & Dainty, J. C. (2003). Estimation of anisoplanatism in adaptive optics by generalized SCIDAR profiling, *Mon. Not. R. Astron. Soc.* 339: 491–494.

Yaglom, A. (1949). On the local structure of the temperature field in a turbulent flow, *IDokl. Akad. Nauk. SSSR* 69: 743–746.

Ziad, A., Aristidi, E., Agabi, A., Borgnino, J., Martin, F. & Fossat, E. (2008). First statistics of the turbulence outer scale at Dome C, *Astron. Astrophys.* 491: 917–921.

2

Atmospheric Turbulence Characterization and Wavefront Sensing by Means of the Moiré Deflectometry

Saifollah Rasouli
Department of Physics, Institute for Advanced Studies in Basic Sciences (IASBS), Zanjan
Optics Research Center, Institute for Advanced Studies in Basic Sciences (IASBS), Zanjan
Iran

1. Introduction

When a light beam propagates through the turbulent atmosphere, the wavefront of the beam is distorted, which affect the image quality of ground based telescopes. Adaptive optics is a means for real time compensation of the wavefront distortions. In an adaptive optics system, wavefront distortions are measured by a wavefront sensor, and then using an active optical element such as a deformable mirror the instantaneous wavefront distortions are corrected. On the other hand, three physical effects are observed when a light beam propagates through a turbulent atmosphere: optical scintillation, beam wandering, and fluctuations in the angle-of-arrival (AA). These effects are used for measuring turbulence characteristic parameters. Fluctuations of light propagation direction, referred to as the fluctuations of AA, are measured by various methods. In wavefront sensing applications the AA fluctuations measurement is a basic step.

Various wavefront sensing techniques have been developed for use in a variety of applications ranging from measuring the wave aberrations of human eyes (Lombardo & Lombardo, 2009) to adaptive optics in astronomy (Roddier, 1999). The most commonly used wavefront sensors are the Shack-Hartmann (Platt & Shack, 2001; Shack & Platt, 1971), curvature sensing (Roddier , 1988), shearing interferometry (Leibbrandt et al., 1996), phase retrieval methods (Gonsalves, 1996) and Pyramid wavefront sensor (Ragazzoni & Farinato, 1999). The Shack-Hartmann (SH) sensor is also the most commonly used technique for measurement of turbulence-induced phase distortions for various applications in atmospheric studies and adaptive optics. But, the dynamic range of the SH sensor is limited by the optical parameters of its microlenses, namely, the spacing and the focal length of the microlens array.

In recent years, some novel methods, based on moiré technique, for the study of atmospheric turbulence have been introduced (Rasouli & Tavassoly, 2006b; 2008; Rasouli, 2010). As a result of these works, due to the magnification of the telescope, the use of moiré technique, and the Talbot effect, measurements of fluctuations in the AA can be up to 2 orders of magnitude more precise than other methods. Also, moiré deflectometry have been used to wavefront sensing in various schemes (Rasouli et al., 2009; 2010). In the recent scheme, an adjustable, high-sensitivity, wide dynamic range two channel wavefront sensor was suggested for measuring distortions of light wavefront transmitted through the atmosphere (Rasouli et al., 2010). In this sensor, a slightly divergent laser beam is passed through the turbulent ground level atmosphere and then a beam-splitter divides it into two beams. The beams pass through a pair of moiré deflectometers which are installed parallel and close together. From deviations in the moiré fringes, two orthogonal components of AA at each location across the wavefront are calculated. The deviations have been deduced in successive frames which allows evolution of the wavefront shape to be determined. In this wavefront sensor the dynamic range and sensitivity of detection are adjustable in a very simple manner. This sensor is more reliable, quite simple, and has many practical applications ranging from wave aberrations of human eyes to adaptive optics in astronomy. Some of the applications, such as measurement of wave aberrations induced by lenses and study of nonlinear optical media, are in progress, now by the author.

At the beginning of the this chapter, moiré pattern, Talbot effect, Talbot interferometry and moiré deflectometry will be briefly reviewed. Also, definition, history and some applications of the moiré technique will be presented. Then, all of the moiré based methods for the atmospheric turbulence study will be reviewed. One of the mean purposes of this chapter is to describe the abilities of the moiré based techniques in the study of atmospheric turbulence with their potentials and limitations. Also, in this chapter a new moiré based wavefront sensing technique that can be used for adaptive optics will be presented. At the end of this chapter, a brief comparison of use of two wavefront sensors, the SH sensor and the two channel moiré deflectometry based wavefront sensor, will be presented.

In addition, a new computationally algorithm for analyzing the moiré fringes will be presented. In this chapter, for the first time, the details of an improved algorithm for processing moiré fringes by means of virtual traces will be presented. By means of the virtual traces one can increase the precision of measurements in all of the moiré based methods, by increasing the moiré fringes spacing, meanwhile at the same time by using a number of virtual traces, the desired spatial resolution is achievable. As a result, the sensitivity of detection is adjustable by merely changing the separation of the gratings and the angle between the rulings of the gratings in moiré deflectometer, and at the same time, the desired spatial resolution is achieved by means of the virtual traces.

2. Moiré technique; definition, history and applications

Generally, superposition of two or more periodic or quasi-periodic structures (such as screens, grids or gratings) leads to a coarser structure, named moiré pattern or moiré fringes. The moiré phenomenon has been known for a long time; it was already used by the Chinese in ancient times for creating an effect of dynamic patterns in silk cloth. However, modern scientific research into the moiré technique and its application started only in the second half of the 19th century. The word moiré seems to be used for the first time in scientific literature by Mulot (Patorski & Kujawinska, 1993).

The moiré technique has been applied widely in different fields of science and engineering, such as metrology and optical testing. It is used to study numerous static physical phenomena such as refractive index gradient (Karny & Kafri, 1982; Ranjbar et al., 2006). In addition, it has a severe potential to study dynamical phenomena such as atmospheric turbulence (Rasouli & Tavassoly, 2006a;b; 2008; Rasouli, 2010), vibrations (Harding & Harris, 1983), nonlinear refractive index measurements (Jamshidi-Ghaleh & Mansour, 2004; Rasouli et al., 2011), displacements and stress (Post et al., 1993; Walker, 2004), velocity measurement

(Tay et al., 2004), acceleration sensing (Oberthaler et al., 1996), etc. The moiré pattern can be created, for example, when two similar grids (or gratings) are overlaid at a small angle, or when they have slightly different mesh sizes. In many applications one of the superposed gratings is the image of a physical grating (Rasouli & Tavassoly, 2005; Ranjbar et al., 2006; Rasouli & Tavassoly, 2006a). When the image forming lights propagate in a perturbed medium, the image grating is distorted and the distortion is magnified by the moiré pattern.

Briefly, moiré technique has diverse applications in the measurements of displacement and light deflection, and it improves the precision of the measurements remarkably. Besides, the required instrumentation is usually simple and inexpensive.

3. Moiré pattern, Talbot effect, Talbot interferometry and moiré deflectometry

As it mentioned, moiré pattern can be created, when two similar straight-line grids (or gratings) are superimposed at a small angle, Fig. 1, or when they have slightly different mesh sizes, Fig. 2. In many applications one of the superimposed gratings is the image of a physical grating or is one of the self-images of the first grating. In applications, the former case is named projection moiré technique and the latter case is called moiré deflectometry or Talbot interferometry.

When a grating is illuminated with a spatially coherent light source, exact images and many other images can be found at finite distances from the grating. This self-imaging phenomenon is named the Talbot effect. By superimposing another grating on one of the self-images of the first grating, moiré fringes are formed. The Talbot interferometry and the moiré deflectometry are not identical, although they seem quite similar at a first glance. In the Talbot interferometry setup, a collimated light beam passes through a grating and then through a distorting phase object. The distorted shadow of the grating forms a moiré pattern with a second grating located at a Talbot plane (also known as Fourier plane). The moiré deflectometry measures ray deflections in the paraxial approximation, provided that the phase object (or the specular object) is placed in front of the two gratings. The resulting fringe pattern, is a map of ray deflections corresponding to the optical properties of the inspected object. Generally, when the image forming lights propagate in a perturbed medium the image grating is distorted and the distortion is magnified by moiré pattern. When the similar gratings are overlaid at a small angle, the moiré magnification is given by (Rasouli & Tavassoly, 2006b)

$$\frac{d_m}{d} = \frac{1}{2sin(\theta/2)} \simeq \frac{1}{\theta}, \tag{1}$$

where d_m, d, and θ stand for moiré fringe spacing, the pitch of the gratings, and gratings' angle.

In case of parallel moiré pattern (Rasouli & Tavassoly, 2008; Rasouli et al., 2011), when the gratings vectors are parallel together and the resulting moiré fringes are parallel to the gratings lines, the moiré magnification is given by

$$\frac{d_m}{d} = \frac{d}{\delta d}, \tag{2}$$

where δd stands for the difference of mesh sizes of the gratings.

Generally, in the moiré technique displacing one of the gratings by l in a direction normal to its rulings leads to a moiré fringe shift s, given by (Rasouli & Tavassoly, 2006b)

$$s = \frac{d}{d_m} l. \tag{3}$$

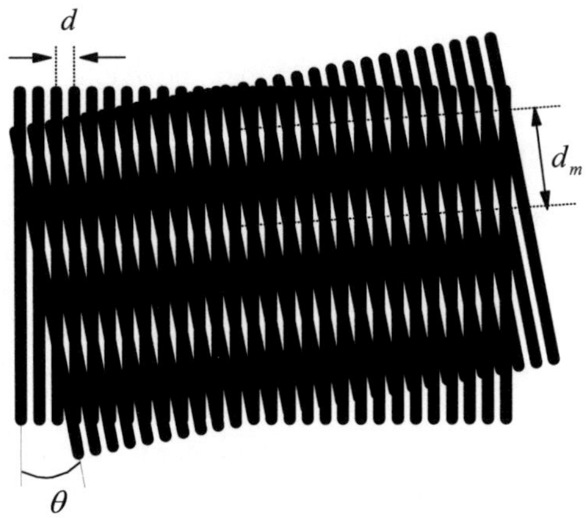

Fig. 1. A moiré pattern, formed by superimposing two sets of parallel lines, one set rotated by angle θ with respect to the other (Rasouli & Tavassoly, 2007).

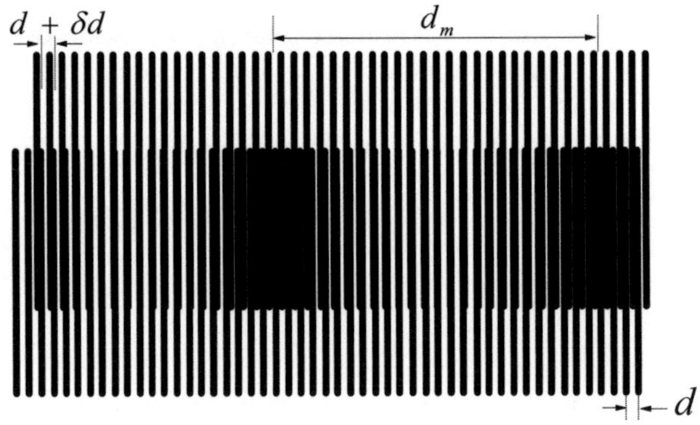

Fig. 2. A moiré pattern, formed by superimposing two sets of parallel lines, when they have slightly different mesh sizes (Rasouli, 2007).

4. Measuring atmospheric turbulence parameters by means of the moiré technique

Changes in ground surface temperature create turbulence in the atmosphere. Optical turbulence is defined as the fluctuations in the index of refraction resulting from small temperature fluctuations. Three physical effects are observed when a light beam propagates through a turbulent atmosphere: optical scintillation, beam wandering, and fluctuations in the AA. These effects are used for measuring turbulence characteristic parameters. Fluctuations of light propagation direction, referred to as the fluctuations of AA, are measured by various methods. In astronomical applications the AA fluctuations measurement is a basic step. Differential image motion monitor (Sarazin, 1990) and generalized seeing monitor systems (Ziad et al., 2000) are based on AA fluctuations. The edge image waviness effect (Belen'kii et al., 2001) is also based on AA fluctuations. In some conventional methods the fluctuations of AA are derived from the displacements of one or two image points on the image of a distant object in a telescope. In other techniques the displacements of the image of an edge are exploited. The precisions of these techniques are limited to the pixel size of the recoding CCD. In following we review some simple but elegant methods that have been presented recently in measuring the AA fluctuations and the related atmospheric turbulence parameters by means of moiré technique.

4.1 Incoherent imaging of a grating in turbulent atmosphere by a telescope

The starting work of the study of atmospheric turbulence by means of moiré technique was published in Rasouli & Tavassoly (2006a). In this work moiré technique have been used in measuring the refractive index structure constant, C_n^2, and its profile in the ground level atmosphere. In this method from a low frequency sinusoidal amplitude grating, installed at certain distance from a telescope, successive images are recorded and stored in a computer. By superimposing the recorded images on one of the images, the moiré patterns are formed. Also, this technique have been used in measuring the modulation transfer functions of the ground-level atmosphere (Rasouli et al., 2006). In the present approach after the filed process, by superimposing the images of the grating the moiré patterns are formed. Thus, observation of the AA fluctuations visually improved by the moiré magnification, but it was not increased precision of the AA fluctuations measurement. Also, this method is not a real-time technique. But, compared to the conventional methods (Belen'kii et al., 2001; Sarazin, 1990; Ziad et al., 2000) in this configuration across a rather large cross section of the atmosphere one can access to large volume of 2-D data.

In this method, when an image point on the focal plane of a telescope objective is displaced by l the AA changes by

$$\alpha = l/f, \tag{4}$$

where f is the objective focal length. Thus, order of measurement precision of the method is similar to the order of measurement precision of the conventional methods like differential image motion monitor (DIMM) (Belen'kii et al., 2001; Sarazin, 1990). Meanwhile, in this method a grating on full size of a CCD's screen are being imaged, but for example in the differential image motion monitor two image points are formed on small section of a CCD's screen.

4.2 Incoherent imaging of a grating on another grating in turbulent atmosphere by a telescope

In 2006 a new technique, based on moiré fringe displacement, for measuring the AA fluctuations have been introduced (Rasouli & Tavassoly, 2006b). This technique have two main advantages over the previous methods. The displacement of the image grating lines can be magnified about ten times, and many lines of the image grating provide large volume of data which lead to very reliable result. Besides, access to the displacement data over a rather large area is very useful for the evaluation of the turbulence parameters depending on correlations of displacements. The brief description of the technique implementation is as follows. A low frequency grating is installed at a suitable distance from a telescope. The image of the grating, practically forms at the focal plane of the telescope objective. Superimposing a physical grating of the same pitch as the image grating onto the latter forms the moiré pattern. Recording the consecutive moiré patterns with a CCD camera connected to a computer and monitoring the traces of the moiré fringes in each pattern yields the AA fluctuations versus time across the grating image. A schematic diagram of the experimental setup is shown in Fig. 3.

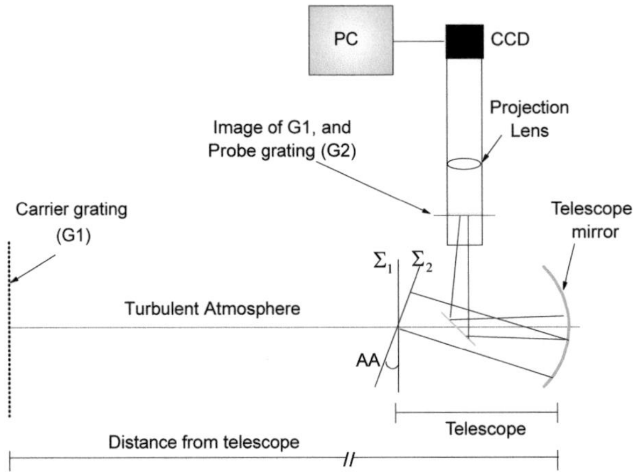

Fig. 3. Schematic diagram of the instrument used for atmosphere turbulence study by projection moiré technique, incoherent imaging of a grating on another grating in turbulent atmosphere by a telescope. (Rasouli & Tavassoly, 2006b; 2007).

The typical real time moiré fringes obtained by the set-up is shown in Fig. 4(a), and its corresponding low frequency illumination after a spatial fast Fourier transform method to low pass filter the data is shown in Fig. 4(b).

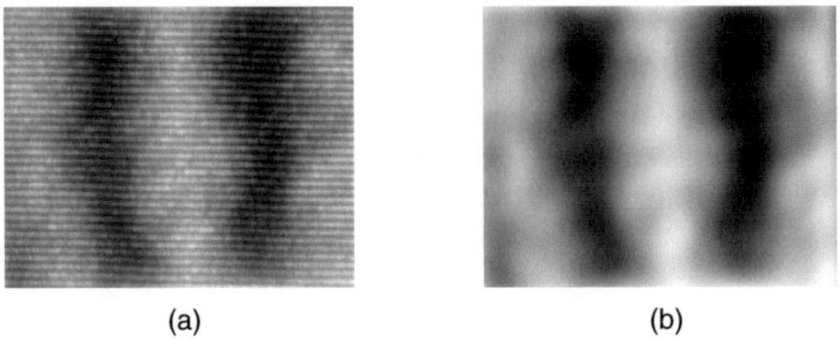

Fig. 4. (a) Typical moiré pattern recorded by the set-up in Fig. 3, (b) the corresponding low frequency illumination (Rasouli & Tavassoly, 2007).

In this method, the component α of the AA fluctuation in the direction perpendicular to the lines of the carrier grating (parallel to the moiré fringes) is given by (Rasouli & Tavassoly, 2006b)

$$\alpha = \frac{1}{f}\frac{d}{d_m}\,s, \tag{5}$$

where f, d, d_m, and s are the telescope focal length, the pitch of the probe gratings, the moiré fringes spacing, and the moiré fringe displacement, respectively. Compared to Eq. (4), here an improving factor $\frac{d}{d_m}$ appears. When the angle between the lines of superimposed gratings is less than 6^0, the magnification is more than ten times. In other word: *"Light-beam deflections due to atmospheric turbulence are one order of magnitude more precise with the aid of moiré patterns (Rasouli & Tavassoly, 2007)."*

4.3 Application of moiré deflectometry in atmospheric turbulence measurements

The next scheme, noteworthy both for its simplicity and its cleverness, illustrates the basic idea (ICO Newsletter, April 2009; Rasouli & Tavassoly, 2008). A monochromatic light wave from a small and distant source is incident on a fine pitch Ronchi ruling. A short distance beyond, a Talbot image of the ruling appears.

With diverging-light illumination of the Ronchi ruling, the Talbot image is slightly larger in scale than the ruling itself. If a duplicate of the ruling is placed in the Talbot image plane, in exactly the same orientation as the original ruling, large fringes result from the moiré effect.

Most importantly, any turbulence-produced local variations in the AA of the incident wave, even if quite small, manifest themselves as easily seen distortions of the moiré fringe pattern. These distortions, captured by a CCD video camera, are analysed by a computer program. The technique have been used to determine parameters that characterize the strength of turbulence measured along horizontal paths. A schematic diagram of the experimental setup is shown in Fig. 5.

In this method the component α of the AA fluctuation in the direction perpendicular to the lines of the gratings is given by (Rasouli & Tavassoly, 2008)

$$\alpha = \frac{d}{d_m}\frac{s}{Z_k}, \tag{6}$$

where Z_k denotes the kth self-image or Talbot's distance is given by (Patorski & Kujawinska, 1993)

$$2k\frac{d^2}{\lambda} = \frac{LZ_k}{L + Z_k}, \tag{7}$$

where λ is the light wavelength and L is the distance between $G1$ and the source.

The implementation of the technique is straightforward, a telescope is not required, fluctuations can be magnified more than ten times, and the precision of the technique can be similar to that reported in the previous work (Rasouli & Tavassoly, 2006b).

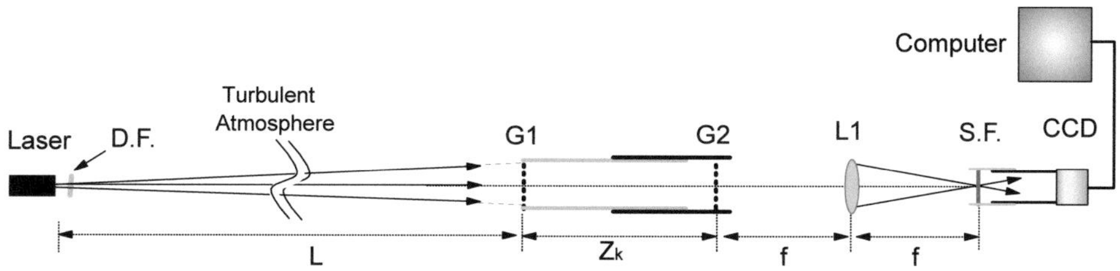

Fig. 5. Schematic diagram of the application of moiré deflectometry in atmospheric turbulence measurements. *D.F.*, *G*1, *G*2, *L*1, and *S.F.*, stand for the neutral density filter, first grating, second grating, Fourier transforming lens, and the spatial filter, respectively (Rasouli & Tavassoly, 2008).

4.4 Use of a moiré deflectometer on a telescope for atmospheric turbulence measurements

Recently, a highly sensitive and high spatial resolution instrument for the study of atmospheric turbulence by measuring the fluctuation of the AA on the telescope aperture plane have been constructed (Rasouli, 2010). A schematic diagram of the instrument is shown in Fig. 6. A slightly divergent laser beam passes through a turbulent ground level atmosphere and enters the telescope aperture. The laser beam is recollimated behind the telescope's focal point by means of a collimator. The collimated beam passes through a moiré deflectometer. Compared to the previous moiré based methods, because of the large area of the telescope aperture, this instrument is more suitable for studying spatial and temporal properties of wavefronts. Because of the magnifications of the telescope and moiré deflectometry, the precision of measurement of the technique is one order of magnitude more precise than previous methods. In other word, the precision of AA fluctuations measurement for the second time have been improved. This instrument has a very good potential for wavefront sensing and adaptive optics applications in astronomy with more sensitivity. Besides, a modified version of this instrument can be used to study other turbulent media such as special fluids and gases. Also, this method is a reliable way to investigate turbulence models experimentally.

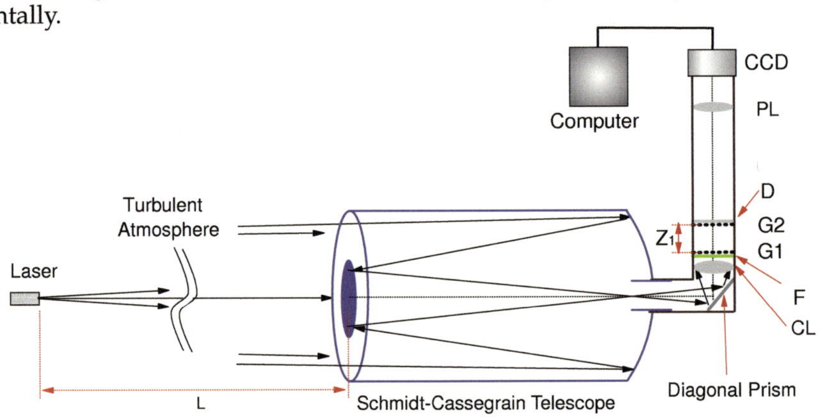

Fig. 6. Schematic diagram of the instrument; use of a moiré deflectometer on a telescope. *CL*, *F*, *G1*, *G2*, and *PL*, stand for the collimating lens, bandpass filter, first grating, second grating, and the lens that projects the moiré pattern produced on the diffuser D on the CCD, respectively. (Rasouli, 2010).

Here, the component α of the AA fluctuation on the telescope aperture plane is given by(Rasouli, 2010):

$$\alpha = \frac{f'}{f}\frac{1}{Z_k}\frac{d}{d_m}s, \tag{8}$$

where f is the telescope focal length and f' is the focal length of the collimating lens. Compared to Eq. (6) here an improving factor f'/f appears. For example, in the work of Rasouli (2010), f=200 cm and $f' = 13.5$ cm have been used, thus the magnification is more than ten times. In other word, the precision of AA fluctuations measurement for second time in this work have been improved. *As a result, due to the magnification of the telescope, the use of Moiré technique, and the Talbot effect, measurements of fluctuations in the AA can be up to 2 orders of magnitude more precise than other methods* (Rasouli & Tavassoly, 2006b; 2008; Rasouli, 2010).

Method	α_{min}	Volume of data	Processing way
DIMM	order of one arc sec	Two image points	Real-time
IIGT	0.5 arc sec	are equal to the CCD pixels number	Non real-time
IIGGT	0.06 arc sec	are equal to the CCD pixels number	Real-time
MD	0.27 arc sec	are equal to the CCD pixels number	Real-time
MDT	0.01 arc sec	are equal to the CCD pixels number	Real-time

Table 1. Comparison of sensitivities and spatial resolutions of different methods; DIMM, IIGT, IIGGT, MD, MDT are stand for the differential image motion monitor, incoherent imaging of a grating by a telescope, incoherent imaging of a grating on another grating by a telescope, moiré deflectometry method, use of a moiré deflectometer on a telescope, respectively. α_{min} stands for the minimum measurable AA fluctuation.

4.5 Comparison of sensitivities and spatial resolutions of different moiré based methods

According to Eqs. (5), (6), and (8), in all of the moiré based methods by increasing the gratings distance, decreasing the pitch of the gratings, or increasing the moiré fringes spacing, the measurement precision is improved. Let us now to compare the sensitivities of all of the reviewed methods by considering typical values that can be used in the works namely: l = 5 μm, d=1/15 mm, f=2 m, f'=10 cm, Z_k =0.5 m, s/d_m= 1/100, the minimum measurable AA fluctuations are obtained using Eqs. (4)-(6), and (8); 2.5×10^{-6}, 3.3×10^{-7}, 1.3×10^{-6}, and 6.6×10^{-8} rad, respectively. More details of the different methods are presented in Table 1.

In comparing implementation of different methods, the implementation of the incoherent imaging of a grating by a telescope is not straightforward. Implementation of the moiré deflectometry method and use of a moiré deflectometer on a telescope are very straightforward. Also, the last method, because of its measurement precision and large area of the telescope aperture has potential applications in diverse fields.

4.6 A brief summary on the study of atmospheric turbulence by means of moiré technique

In brief incorporation of moiré technique in the study of atmospheric turbulence provides the following advantages:

- Access to large volume of data in 2-Ds
- Correlations calculations in 2-Ds at desired scale
- The required instrumentation is usually simple and inexpensive
- The presented techniques usually are very flexible and can be applied in a wide range of turbulence strengths, by choosing gratings of adequate pitch, size, and separation.
- Improvement of measurement precision; as a result of the works, measurements of fluctuations in the AA can be up to 2 orders of magnitude more precise than other methods.

5. Wavefront sensing based on moiré deflectometry

Recently, a wide dynamic range two channel wavefront sensor based on moiré deflectometry has been constructed for measuring atmospheric distortions of wavefront (Rasouli et al., 2009; 2010). Schematic diagram of the sensor is shown in Fig. 7. In this sensor, a collimated laser beam passes through a time-varying refractive index field, like a turbulent medium, and then a beam-splitter divides it into two beams. A mirror reflects the second beam into a direction parallel to the first beam propagation direction, and the beams pass through a pair of moiré deflectometers. The moiré deflectometers are installed parallel and close together. The gratings' rulings are roughly parallel in each moiré deflectometer but are perpendicular in the two beams. Moiré patterns are formed on a plane where the second gratings of the moiré deflectometers and a diffuser are installed. The moiré patterns are projected on a CCD camera. Using moiré fringes fluctuations two orthogonal components of the AA across the wavefront have been calculated. The fluctuations have been deduced in successive frames, and then evolution of the wavefront shape is determined. The dynamic range and sensitivity of detection are adjustable by merely changing the distance between two gratings in both moiré deflectometers and relative grating ruling orientation. The spatial resolution of the method is also adjustable by means of bright, dark, and virtual traces for given moiré fringes without paying a toll in the measurement precision. The implementation of the technique is straightforward. The measurement is relatively insensitive to the alignment of the beam into the sensor. This sensor has many practical applications ranging from wave aberrations of human eyes to adaptive optics in astronomy.

In this sensor, the incident wavefront gradients in x and y-directions at a point (x, y) are determined by (Rasouli et al., 2010)

$$\left[\frac{\partial U(x,y)}{\partial x}, \frac{\partial U(x,y)}{\partial y}\right] = \frac{d}{Z_k}\left[\frac{s_y}{d'_m}, \frac{s_x}{d_m}\right]. \tag{9}$$

where, d_m, d'_m, s_y, and s_x are the moiré fringe spacing in the first and second channels, and the moiré fringes shifts in the first and second channels, respectively.

Typical reconstructed wavefront surface for a collimated laser beam passes through a turbulent column of hot water vapor rising from a small cup using this wavefront sensor in a region of $20mm \times 20mm$ are shown in Fig. 8.

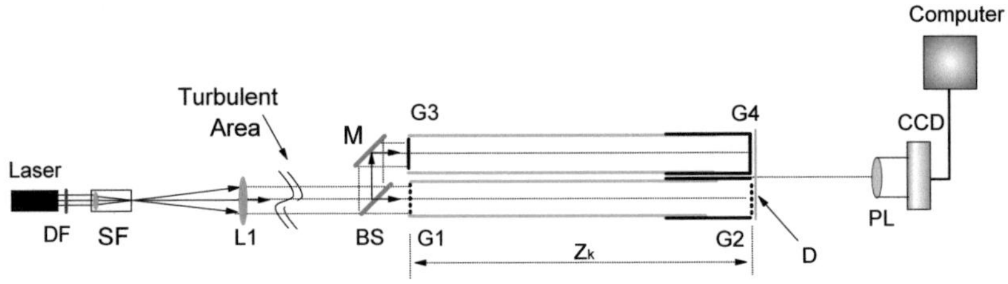

Fig. 7. Schematic diagram of the experimental setup of two channel wavefront sensor. G, L, M, and $S.F.$ stand for the gratings, lenses, mirrors, and spatial filters respectively. $D.F.$, $B.S.$ and Z_k stand for the neutral density filter, beam splitter, and talbot distance, respectively.

5.1 An improved algorithm for processing moiré fringes by means of virtual traces

The most commonly approach in the moiré fringes processing are based on measurement of the displacements of the bright or dark moiré fringes. In this approach, according to the Eqs. (3), (5), (6), (8) and (9), by increasing the moiré fringe spacing the precision of measurements can be improved. But, then the number of moiré fringes in the field of view is decreased and the spatial resolution of the method is decreased. Recently, an improved algorithm have been used for processing the moiré fringes to overcome this limitation (Rasouli et al., 2010; Rasouli & Dashti, 2011). A new concept that is called virtual traces in the moiré patterns have been introduced . In this approach, the traces of bright moiré fringes, dark moiré fringes, and of points with intensities equal to the mean intensity of the adjacent bright and dark traces (first order virtual traces) were determined. One can potentially produce a large number of virtual traces between two adjacent bright and dark traces by using their intensities and

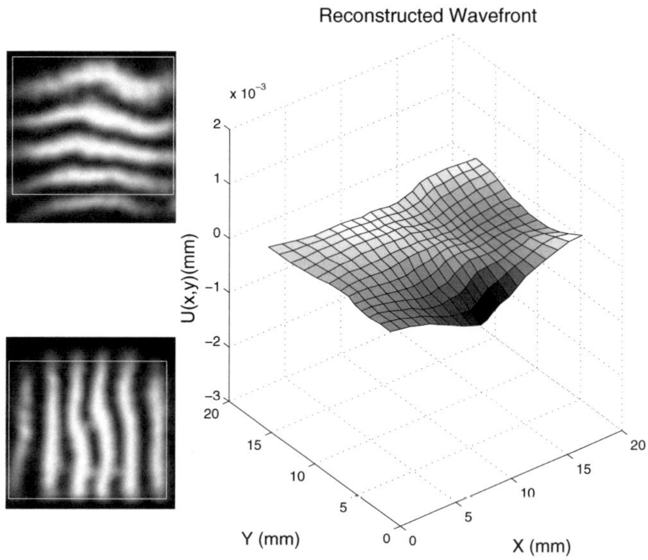

Fig. 8. Moiré fringes in the horizontal and vertical directions and the reconstructed wavefront, surface plot, corresponding to distortions generated by a turbulent column of hot water vapor rising from a small cup in a region of 20×20 mm^2.

locations. One can increase the precision of measurements by increasing the moiré fringes spacing, meanwhile at the same time by using a number of virtual traces, the desired spatial resolution is achievable. Thus, the sensitivity of detection is adjustable by merely changing the separation of the gratings and the angle between the rulings of the gratings in moiré deflectometer, and at the same time, the desired spatial resolution is achieved by means of the virtual traces.

Here, for the first time, the details of the mentioned improved algorithm for processing moiré fringes by means of virtual traces are presented. We use the algorithm on one of the moiré fringes of Fig. 8. As it previously mentioned, the distortions on the fringes correspond to a turbulent column of hot water vapor rising from a small cup. Low-frequency illumination distribution of the first bright and dark moiré fringes in the horizontal direction are shown in Fig. 9. The corresponding derived bright (I_b) and dark (I_d) traces are shown in Fig. 10.

Mathematically, the moiré fringes intensity profile in direction perpendicular to the moiré fringes, after a spatial fast Fourier transform method to low pass filter the data, can be written as

$$I(y) = \left[(\frac{I_b + I_d}{2}) + (\frac{I_b - I_d}{2}) cos \frac{2\pi}{d_m}(y + y_{0b}) \right], \tag{10}$$

where d_m, I_b, I_d, and y_{0b} are the moiré fringes spacing, the intensity of bright and dark traces, and the position of the reference bright trace, respectively. Here, we have used d_m instead of previously used $d_m^{'}$.

It should be mentioned that, due to the presence of air turbulence in the path, the image grating - one of the superimposed gratings for generation the moiré pattern - is distorted and the distortion is magnified by the moiré pattern. As a result, the moiré fringes intensity profile a little differs from Eq. (10). We will show the distorted moiré fringes intensity profile by $I'(y)$.

From Eq. (10), mid points between the adjacent bright and dark traces have an intensity equal to $\frac{I_b+I_d}{2}$. Now, for the case of distorted moiré pattern, the traces of points with intensities equal to the mean intensity of the adjacent bright and dark traces ($I_{vir}^{(1)} = \frac{I'_b+I'_d}{2}$) to be determined the first order virtual trace. By use of following equation in all of the columns of the intensity distribution of the moiré pattern, one can find the first order virtual trace

$$I_{vir}^{(1)} = \frac{I'_b + I'_d}{2} = I'(y_{vir}^{(1)}) \rightarrow y_{vir}^{(1)} = y\left(I' = \frac{I'_b + I'_d}{2}\right). \quad (11)$$

Intensity profile of moiré fringes in the direction perpendicular to the moiré fringes and the corresponding point on the first order virtual trace, $(y_{vir}^{(1)}, I_{vir}^{(1)})$, are shown in Fig. 11.

One can potentially produce a large number of virtual traces between two adjacent bright and dark traces by using their intensities and locations. In the non-distorted moiré pattern, by finding the intensity of the mid points between the first order virtual trace and the adjacent bright and dark traces using Eq. (10), one can produce two second order virtual traces, that are named $I_{vir}^{(2b)}$ and $I_{vir}^{(2d)}$, respectively. For non-distorted moiré pattern, their intensities are given by

$$I_{vir}^{(2b)} = \left[\left(\frac{I_b + I_d}{2}\right) + \left(\frac{I_b - I_d}{2}\right)cos\frac{2\pi}{d_m}(y_{vir}^{(2b)} + y_{0b})\right], \quad (12)$$

$$I_{vir}^{(2d)} = \left[\left(\frac{I_b + I_d}{2}\right) + \left(\frac{I_b - I_d}{2}\right)cos\frac{2\pi}{d_m}(y_{vir}^{(2d)} + y_{0b})\right], \quad (13)$$

where, $y_{vir}^{(2b)} = \frac{y_b+y_{vir}^{(1)}}{2}$ and $y_{vir}^{(2b)} = \frac{y_d+y_{vir}^{(1)}}{2}$. Now, for the case of distorted moiré pattern, the traces of points with intensities equal to the intensity of points with $y = \frac{y_b+y_{vir}^{(1)}}{2}$ and $y = \frac{y_d+y_{vir}^{(1)}}{2}$, were obtained from Eqs. (12) and (13), to be determined the second order virtual traces, $I_{vir}^{(2b)}$ and $I_{vir}^{(2b)}$, respectively. By use of following equations in all of the columns of the intensity distribution of moiré pattern, one can find the second order virtual traces

$$I_{vir}^{(2b)} = I'\left(y = \frac{y_b + y_{vir}^{(1)}}{2}\right) \rightarrow y_{vir}^{(2b)} = y\left(I_{vir}^{(2b)}\right), \quad (14)$$

$$I_{vir}^{(2d)} = I'\left(y = \frac{y_d + y_{vir}^{(1)}}{2}\right) \rightarrow y_{vir}^{(2d)} = y\left(I_{vir}^{(2d)}\right). \quad (15)$$

The procedure to produce higher order virtual traces is similar to that one were used for the second order virtual traces. In Fig. 12, three virtual traces, $I_{vir}^{(1)}$, $I_{vir}^{(2b)}$, and $I_{vir}^{(2d)}$ are produced between two adjacent bright and dark traces by using their intensities and locations.

5.2 Wavefront reconstruction

Another major part of a wavefront sensor is a software to convert the 2-D wavefront gradients data into 2-D wavefront phase data. In order to perform the wavefront reconstruction from

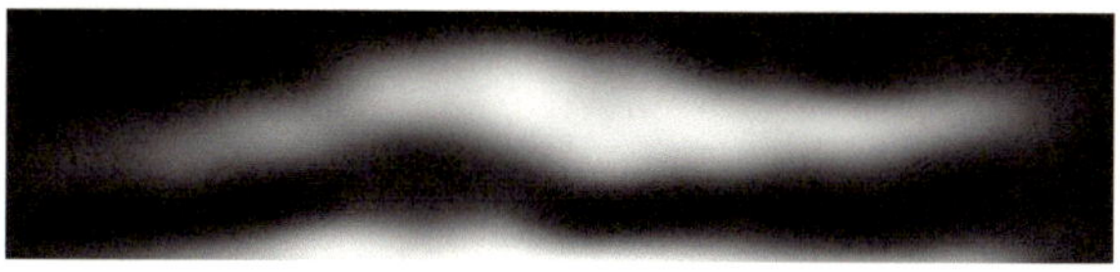

Fig. 9. Low-frequency illumination distribution of a typical bright and dark moiré fringes.

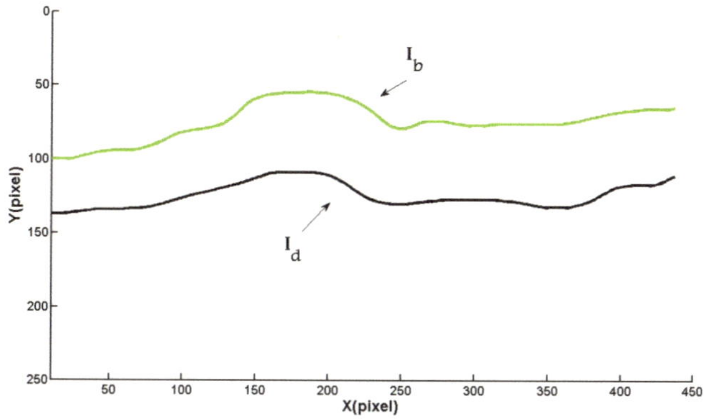

Fig. 10. The derived traces of bright (I_b) and dark (I_d) moiré fringes corresponding to the fringes are shown in Fig. 9.

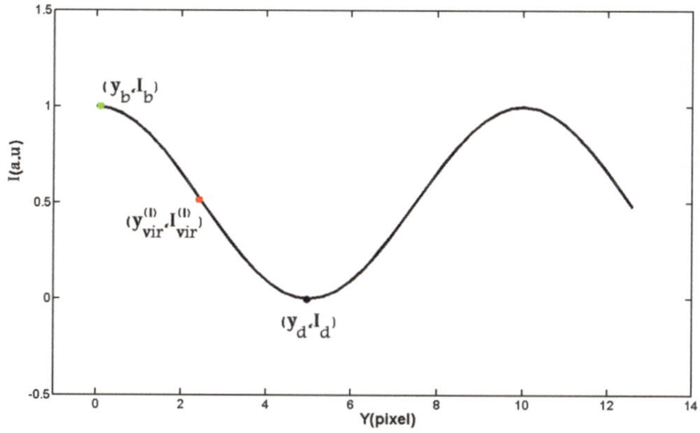

Fig. 11. Typical intensity profile of non-distorted moiré fringes in the direction perpendicular to the moiré fringes and the corresponding point on the first order virtual trace.

the measured moiré patterns, we consider the displacements of the bright, dark, and the first order virtual traces with respect to their reference positions, which represents an estimate of the local x-gradients or y-gradients of the wavefront phase. The reference positions of the traces are determined from a long-exposure frame. In practice, by considering two sets of vertical and horizontal moiré traces of a frame in a x-y coordinate system (the vertical and horizontal traces are overlapped in the x-y coordinate system), the intersection points of the vertical and horizontal bright, dark, and the first order virtual traces are determined. x-gradients and y-gradients of the wavefront are deduced from the displacements of the intersection points in successive frames. From this 2-D gradient field we performed an estimate of the wavefront.

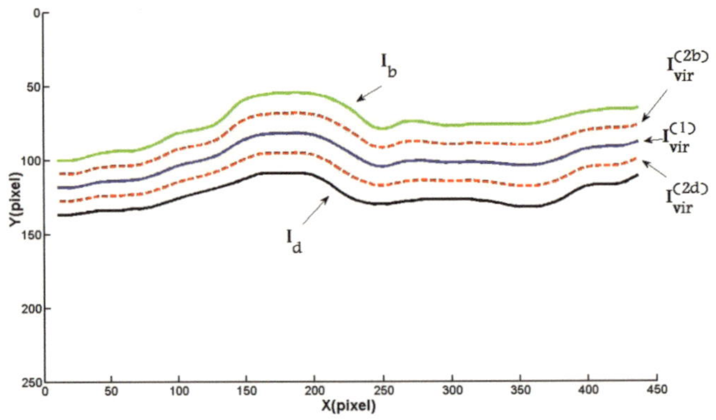

Fig. 12. Three virtual traces are produced between two adjacent bright and dark traces by using their intensities and locations.

Algorithmically, this involves the calculation of a surface by an integration-like process. The reconstruction problem can be expressed in a matrix-algebra framework. The unknowns, a vector Φ of N phase values over a grid, must be calculated from the data, from a measurement vector **S** of M elements of wavefront gradients in two directions. In the context of wavefront reconstruction, a general linear relation like $\Phi = \mathbf{BS}$ is used, where **B** is the so-called reconstruction matrix (Roddier, 1999). A number of techniques are available to derive reconstruction matrix (Fried, 1977; Herrmann, 1980; Hunt, 1979; Southwell, 1980). A linear model of wavefront sensor allows the linking of the measurements **S** to the incoming wavefront or its phase. The matrix equation between **S** and Φ reads as $\mathbf{S} = \mathbf{A}\Phi$, where **A** is called the interaction matrix. For the geometry of discretization, the 2-D map of intersection points of the traces, the interaction matrix is determined, then the reconstruction matrix is obtained. In the mentioned work Hudgin's and Frid's discretization have been examined (Herrmann, 1980; Hunt, 1979).

5.3 Comparison of SH method and moiré deflectometry based two channel wavefront sensor

In this section, an adjustable, high sensitivity, wide dynamic range two channel wavefront sensor based on moiré deflectometry has been reviewed. In this sensor the dynamic range and sensitivity of detection are adjustable by merely changing the separation of the gratings and the angle between the rulings of the gratings in both the moiré deflectometers. This overcomes the deficiency of the Shack-Hartman sensors in that these require expensive retrofitting to change sensitivity. The spatial resolution of the method is also adjustable by means of bright, dark, and virtual traces for a given set of moiré fringes without paying a toll in the measurement precision. By this method discontinuous steps in the wavefront are detectable, because AA fluctuations are measured across the wavefront. Also, unlike the SH sensor, in this sensor the measurement is relatively insensitive to the alignment of the beam into the sensor. The implementation of the technique is straightforward and it overcomes some of the technical difficulties of the SH technique. The required instrumentation for this sensor is usually simple and inexpensive. This sensor has many practical applications ranging from wave aberrations of human eyes to adaptive optics in astronomy.

Finally, for low light applications as one would normally expect in astronomy (to work with stars), the sensor can be performed with phase gratings on a large-sized telescope in conjunction by use of a highly sensitive CCD. Also, it seems that using a laser guide star one can overcome to this limitation.

6. References

Belen'kii, M. S.; Stewart, J. M. & Gillespie, P. (2001). Turbulence-induced edge image waviness: theory and experiment, *Appl. Opt.*, Vol. 40: 1321–1328.

Fried, D. L. (1977). Least-square fitting a wavefront distortion estimate to an array of phase-difference measurements, *J. Opt. Soc. Am.*, Vol. 67: 370–375.

Gonsalves, R. A. (1982). Phase retrieval and diversity in adaptive optics, *Opt. Eng.*, Vol. 21: 829–832.

Harding, K. G. & Harris, J. S. (1983). Projection moiré interferometer for vibration analysis, *Appl. Opt.*, Vol. 22: 856–861.

Herrmann, J. (1980). Least-squares wavefront errors of minimum norm, *J. Opt. Soc. Am.*, Vol. 70: 28–35.

Hunt, B. R. (1979). Matrix formulation of the reconstruction of phase values from phase differences, *J. Opt. Soc. Am.*, Vol. 69: 393–399.

ICO Newsletter, (2009). *ICO Newsletter*, April 2009, Number 79.

Jamshidi-Ghaleh, K. & Mansour, N. (2004). Nonlinear refraction measurements of materials using the moiré deflectometry, *Optics Communications*, Vol. 234: 419–425.

Karny, Z. & Kafri, O. (1982). Refractive-index measurements by moiré deflectometry, *Appl. Opt.*, Vol. 21(18): 3326–3328.

Leibbrandt, G. W. R.; Harbers, G. & Kunst, P. J. (1996). Wavefront analysis with high accuracy by use of a double-grating lateral shearing interferometer, *Appl. Opt.*, Vol. 35: 6151–6161.

Lombardo, M. & Lombardo, G. (2009). New methods and techniques for sensing the wave aberrations of human eyes, *Clinical and Experimental Optometry*, Vol. 92: 176–186.

Nishijima, Y. & Oster, G. (1983). Moiré patterns: Their application to refractive index gradient measurements, *J. Opt. Soc. Am.*, Vol. 54: 1–5.

Oberthaler, M. K.; Bernet, S.; Rasel, Et. M.; Schmiedmayer, J. & Zeilinger, A. (1996). Inertial sensing with classical atomic beams, *Phys. Rev. A*, Vol. 54(4): 3165–3176.

Patorski, K. & Kujawinska, M. (1993). *Handbook of the moiré fringe technique*, Elsevier, Amsterdam.

Platt, B. C. & Shack, R. V. (2001). History and principles of Shack-Hartmann wavefront sensing, *Journal of Refractive Surgery*, Vol. 17: S573–S577.

Post, D.; Han, B. & Ifju, P. (1993). *High Sensitivity moiré: experimental analysis for mechanics and materials*, Springer, Berlin, Germany.

Ragazzoni, R. & Farinato, J. (1999). Phase retrieval and diversity in adaptive optics, *Opt. Eng.*, Vol. 350: L23–L26.

Rasouli, S. & Tavassoly, M. T. (2005). Moiré deflectometer for measuring distortion in sheet glasses, *Proceedings of SPIE, ICO20: Optical Devices and Instruments*, pp. 6024, doi: 10.1117/12.666818, SPIE.

Ranjbar, S.; Khalesifard, H. R.; & Rasouli, S. (2006). Nondestructive measurement of refractive index profile of optical fiber preforms using moiré technique and phase shift method, *Proceedings of SPIE, ICO20: Optical Communication*, pp. 602520, doi: 10.1117/12.667094, SPIE.

Rasouli, S. & Tavassoly, M. T. (2006). Measurement of the refractive-index structure constant, Cn2, and its profile in the ground level atmosphere by moiré technique, *Proceedings of SPIE, Optics in Atmospheric Propagation and Adaptive Systems IX*, pp. 63640G, doi: 10.1117/12.683873, SPIE.

Rasouli, S.; Madanipour, K. & Tavassoly, M. T. (2006). Measurement of modulation transfer function of the atmosphere in the surface layer by moiré technique, *Proceedings of SPIE, Optics in Atmospheric Propagation and Adaptive Systems IX*, pp. 63640K, doi: 10.1117/12.687614, SPIE.

Rasouli, S. & Tavassoly, M. T. (2006). Application of moiré technique to the measurement of the atmospheric turbulence parameters related to the angle of arrival fluctuations, *Opt. Lett.*, Vol. 31(22): 3276 – 3278, ISSN 0146-9592.

Rasouli, S. & Tavassoly, M. T. (2007). Moiré technique improves the measurement of atmospheric turbulence parameters. *SPIE Newsroom*, DOI 10.1117/2.1200702.0569. http://spie.org/documents/Newsroom/Imported/0569/0569-2007-02-20.pdf.

Rasouli, S. (2007). *Ph. D. Thesis: Study of the atmosphere turbulence parameters and large scale structure vibrations using moiré technique*, IASBS, Zanjan, IRAN.

Rasouli, S. & Tavassoly, M. T. (2008). Application of the moiré deflectometry on divergent laser beam to the measurement of the angle of arrival fluctuations and the refractive index structure constant in the turbulent atmosphere, *Opt. Lett.*, Vol. 33(9): 980 – 982, ISSN 0146-9592

Rasouli, S.; Ramaprakash, A. N.; Das, H. K.; Rajarshi, C. V.; Y. Rajabi; & Dashti, M. (2009). Twochannel wavefront sensor arrangement employing moiré deflectometry, *Proceedings of SPIE, Optics in Atmospheric Propagation and Adaptive Systems XII*, pp. 74760K-1, doi: 10.1117/12.829962, SPIE.

Rasouli, S. (2010). Use of a moiré deflectometer on a telescope for atmospheric turbulence measurements, *Opt. Lett.*, Vol. 35(9): 1470 – 1472, ISSN 0146-9592.

Rasouli, S.; Dashti, M.; & Ramaprakash, A. N. (2010). An adjustable, high-sensitivity, wide dynamic range two channel wavefront sensor based on the moiré deflectometry, *Opt. Exp.*, Vol. 18(23): 23906 – 23915.

Rasouli, S. & Dashti, M. (2010). An improved algorithm for processing moiré fringes by means of virtual traces, *Proceeding of 17th Iranian Conference on Optics and Photonics*, pp. 252-255, 2011, (in Persian).

Rasouli, S.; Ghasemi, H.; Tavassoly, M. T. & Khalesifard, H. R. (2011). Application of "parallel" moiré deflectometry and the single beam Z-scan technique in the measurement of the nonlinear refractive index, *Appl. Opt.*, Vol. 50(16): 2356–2360.

Roddier, E. (1988). Curvature sensing and compensation: a new concept in adaptive optics, *Appl. Opt.*, Vol. 27: 1223–1225.

Roddier, F. (1999). *Adaptive optics in astronomy*, Cambridge university press, Cambridge, United Kingdom.

Sarazin, M. (1990). The ESO differential image motion monitor. *Astron. Astrophys.*, Vol. 227: 294–300.

Shack, R. V. & Platt, B. C. (1971). Production and use of a lenticular Hartmann screen, *J. Opt. Soc. Am.*, Vol. 61: 656.

Southwell, W. H. (1980). Wavefront estimation from wavefront slope measurements, *J. Opt. Soc. Am.*, Vol. 70(8): 998–1006.

Tay, C. G.; Quan, C.; Fu, Y. & Huang, Y. (2004). Instantaneous velocity, displacement, and contour measurement by use of shadow moiré and temporal wavelet analysis, *Appl. Opt.*, Vol. 43: 4164–4171.

Ziad, A.; Conan, R.; Tokovinin, A.; Martin, F. & Borgnino, J. (2000). From the grating scale monitor to the generalized seeing monitor, *Appl. Opt.*, Vol. 39: 5415–5425.

Walker, C. A. (2004). *Handbook of moiré measurement*, Institute of Physics, London.

Part 2

Imaging and Laser Propagation Systems

3

Direct Imaging of Extra-Solar Planets – Homogeneous Comparison of Detected Planets and Candidates

Ralph Neuhäuser and Tobias Schmidt
Astrophysical Institute and University Observatory, Friedrich Schiller University Jena, Jena Germany

1. Introduction

Planets orbiting stars other than the Sun are called *extra-solar planets* or *exo-planets*. Since about 1989, several hundred such objects were detected using various techniques. The first companions with planetary masses orbiting another star were found around a pulsar, a fast rotating neutron star, by variations of the otherwise very stable radio pulses (Wolszczan & Frail, 1992). With the radial velocity technique, one can detect the motion of the star in radial direction towards and away from us due to the fact that both a star and a planet orbit their common center of mass (first successfully done on HD 114762 and 51 Peg, Latham et al. (1989) and Mayor & Queloz (1995), respectively). This one-dimensional technique yields the lower mass limit $m \cdot \sin i$ of the companion mass m due to the unknown orbit inclination i, so that planets detected only by the radial velocity method are *planet candidates*, they could also be higher mass brown dwarfs or low-mass stars. Several hundred planet candidates were detected with this method. The reflex motion of the star in the two other dimensions due to the orbiting companion can be measured with the astrometry method (e.g. Benedict et al. (2002)), but no new planets were found with this technique so far. If the orbital plane of a planet is in the line of sight towards the Earth, then the planet will move in front of the star once per orbital period, which can be observed as transit, i.e. as small drop in the brightness of the star. This determines the inclination i and, for radial velocity planet candidates, can confirm candidates to be true planets. The transit method could confirm almost one dozen planet candidates previously detected by the radial velocity method (the first was HD 209458 b, Charbonneau et al. (2000)); in addition, more than 100 planets were originally discovered as candidates by the transit method and then confirmed by radial velocity. All the techniques mentioned above are *in*direct techniques, i.e. they all observe only the star, but not the planet or planet candidate, i.e. it is never known which photons were emitted by the star (most) and which by the planets (negligibly few). This is different only in the direct imaging technique, which we discuss below in detail. For number of planets, lists, properties, and references, see Schneider et al. (2011) with updates on www.exoplanet.eu or Wright et al. (2011) with updates on www.exoplanets.org.

For all techniques, it is also relevant to define what is called a *planet*. In the case of extra-solar planet, what is relevant is to define the upper mass limit of planets and the distinction from brown dwarfs. For the Solar System, the International Astronomical Union (IAU) has defined the lower mass limit for planets: *A celestial body that (a) is in orbit around the Sun, (b) has sufficient mass for its self-gravity to overcome rigid body forces, so that it assumes a hydrostatic equilibrium (nearly round) shape, and (c) has cleared the neighbourhood around its orbit* (www.iau.org).

For the upper mass limit of planets, there are still several suggestions:

- The IAU *Working Group on Extrasolar Planets* has agreed on the following preliminary working definition: *Objects with true masses below the limiting mass for thermonuclear fusion of deuterium (currently calculated to be 13 Jupiter masses for objects of solar metallicity) that orbit stars or stellar remnants are planets (no matter how they formed). The minimum mass/size required for an extra-solar object to be considered a planet should be the same as that used in our Solar System. Sub-stellar objects with true masses above the limiting mass for thermonuclear fusion of deuterium are brown dwarfs, no matter how they formed nor where they are located. Free-floating objects in young star clusters with masses below the limiting mass for thermonuclear fusion of deuterium are not planets, but are sub-brown dwarfs* (www.iau.org).
- The mass (or $m \cdot \sin i$) distribution of sub-stellar companions is bi-modal, which may indicate that the two populations formed differently; the deviding mass can be used to define planets as those below the so-called *brown dwarf desert*, which lies at around $\sim 25\ M_{Jup}$ (Grether & Lineweaver, 2006; Sahlmann et al., 2010; Schneider et al., 2011; Udry, 2010); Schneider et al. (2011) in their catalog on www.exoplanet.eu now include all those companions with mass below $\sim 25\ M_{Jup}$ within a 1 σ error.
- One can try to distinuish between planets and brown dwarfs by formation, e.g. that planets are those formed in circumstellar disks with solid or fluid cores and brown dwarfs being those formed star-like by direct gravitational contraction. In such a case, the mass ranges may overlap.

There is still no consensus on the definition of planets and their upper mass limit. The second and third suggestions above, however, may be consistent with each other, because the bi-modal distribution in masses may just be a consequence of different formation mechanism. We will use $\sim 25\ M_{Jup}$ within a 1 σ error as upper mass limit for this paper.

For a direct detection of a planet close to its host star, one has to overcome the large dynamical range problem (see Fig. 1 and 4): The planet is much fainter than its host star and very close to its bright host star. Normal Jupiter-like planets around low-mass stars ($\sim 0.1\ M_{\odot}$) with one to few Gyr age are 6 orders of magnitude fainter than their host stars (Burrows et al., 1997) - unless the planet would have a large albedo and would be very close to the star and, hence, would reflect a significant amount of star light, but then it is too close to the star for direct detection. Another exception are young planets, which are self-luminous due to ongoing contraction and maybe accretion, so that they are only 2 to 4 orders of magnitude fainter (for 13 to 1 Jup mass planets, respectively) than their (young) host stars, again for $0.1\ M_{\odot}$ stars (Burrows et al. (1997),Baraffe et al. (1998)). Hence, direct imaging of planets is less difficult around young stars with ages up to a few hundred Myr.

In this article, we will compile the planets and candidates imaged directly so far: We will compile all their relevant properties to compare them in a homogeneous way, i.e. to estimate their luminosities, temperatures, and masses homogeneously. So far, the different teams, who have found and published the objects, use different methods to estimate the companion mass, which is the most critical parameter to decide about the nature of the object as either a planet or a brown dwarf. We will then also discuss each object individually.

2. Adaptive Optics observations to detect candidates

Given the problem of dynamical range mentioned above, i.e. that planets are much fainter than stars and very close to stars, one has to use Adaptive Optics (AO) imaging in the near-infrared JHKL bands (1 to 3.5 μm), in order to directly detect a planet, i.e. to resolve it from the star. The infrared (IR) is best, because planets are cool and therefore brightest in the near- to mid-IR, while normal stars are brighter in the optical than in the IR. Two example images are given in Fig. 1.

Before any planets or planet candidates became detectable by ground-based AO observations, brown dwarfs as companions to normal stars were detected, because brown dwarfs are more massive and, hence, brighter, Gl 229 B being the first one (Nakajima et al. (1995), Oppenheimer et al. (1995)).

We will now present briefly the different observational techniques.

In *normal near-IR imaging* observations, even without AO, one would also take many short exposures in order not to saturate the bright host star (typically on the order of one second), with some offset either after each image or after about one minute (the time-scale after which the Earth atmosphere changes significantly) - called jitter or dither pattern. One can then subtract each image from the next or previous image (or a median of recent images) in order to subtract the background, which is actually foreground emission from Earth atmosphere etc. Then, one can add up or median all images, a procedure called shift+add. Without AO and/or with exposure times much longer than the correlation time of the atmosphere, such images will be far from the diffraction limit. Objects like TWA 5 B (Fig. 1) or HR 7329 B, also discussed below, were detected by this normal IR imaging with the 3.5m ESO NTT/SofI (Neuhäuser, Guenther, Petr, Brandner, Huélamo & Alves (2000),Guenther et al. (2001)).

In *speckle imaging*, also without AO, each single exposure should be as short as the correlation time of the atmosphere (at the given wavelength), so that each image can be diffraction-limited. Then, one also applies the shift+add technique. A faint planet candidate near TWA 7 was detected in this way with the 3.5m ESO NTT/Sharp (Neuhäuser, Brandner, Eckart, Guenther, Alves, Ott, Huélamo & Fernández, 2000), but later rejected by spectroscopy (Neuhäuser et al., 2002).

In *Adaptive Optics* (AO) IR imaging, each single exposure should also be short enough, in order not to saturate on the bright host star. If the host star image would be saturated, one cannot measure well the position of the photocenter of its PSF, so that the astrometric precision for the common proper motion test would be low. One also applies the shift+add technique. Most planets and candidates imaged directly were detected by normal AO imaging, see e.g. Fig. 1 (TWA 5), but are also limited regarding the so-called *inner working angle*, i.e. the lowest possible separation (e.g. the diffraction limit), at which a faint planet can be detected. The diffraction limit ($\sim \lambda/D$) at D=8 to 10 meter telescopes in the K-band ($\lambda = 2.2\,\mu$m) is 0.045 to 0.057 arc sec; one cannot improve the image quality (i.e. obtain a smaller diffraction limit) by always increasing the telescope size because of the seeing, the turbulence in the Earth atmosphere, hence AO corrections. One can combine the advantages of speckle and AO, if the individual exposures are very short and if one then saves all exposures (so-called *cube mode* at ESO VLT NACO AO instrument).

If the host star nor any other star nearby (in the isoplanatic patch) is bright enough as AO guide star, then one can use a *Laser Guide Star*, as e.g. in the Keck AO observations of Wolf 940 A and B, a planet candidate (Burningham et al., 2009), see below.

One can also place the bright host star behind a coronagraph, so that the magnitude limit will be larger, i.e. fainter companions would be detectable. However, one then cannot measure the photocenter position of the host star, so that the astrometric precision for the common proper motion test would be low. One can use a *semi-transparent coronagraph*, so that both star and companion are detected. We show an example in Fig. 1, the star ϵ Eri, where one close-in planet may have been detected by radial velocity and/or astrometry (Hatzes et al. (2000),Benedict et al. (2006)) and where there are also asymmetries in the circumstellar debris disk, which could be due to a much wider planet; such a wide planet might be detectable with AO imaging, but is not yet detected - neither in Janson et al. (2007) nor in our even deeper imaging observation shown in Fig. 1.

For any AO images with simple imaging (shift+add), or also when using a semi-transparent coronagraph and/or a Laser Guide Star, one can then also subtract the Point Spread Function (PSF) of the bright host star after the shift+add procedure, in order to improve the dynamic range, i.e. to improve the detection capability for very small separations (see Fig. 4). For *PSF subtraction*, one can either use another similar nearby star observed just before or after the target (as done e.g. in the detection of β Pic b, Lagrange et al. (2009)) or one can measure the actual PSF of the host star in the shift+add image and then subtract it.

Moreover one can obtain the very highest angular resolutions at the diffraction limit by using sparse aperture interferometric masks in addition to AO. While very good dynamic ranges can be achieved very close to stars, the size of the apertures in masking interferometry is limited by the number of holes which are needed, in order to preserve non-redundancy, thus limiting the total reachable dynamic range. Currently reached detection limits at VLT can be found e.g. in Lacour et al. (2011), beginning to reach the upper mass limits for planets given above.

In order to reduce present quasistatic PSF noise further one can use another technique called Angular Differential Imaging (ADI) (Marois et al., 2006). Using this method a sequence of images is acquired with an altitude/azimuth telescope, while the instrument field derotator is switched off, being the reason for the technique's alias name Pupil Tracking (PT). This keeps the instrument and telescope optics aligned and allows the field of view to rotate with respect to the instrument. For each image, a reference PSF is constructed from other appropriately selected images of the same sequence and subtracted before all residual images are then rotated to align the field and are combined.
This technique was further improved by introducing an improved algorithm for PSF subtraction in combination with the ADI. Lafrenière, Marois, Doyon, Nadeau & Artigau (2007) present this new algorithm called 'locally optimized combination of images' (LOCI).

While the ADI is inefficient at small angular separations, the simultaneous Spectral Differential Imaging (SDI) technique offers a high efficiency at all angular separations. It consists in the simultaneous acquisition of images in adjacent narrow spectral bands within a spectral range where the stellar and planetary spectra differ appreciably (see Lafrenière, Doyon, Nadeau, Artigau, Marois & Beaulieu, 2007, and references therein).

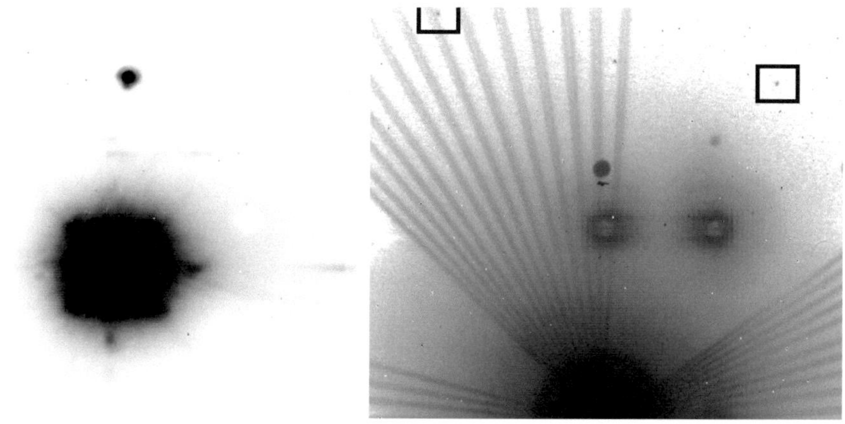

Fig. 1. Left: Our latest AO image of TWA 5 A (center) and B (2 arc sec north of A) obtained with VLT/NACO on 2008 June 13 in the K band. The mass of the companion is in the range of 17 to 45 M_{Jup} according to the Burrows et al. (1997) model (Table 2). Right: Our recent AO image of ϵ Eri obtained with VLT/NACO with the star located behind a semi-transparent coronagraph. Due to the large brightness of the star, reflection effects are also strong. Several very faint objects are detected around ϵ Eri (see boxes); however, they are all non-moving background objects as found after two epochs of observations; there is no planet nor planet candidate detected, yet, nor any additional faint object with only one epoch of imaging observation. Asymmetries in the debris disk around ϵ Eri might be due to a wide planet.

Moreover different kinds of phase masks are in use and are especially effective in combination with Adaptive Optics and a coronagraph. Recently Quanz et al. (2010) presented first scientific results using the Apodizing Phase Plate coronagraph (APP) on VLT NACO to detect β Pic b at 4 μm.

One can also detect planets as companions to normal stars with optical imaging from a *space telescope* like the Hubbe Space Telescope, see e.g. Kalas et al. (2008) for the images of the planet Fomalhaut b. From outside the Earth atmosphere, there is no atmospheric seeing, so that one can always reach the diffraction limit.

Previous reviews of AO imaging of planets were published in Duchêne (2008), Oppenheimer & Hinkley (2009), and Absil & Mawet (2010). Previous homogeneous mass determinations of planets and candidates imaged directly were given in Neuhäuser (2008) and Schmidt et al. (2009).

3. Proper motion confirmation of candidates

Once a faint object is directly detected close to a star, one can consider it a planet *candidate*, which needs to be confirmed. Two common tests can be performed on such candidates:
(a) Common proper motion test: Both the star and the planet have to show the same (or at least very similar) proper motion. The host star is normally a relatively nearby star (up to a few hundred pc, otherwise the planet would be too faint, i.e. not detectable), so that its proper motion is normally known. If the faint object would be a background star, it would be 1 to several kpc distant, so that its proper motion should be negligible compared to the star. Hence, if both the star and the faint object show the same proper motion, then the companion is not

a non-moving background star, but a co-moving companion. Given the orbital motion of the star and its companion, depending on the inclination and eccentricity, one would of course expect that their proper motions are not identical, but the differences (typically few milli arc sec per year, mas/yr) are negligible compared to the typical proper motions. Instead of (or best in addition to) common proper motion, it is also sufficient to show that both objects (primary and companion candidate) show the same radial velocity, and that the secular evolution of the radial velocity is consistent with orbital motion and inconsistent with the background hypothesis.
(b) Spectrum: If the faint object next to the star would be a planet, its mass and temperature should be much smaller than for the star. This can be shown by a spectrum. Once a spectrum is available, one can determine the spectral type and temperature of the companion. If those values are consistent with planetary masses, then the faint object is most certainly a planet orbiting that star. However, it could still be a very low-mass cool background object (very low-mass L-type star or L- or T-type brown dwarf). In cases where the companion is too faint and/or too close to the star, a spectrum might not be possible, yet, so that one should try to detect the companion in different bands to measure color indices, which can also yield (less precise) temperature or spectral type; then, however, one has the problem to distinguish between a reddened background object and the truely red (i.e. cool, e.g. planetary) companion.
The case of the ScoPMS 214 companion candidate (no. 1 or B) has shown that both tests are neccessary for a convincing case: The young K2 T Tauri star ScoPMS 214, member of the Scorpius T association, shares apparently common proper motion with a faint object nearby (3 arc sec separation) over five years; however, a spectrum of this companion candidate has shown that it is a foreground M dwarf (Metchev & Hillenbrand, 2009). Hence, the spectroscopic test is indeed necessary. Also, red colors alone (even if together with common proper motion) is not convincing, because a faint object near a star could just be reddened by extinction (background) instead of being intrinsically red, i.e. cool.

It is not sufficient to show that a star and a faint object nearby show the same proper motion (or proper motion consistent within 1 to 3 σ), one also has to show that the data are inconsistent with the background hypothesis. Common proper motion can be shown with two imaging detections with an epoch difference large enough to significantly reject the background hypothesis, namely that the faint object would be an unrelated non-moving background object. The epoch difference needed depends on the astrometric precision of the detector(s) used and the proper motion of the host star. We show an example in Fig. 2 and 3. Spectra are usually taken with an infrared spectrograph with a large telescope and AO.

Fig. 2 shows the change in separation between TWA 5 A and B with time, Fig. 3 shows the position angle (PA) of TWA 5 B with respect to TWA 5 A. In Fig. 3, the expectation for the background hypothesis is also plotted and clearly rejected by many σ. All data points are consistent with TWA 5 A and B being a common proper motion pair. Both the PA and the separation values decrease since the first detection in 1999. Such a (small) change in separation and/or PA can be interpreted as evidence of orbital motion, which of course has to be expected. One can conclude from these data that the orbit is eccentric and/or inclined.

A detection of a change in either separation or PA is actually a detection of a difference in the proper motions of A and B. It can be interpreted as evidence for slightly different proper motion. Given that most directly imaged planets (or candidates) are detected close

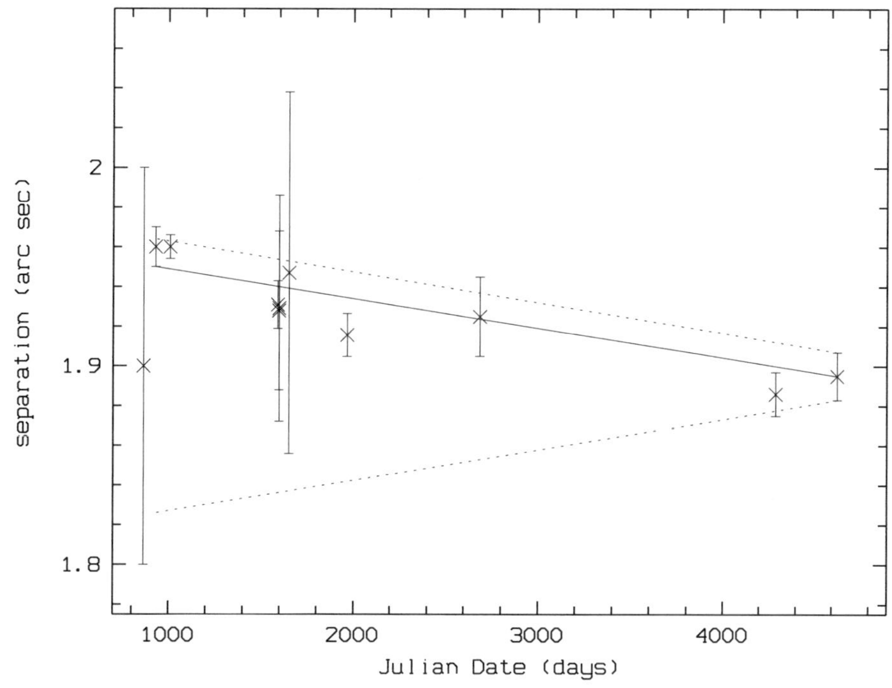

Fig. 2. Separation (in arc sec) versus observing epoch (JD - 2450000 in days) between the host star TWA 5 A (actually the photo-center of close Aa+B pair) and the sub-stellar companion TWA 5 B using data from Neuhäuser et al. (2010). The dotted lines (starting from the 2008 data point opening to the past) indicate maximum possible separation change due to orbital motion for a circular edge-on orbit. The expectation for the background hypothesis is not shown for clarity (and is rejected in Fig. 3). All data points are fully consistent with common proper motion, but not exactly identical proper motion (constant separation). Instead, the data are fully consistent with orbital motion: The separation decreases on average by $\sim$ 5.4 mas per year, as shown by the full line, which is the best fit. The figure is adapted from Neuhäuser et al. (2010).

to members of young associations (like TWA, Lupus, β Pic moving group etc.), it is therefore also possible that both the host star A and the faint object nearby (B or b) are both independent members of that association, not orbiting each other. Such an association is partly defined by the way that all or most members show a similar proper motion. In such cases, it might be necessary to show not only common proper motion (i.e. similar proper motion within the errors) or slightly different proper motion (consistent with orbital motion), but also curvature in the orbital motion. Such curvature would be due to acceleration (or deceleration) in case of a non-circular orbit. It could also be due to apparent acceleration (or deceleration) in the 2-dimensional apparent orbit on the plane of the sky for an orbit that is inclined towards the plane of the sky. Curvature can also be detected if the faint object is not anymore a bound

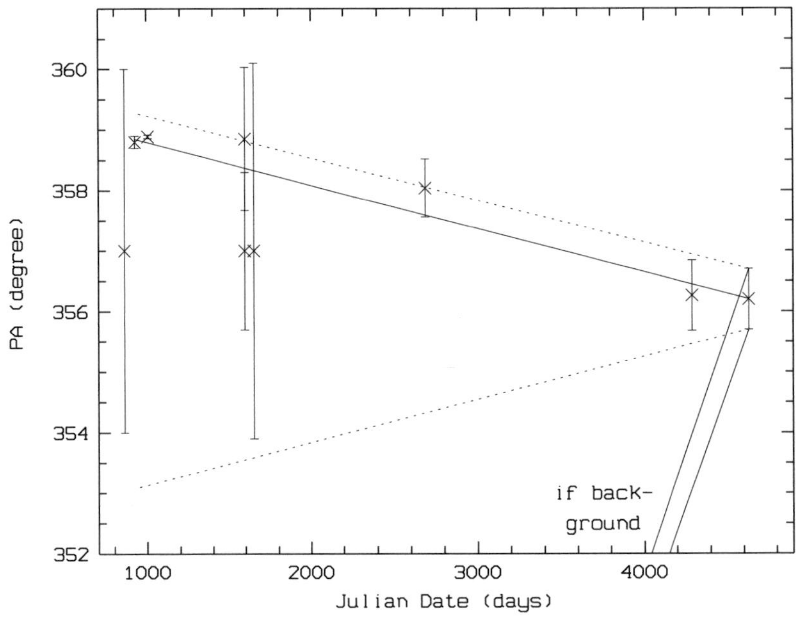

Fig. 3. Position angle PA (in degrees) versus observing epoch (JD - 2450000 in days) for TWA 5 B with respect to TWA 5 A (actually the photo-center of close Aa+B pair) using data from Neuhäuser et al. (2010). The dotted lines (starting from the 2008 data point opening to the past) indicate maximum possible PA change due to orbital motion for a circular pole-on orbit. The full lines with strong positive slope in the lower right corner are for the background hypothesis, if the bright central star (TWA 5 A) moved according to its known proper motion, while the fainter northern object (now known as B) would be a non-moving background object; the data points are inconsistent with the background hypothesis by many σ. All data points are fully consistent with common proper motion, but not exactly identical proper motion (constant PA). Instead, the data are fully consistent with orbital motion: The PA appears to decrease by $\sim 0.26^{\circ}$ per year, as shown by the full line, which is the best fit. The figure is adapted from Neuhäuser et al. (2010).

companion, but is currently been ejected, i.e. on a hyperbolic orbit. Hence, to convincingly prove that a faint object is a bound companion, one has to show curvature that is not consistent with a hyperbolic orbit.

For all of the directly imaged planets and candidates listed below, common proper motion between the candidate and the host star has been shown. For only few of them, evidence for orbital motion is shown, e.g. DH Tau (Itoh et al., 2005) GQ Lup (Neuhäuser et al., 2008), Fomalhaut (Kalas et al., 2008), HR 8799 bcde (Marois et al., 2008; 2010), TWA 5 (Neuhäuser et al., 2010), PZ Tel (Mugrauer et al., 2010), β Pic (Lagrange et al., 2010), and HR 7329 (Neuhäuser et al., 2011). For only two of them, curvature in the orbital motion was detected, namely in PZ Tel (Mugrauer et al., 2010) and TWA 5 (Neuhäuser et al., 2010).

4. Data base: Planets and candidates imaged directly

Searching the literature, we found 25 stars with directly imaged planets and candidates. We add 2M1207, a planet candidate imaged directly, whose primary would be a brown dwarf. In two cases, there is more than one planet (or candidate) detected directly to orbit the star: 4 planets around HR 8799 and two candidates in the GJ 417 system, see below. Most planets and candidates orbit single stars, but there are some as members of hierachical systems with three or more objects (one planet candidate plus two or more stars, such as TWA 5 or Ross 458). We gathered photometric and spectral information for all these objects, to derived their luminosities in a homogeneous way, taking a bolometric correction into account (Table 1). According to the mass estimate in Table 2, all of them can have a mass below 25 M_{Jup}, so that they are considered as planets.

The masses of such companions can be determined in different ways, the first two of which are usually used:

- Given the direct detection of the companion, its brightness is measured. If the companionship to the star is shown, e.g. by common proper motion, then one can assume that the companion has the same distance and age as its host star. If either a spectrum or color index is also observed, one can estimate the temperature of the companion, so that the bolometric correction can be determined; if neither color nor spectrum is available, one can often roughly estimate the temperature from the brightness difference, assuming companionship. From brightness, bolometric correction, and distance, one can estimate the luminosity. Using theoretical evolutionary models, one can then estimate the mass from luminosity, temperature, and age. However, those models are uncertain due to unknown initial conditions and assumptions. In particular for the youngest systems, below 10 Myr, the values from the models are most uncertain. Masses derived in this way are listed in Table 2.
- If a good S/N spectrum with sufficient resolution is obtained, one can also measure the effective temperature and surface gravity of the companion. Then, from temperature and luminosity, one can estimate the companion radius. Then, from radius and gravity, one can estimate the companion mass. This technique is independent of the uncertain models, but needs both distance and gravity with good precision. Since gravities (and sometimes also distances) cannot be measured precisely, yet, the masses derived in this way typically have a very large possible range.
- In the case of the directly imaged planet around the star Fomalhaut, an upper mass limit of $\sim$ 3 M_{Jup} for the companion could be determined by the fact that a dust debris ring is located just a few AU outside the companion orbit (Kalas et al., 2008). In other planet candidates also orbiting host stars with debris disks, such an upper mass limit estimate should also be possible, e.g. in HR 8799 (Marois et al. (2008),Reidemeister et al. (2009)), β Pic (see Freistetter et al., 2007), PZ Tel (Biller et al. (2010),Mugrauer et al. (2010)), and HR 7329 (Neuhäuser et al., 2011).
- If there are several planets or candidates imaged around the same star, then one can also try to determine masses or limits by stability arguments, see e.g. HR 8799 (Marois et al. (2008),Reidemeister et al. (2009)).
- If there are other sub-stellar objects with very similar values regarding temperature, luminosity, and age, for which there is also a direct mass estimate, e.g. directly obtained

in an eclipsing double-lined binary such as 2M0535 (Stassun et al., 2006) or in a visually resolved system with full orbit determination such as HD 130948 BC (Dupuy et al., 2009), one can conclude that the sub-stellar companion in question also has a similar mass. If a sub-stellar companion has temperature and luminosity smaller than another sub-stellar object with a direct mass estimate, but the same age, then the sub-stellar companion in question should have a smaller mass. For HD 130948 BC, there is only an estimate for the total mass being 114 ± 3 M_{Jup} (Dupuy et al., 2009), i.e. somewhat too large for comparison to planet candidates. The object 2M0535 is an eclipsing double-lined spectroscopic binary comprising of two brown dwarfs, member of the Orion star forming region, hence not older than a few Myr, maybe below 1 Myr. For a double-lined spectroscopic binary, one can determine brigtness, temperatures, luminosities, and lower mass limits $m \cdot \sin i$ for both objects individually. The orbital inclination i can then be obtained from the eclipse light curve. Hence, both masses are determined dynamically without model assumptions, the masses are 60 ± 5 M_{Jup} for A and 38 ± 3 M_{Jup} for B (Stassun et al., 2007). Given that several of the sub-stellar companions discussed here have a very similar age, we can compare them with 2M0535 A and B. If all parameters are similar, than the masses should also be similar. If a companion has lower values (at a similar age), i.e. being cooler and fainter, then it will be lower in mass. See Table 1 and 2 for the values and the comparison. Such a comparison should also be done with great care, because also other properties like magnetic field strength, spots on the surface, and chemical composition (metallicity) affect the analysis.

- If one could determine a full orbit of two objects around each other, one can then estimate the masses of both the host star and the companion using Kepler's 3rd law as generalized by Newton. However, since all planets and planet candidates imaged directly so far have large separations ($\geq$ 8.5 AU) from their host star (otherwise, they would not have been detected directly), the orbital periods are typically tens to hundreds of years, so that full orbits are not yet observed.

5. Comments on individual objects

Here, we list data, arguments, and problems related to the classification of the companions as planets. We include those sub-stellar companions, where the common proper motion with a *stellar* primary host star has been shown with at least 3 σ and where the possibly planetary nature, i.e. very low mass and/or cool temperature, has been shown by a spectrum - or at least a very red color with known (small) extinction, in order to exclude reddened background objects. We also include the brown dwarf 2M1207 with its fainter and lower-mass sub-stellar companion, even though the primary object is not a star, but we do not list other brown dwarfs with possibly planetary mass companions. We include only those systems in our list for new and homogeneous mass determination, where the age is considered to be possibly below $\sim$ 500 Myr, otherwise age and sub-stellar companion mass is probably too large; however, we do include those older systems, where the mass of the sub-stellar companion has already been published and estimated to be near or below $\sim$ 25 M_{Jup}, e.g. WD 0806-661 (Luhman et al., 2011) and Wolf 940 B (Burningham et al., 2009). We also exclude those systems, however, where the age is completely unconstrained, e.g. HD 3651 (Mugrauer et al., 2006), GJ 758 (Thalmann et al., 2009), and several others listed, e.g., in Faherty et al. (2010). This compilation is the 3rd version (after Neuhäuser (2008) and Schmidt et al. (2009)) and we do plan to renew and enlarge the catalog later; then, we will consider to also include possibly planetary mass companions imaged directly around sub-stellar primaries and around old stars or stars with unconstrained age.

We list the objects in the chronologic order of the publications of the common proper motion confirmations; if the significant confirmation were published later than the discovery, we list the object at the later date.

GG Tau Bb: The T Tauri star GG Tau in the Taurus T association (hence 0.1 to 2 Myr young at $\sim$ 140 pc distance) is a quadruple system with two close pairs GG Tau Aa+Ab (separation $\sim 0.25'' \simeq 35$ AU, the fainter component in this northern binary is sometimes called GG Tau a) and GG Tau Ba+Bb ($\sim 1.48'' \simeq 207$ AU), the separation between A and B is $10''$. The system has been studied in detail by White et al. (1999) using the HST as well as HIRES and LRIS at Keck. The object of interest here is GG Tau Bb, also called GG Tau/c. White et al. (1999) determine a spectral type of M7, zero extinction, Lithium absorption, and Hα emission (hence, young, not a reddened background object). According to Baraffe et al. (1998) models, at the age and distance of the star GG Tau Ba, the sub-stellar object GG Tab Bb has a mass of 40 to 60 M_{Jup}, or only 20 to 50 M_{Jup} according to D'Antona & Mazzitelli (1994; 1997). Also Woitas et al. (2001) give 20 to 40 M_{Jup} for GG Tau Bb using the D'Antona & Mazzitelli (1998) and Baraffe et al. (1998) models. We compared its properties with the Burrows et al. (1997) models, where the mass can be as low as ~ 23 M_{Jup}, so that we include the object in this study. Kinematic confirmation was done by White et al. (1999), who could determine the radial velocities of both GG Tau Ba and Bb to be 16.8 ± 0.7 km/s and 17.1 ± 1.0 km/s, i.e. consistent with each other and also with GG Tau A. The pair GG Tau A+B is also a common proper motion pair according to the NOMAD (Zacharias et al., 2005) and WDS catalogs (Mason et al., 2001). If accepted as planet, it would be the first planet imaged directly and confirmed by both common proper motion and spectroscopy. Orbital motion or curvature in orbital motion of GG Tau Bb around Ba were not yet reported.

TWA 5: The first direct imaging detection of the companion $1.960 \pm 0.006''$ (86.2 ± 4.0 AU) off TWA 5 (5 to 12 Myr as member of the TW Hya association, 44 ± 4 pc, M1.5) was done with NASA/IRTF and Keck/LRIS by Webb et al. (1999) and with HST/Nicmos and Keck/NIRC by Lowrance et al. (1999). The common proper motion and spectral (M8.5-9) confirmation was given in Neuhäuser, Guenther, Petr, Brandner, Huélamo & Alves (2000), who derived a mass of 15-40 M_{Jup} from formation models at the age and distance of the star. The mass lies anywhere between 4 and 145 M_{Jup}, if calculated from temperature (2800 ± 100 K), luminosity ($\log(L_{bol}/L_{\odot}) = -2.62 \pm 0.30$ at 44 ± 4 pc), and gravity ($\log g = 4.0 \pm 0.5$ cgs), as obtained by comparison of a VLT/Sinfoni K-band spectrum with Drift-Phoenix model atmospheres (Neuhäuser et al., 2009). The temperature error given in Neuhäuser et al. (2009) may be underestimated, because it is only from the K-band, so that we use a conservative, larger error here in Table 2. In our Table 2 below, we give 17 to 50 M_{Jup} as possible mass range. For any of those mass ranges, it might well be below 25 M_{Jup}, hence it is also a planet candidates imaged directly (called here TWA 5 B, but not b, in order not to confuse with TWA 5 Ab). Our latest image of TWA 5 A+B is shown in Fig. 1. Orbital motion of B around A was shown in Neuhäuser et al. (2010); the host star A is actually a very close 55 mas binary star. The data for separation and position angle plotted in Fig. 2 and 3 here are corrected for the binarity of the host star, i.e. are the values between the companion B and the photocenter of Aa+Ab using the orbit of Aa+b from Konopacky et al. (2007). What was plotted as orbital motion in figures 1 and 2 in Neuhäuser et al. (2010), actually is a small difference in proper motions of TWA

Object name	Luminosity $\log(L_{bol}/L_{\odot})$	Magnitude M_K [mag]	Temp. T_{eff} [K]	Age [Myr]	Comments
Reference object (eSB2 brown dwarf - brown dwarf binary 2M0535):					
2M0535 A	-1.65 ± 0.07	5.29 ± 0.16	2715 ± 100	1 (0-3)	Stassun et al. (2007)
B	-1.83 ± 0.07	5.29 ± 0.16	2820 ± 105	1 (0-3)	Stassun et al. (2007)
Directly detected planet candidates:					
GG Tau Bb	-1.84 ± 0.32	6.28 ± 0.79	2880 ± 150	0.1-2	(1), (2)
TWA 5 B	-2.62 ± 0.30	8.18 ± 0.28	2800 ± 450	5-12	
GJ 417 B & C	-4.14 ± 0.06	11.74 ± 0.05	1600 ± 300	80-300	each object (1), (3)
GSC 08047 B/b	-3.58 ± 0.28	10.75 ± 0.60	2225 ± 325	25-40	
DH Tau B/b	-2.81 ± 0.32	8.46 ± 0.78	2750 ± 50	0.1-10	
GQ Lup b	-2.25 ± 0.24	7.37 ± 0.78	2650 ± 100	0.1-2	
2M1207 b	-4.74 ± 0.06	13.33 ± 0.12	1590 ± 280	5-12	
AB Pic B/b	-3.73 ± 0.09	10.82 ± 0.11	2000^{+100}_{-300}	25-40	(1)
LP 261-75 B/b	-3.87 ± 0.54	11.18 ± 1.34	1500 ± 150	100-200	(1)
HD 203030 B/b	-4.64 ± 0.07	13.14 ± 0.12	1440 ± 350	130-400	T_{eff} error (3)
HN Peg B/b	-4.93 ± 0.16	14.37 ± 0.25	1450 ± 300	200-300	(1), (3)
CT Cha b	-2.68 ± 0.21	8.86 ± 0.50	2600 ± 250	0.1-4	(7)
Fomalhaut b	≤ -6.5	$M_H \geq 23.5$		100-300	no colors/spectra
HR 8799 b	-5.1 ± 0.1	14.05 ± 0.08	1300 ± 400	20-1100	
c	-4.7 ± 0.1	13.13 ± 0.08	~ 1100	20-1100	
d	-4.7 ± 0.1	13.11 ± 0.12		20-1100	
e	-4.7 ± 0.2	12.93 ± 0.22		20-1100	
Wolf 940 B/b	-6.07 ± 0.04	18.36 ± 0.16	600 ± 100	3.5-6 Gyr	
G 196-3 B/b	$-3.8^{+0.2}_{-0.3}$	11.17 ± 0.62	1870 ± 100	20-600	
β Pic b	$-3.903^{+0.074}_{-0.402}$	11.20 ± 0.11	1700 ± 300	8-20	no spectra, (1), (4)
RXJ1609 B/b	-3.55 ± 0.20	10.36 ± 0.35	1800^{+200}_{-100}	5 (1-10)	
PZ Tel B/b	-2.58 ± 0.08	8.14 ± 0.15	2600 ± 100	8-20	(5), (1)
Ross 458 C	-5.62 ± 0.03	16.11 ± 0.05	650 ± 25	150-800	
GSC 06214 B/b	-3.09 ± 0.12	9.17 ± 0.23	2050 ± 450	5 (1-10)	(1), (3)
CD-35 2722 B/b	-3.59 ± 0.09	10.37 ± 0.16	1800 ± 100	50-150	(1)
HIP 78530 B/b	-2.55 ± 0.13	8.19 ± 0.18	2800 ± 200	5 (1-10)	
WD 0806-661 B/b		$M_J \geq 21.7$	300	1.5-2.7 Gyr	(6)
SR 12 C	-2.87 ± 0.20	9.09 ± 0.44	2400^{+155}_{-100}	0.3-10	
HR 7329 B/b	-2.63 ± 0.09	8.21 ± 0.12	2650 ± 150	8-20	

Table 1. Observed properties of the directly imaged planet candidates. References given in Sect. 5 (1) luminosity from spectral type (BC from Golimowski et al. (2004)), K magnitude and distance (2) temperature from spectral type using Luhman et al. (2003) (3) temperature from spectral type using Golimowski et al. (2004) (4) temperature from Ks - L′ color (5) temperature from JHK colors (6) detected at 4.5 μm (7) Only for CT Cha b extinction was taken into account as given in Schmidt et al. (2008)

5 A and B, so small, that it is consistent with the expected orbital motion; figures 1 and 2 in Neuhäuser et al. (2010) only show a linear fit. However, true orbital motion cannot be linear, but always shows some curvature. In figure 3 in Neuhäuser et al. (2010), it is shown from geometric fits that an eccentric orbit of B around A is most likely, hence curvature is detected (with low significance). Curvature is not yet a final proof for being bound, because it would also be expected for an hyperblic orbit. The orbital period is $\sim$ 950 yr (for a circular orbit) or $\sim$ 1200 yr for an eccentric orbit with the best fit eccentricity $e = 0.45$ (Neuhäuser et al., 2010).

GJ 417: Common proper motion between the binary star GJ 417 (or Gl 417, CCDM J11126+3549AB, WDS J11125+3549AB, called primary A in Bouy et al. (2003)) and the companion 2MASS J1112256+354813 was noticed by Bouy et al. (2003). The primary, GJ 417, is comprised by two stars both with spectral type G0 (Simbad). This primary GJ 417 A (or GJ 417 Aa+b) has common proper motion with 2MASS J1112256+354813 at a wide separation of $\sim 90''$ (or $\sim$ 1953 AU at $\sim$ 21.7 pc, Bouy et al. (2003)). The secondary, 2MASS J1112256+354813, is a close binary itself (called B and C in Bouy et al. (2003)) with $\sim 0.0700''$ separation with almost equal magnitudes and a combined spectral type of L4.5 (Bouy et al., 2003). At the distance of GJ 417, this separation corresponds to only $\sim$ 1.5 AU projected separation and a very short period of few years. The age of the system is given to be only 80 to 300 Myr (Faherty et al., 2010), so that the mass of B and C can be near the planetary mass regime (Faherty et al., 2010).

GSC 08047 (GSC 08047-00232): The first direct imaging detections of this companion $3.238 \pm 0.022''$ (219 ± 59 AU) off GSC 08047 (K2, 50-85 pc, 25 to 40 Myr as member of the TucHor Association) was shown in Neuhäuser et al. (2003) using simple IR imaging with NTT/SofI and IR speckle imaging with NTT/Sharp as well as in Chauvin et al. (2003) with AO imaging using NTT/Adonis. Neuhäuser & Guenther (2004) could show common proper motion, while both Neuhäuser & Guenther (2004) and Chauvin, Lagrange, Lacombe, Dumas, Mouillet, Zuckerman, Gendron, Song, Beuzit, Lowrance & Fusco (2005) presented spectra (M6-9.5). Based on formation models, Neuhäuser & Guenther (2004) derived the mass of the companion to be 7-50 M_{Jup} at the age and distance of the star.

DH Tau: Direct imaging AO detection with Subaru/CIAO of the companion $2.351 \pm 0.001''$ ($\sim$ 329 AU) off DH Tau (0.1 to 10 Myr and at $\sim$ 140 pc as member of the Taurus T association, M0.5) were given in Itoh et al. (2005), who also could show common proper motion and a high-resolution spectrum with Subaru/CISCO giving temperature and gravity yielding a mass of 30-50 M_{Jup}. A small difference observed in the position angle is consistent with orbital motion (Itoh et al., 2005).

GQ Lup: Direct imaging AO detection with VLT/NACO of the companion $0.7347 \pm 0.0031''$ ($\sim$ 100 AU) off GQ Lup (0.1 to 2 Myr and at $\sim$ 140 pc as member of the Lupus-I T association, K7) together with common proper motion, spectral classification (late-M to early-L), and a mass estimate of 1-42 M_{Jup} (Neuhäuser et al., 2005) were confirmed by Janson et al. (2006) giving a mass of 3-20 M_{Jup}, Marois et al. (2007) deriving 10-20 M_{Jup}, and McElwain et al. (2007) listing 10-40 M_{Jup}, all from photometry and temperature of the companion, age, and distance of the star, together with formation models. Seifahrt et al. (2007) obtained higher-resolution VLT/Sinfoni spectra to derive the gravity of the companion and determined the mass model-independant to be 4-155 M_{Jup}. Evidence for a few mas/yr orbital motion (Neuhäuser et al., 2008) does not yet show curvature. Hence, the companion to GQ Lup can be a massive planet or a low-mass brown dwarf.

2M1207 (2MASSWJ1207334-393254): While the first direct imaging AO detection with VLT/NACO of this companion $0.7737 \pm 0.0022''$ (40.5 ± 1 AU) off the brown dwarf 2M1207 A (5 to 12 Myr as member of the TW Hya Association, 52.4 ± 1.1 pc, M8 brown dwarf) was published in Chauvin et al. (2004), the proper motion confirmation was given in Chauvin, Lagrange, Dumas, Zuckerman, Mouillet, Song, Beuzit & Lowrance (2005). Mugrauer et al. (2005) and Close et al. (2007) noticed that the binding energy between 2M1207 and its companion may not be sufficient for being bound or for staying bound for a long time: The total mass is too low for the large separation. Orbital motion or curvature in the orbital motion were not yet shown. The companion also appears to be too faint given its L5-L9.5 spectral type and 5 to 12 Myr age (Mohanty et al., 2007).

AB Pic: The first direct imaging AO detection with VLT/NACO of this companion $5.460 \pm 0.014''$ (251.7 ± 8.9 AU) off AB Pic (46.1 ± 1.5 pc, K1, 25 to 40 Myr as member of the TucHor Association) together with spectrum (L0-2), proper motion confirmation, and a mass estimate based on formation models to be 13-14 M_{Jup} were published in Chauvin, Lagrange, Zuckerman, Dumas, Mouillet, Song, Beuzit, Lowrance & Bessell (2005). Bonnefoy et al. (2010) obtained temperature and gravity with VLT/Sinfoni spectra and estimated the mass to be 1 to 45 M_{Jup}.

LP 261-75: Reid & Walkowicz (2006) noticed this common proper motion pair: The companion 2MASSW J09510549+3558021 (or LP 261-75 B) has a separation of $\sim 12''$ (~ 744 AU) off LP 261-75 (62 ± 38 pc, M4.5, 100 to 200 Myr due to its activity). Reid & Walkowicz (2006) presented a spectrum of the companion (L6), showed proper motion confirmation, and estimated the mass based on formation models to be 15-30 M_{Jup}.

HD 203030: The first direct imaging AO detection with Palomar of this companion $11.923 \pm 0.021''$ (503 ± 15 AU) off HD 203030 (42.2 ± 1.2 pc, G8, 130 to 400 Myr due to its activity) together with spectrum (L7-8), proper motion confirmation, and a mass estimate based on formation models to be 12-31 M_{Jup} were published in Metchev & Hillenbrand (2006).

HN Peg: First direct imaging detection with 2MASS and Spitzer/IRAC of this companion $43.2 \pm 0.4''$ (773 ± 13 AU) off HN Peg (17.89 ± 0.14 pc, G0, 200 to 300 Myr, if a member of the Her-Lyr group) together with common proper motion confirmation were shown in Luhman et al. (2007). Given the NASA/IRTF/SpecX spectral detection of methane, the companion can be classified T2.5 ± 0.5, so that it has 12-30 M_{Jup} at the age and distance of the star. The large projected separation of ~ 773 AU may favor the brown dwarf interpretation.

CT Cha: First direct imaging AO detection with VLT/NACO of this companion $2.670 \pm 0.036''$ (441 ± 87 AU) off CT Cha (165 ± 30 pc and 0.1 to 4 Myr as member of the Cha I T association, K7) together with VLT/Sinfoni JHK spectra (M8-L0), common proper motion confirmation, and a mass estimate based on formation models to be 11-23 M_{Jup} at the age and distance of the star were published in Schmidt et al. (2008), so that this companion can also be a high-mass planet or low-mass brown dwarf.

Fomalhaut b: Direct imaging detection in the red optical with the Hubble Space Telescope (HST) of this companion $12.7''$ (~ 100 AU) off Fomalhaut (7.704 ± 0.028 pc, A4, 100 to 300 Myr old) together with common proper motion confirmation were published in Kalas et al. (2008). They also estimated the mass of the companion to be below ~ 3 M_{Jup} due to its location close to the dusty debris disk seen in reflected optical light. Kalas et al. (2008) also obtained a few imaging photometric points and upper limits; those data points were not consistent with the expected spectrum of a low-mass cool planet, so that they could not exclude that the emission is reflected light from the small cloud-let. A spectrum or an IR detection of the companion could not yet be obtained. After two imaging epochs, the slightly different positions of the companion with respect to the star are consistent with orbital motion (Kalas et al., 2008), but curvature in the orbital motion was not yet shown.

HR 8799: Direct imaging AO detection with Keck/NIRC2 and Gemini North/NIRI of the companions b, c, and d was shown by Marois et al. (2008) together with common proper motion confirmation for all three companions, while orbital motion could be shown for companions b and c. The orbital motion confirmation of the third candidate, d, was later given with higher significance by Metchev et al. (2009). A fourth candidate, HR 8799 e, was later detected by Marois et al. (2010), whose orbital (and common proper motion, however no significance is explicitely given for this) with the star was shown within this publication. The companions b, c, d, and e have separations of $1.713 \pm 0.006''$, $0.952 \pm 0.011''$, $0.613 \pm 0.026''$, and $0.362 \pm 0.033''$, respectively, which correspond to 14 to 67 AU at the distance of the star being 39.4 ± 1.0 pc, similar as the solar system dimension. Spectra were taken for HR 8799 b by Bowler et al. (2010) and Barman et al. (2011) and for HR 8799 c by Janson et al. (2010), showing tempartures of 1300 – 1700 K and 1100 ± 100 K for HR 8799 b and $\sim$ 1100 K for HR 8799 c, respectively. The age of the star is somewhat uncertain: Given its bright debris disk, it might be as young as 20 Myr, then the companions are certainly below 13 M_{Jup}; astroseismology, however, seem to indicate that the star can be as old as $\sim$ 1.1 Gyr, then the companions would be brown dwarfs.

Wolf 940: The first imaging detection in UKIDSS, later also detected with Keck AO and laser guide star, of this companion $32''$ ($\sim$ 400 AU) off Wolf 490 (12.53 ± 0.71 pc, 3.5-6 Gyr due to activity, M4) was presented by Burningham et al. (2009) together with common proper motion and spectra (T8.5). Comparison of their spectra with BT Settl models yields 500 to 700 K, while the temperature and gravity estimates also given in Burningham et al. (2009), 570 ± 25 K, are obtained using the radius from cooling models for the age range from the stellar activity, hence possibly less reliable. A mass was also obtained from models for the given age range, namely 20-32 M_{Jup} (Burningham et al., 2009).

G 196-3: The first direct imaging detection of the companion $\sim 16.2''$ (243-437 AU) off G186-3 (host star: M2.5V, 20 to 600 Myr, 15-27 pc) was published by Rebolo et al. (1998) using the 1m NOT/ALFOSC and HiRAC instruments in the red optical and IR without AO, together with common proper motion (2 σ only) and spectroscopic (L3) confirmation. The mass was estimated to be 15-40 M_{Jup} or 12-25 M_{Jup} from cooling models for 20 to 600 Myr (Rebolo et al., 1998) or 20 to 300 Myr (Zapatero Osorio et al., 2010). if the 2 σ common proper motion is accepted as confirmation, then is would actually be the first imaged planet (candidate) confirmed as co-moving companion by proper motion (Rebolo et al., 1998); higher significance for common proper motion was presented by Zapatero Osorio et al. (2010).

β Pic: The first direct imaging AO detection with VLT/NACO of this companion $0.441 \pm 0.008''$ (8.57 ± 0.18 AU) off β Pic (19.440 ± 0.045 pc, 8 to 20 Myr, A6) was presented in Lagrange et al. (2009), while the 2nd epoch image with common proper motion confirmation was presented in Lagrange et al. (2010). This planet detected at $\sim$ 10 AU projected separation was predicted before by Freistetter et al. (2007) at $\sim$ 12 AU semi-major axis with $\sim$ 2-5 M_{Jup} to account for the main warp, the two inner belts, and the falling evaporaing bodies in the β Pic debris disk. Lagrange et al. (2009) estimated the mass of the detected object to be $\sim$ 6-13 M_{Jup}

based on uncertain formation models at the age and distance of the star. A spectrum of the companion could not yet be obtained. After two imaging epochs, the different positions of the companion with respect to the star are consistent with orbital motion (Lagrange et al., 2010), but the object can still be a moving background object.

RXJ 1609 (1RXS J160929.1-210524): The first direct imaging AO detection with Gemini of this companion $2.219 \pm 0.006''$ ($\sim$ 311 AU) off RXJ1609 ($\sim$ 145 pc and 1 to 10 Myr as member of the Sco-Cen T association, K7-M0) was presented by Lafrenière et al. (2008), while the common proper motion confirmation was published by Lafrenière et al. (2010) and also confirmed by Ireland et al. (2011). From the JHK Gemini/NIRI spectra (with the AO system Altair), Lafrenière et al. (2008; 2010) obtained a spectral type (L2-5) and temperature, and with age and distance of the star, a mass estimate of 6-11 M_{Jup} from uncertain formation models.

PZ Tel: The first direct imaging AO detections of this companion $0.3563 \pm 0.0011''$ (18.35 ± 0.99 AU) off the star PZ Tel (51.5 ± 2.6 pc, 8 to 20 Myr as member of the β Pic moving group, G9) were obtained with Gemini/NICI by Biller et al. (2010) and with VLT/NACO by Mugrauer et al. (2010). They both could also confirm common proper motion. They estimate the mass of PZ Tel B to be 30-42 or 24-40 M_{Jup}, respectively, again possibly below 25 M_{Jup}, so that this companion could be classified as planetary companion imaged directly. Mugrauer et al. (2010) not only show common proper motion between A and B with $\geq 39\,\sigma$, but also orbital motion of B around A with $\geq 37\,\sigma$ including curvature of orbital motion at 2 σ significance.

Ross 458 C: The imaging detection of the companion C (possibly also to be called *b* as planet candidate) to the host star binary Ross 458 A+B (12 pc, M0+M7 pair, 150 to 800 Myr) was published in Goldman et al. (2010) and Scholz (2010) using the 3.5m UKIRT Infrared Deep Sky Survey together with common proper motion confirmation, with the separation of Ross C being $102''$ or 1100 AU. The spectroscopic (T8) confirmation was presented in Burgasser et al. (2010) using the 6.5m Magellan/FIRE. The mass of C is estimated to be $\sim$ 14 M_{Jup} from magnitude, temperature, and gravity of the companion at the distance of the host star (Burgasser et al., 2010).

GSC 06214 (GSC 06214-00210): The first direct imaging detections with the 200-inch Palomar/PHARO and the 10-m Keck/NIRC2 detectors, using a combination of conventional AO imaging and non-redundant mask interferometry (sparse aperture mask with AO) of this companion $2.2033 \pm 0.0015''$ (i.e. 319 ± 31 AU) off the star GSC 06214 (1 to 10 Myr and 145 ± 14 pc as member of ScoCen T association, M1) together with common proper motion confirmation was published by Ireland et al. (2011), who estimate the mass of the companion from its colors (neglecting extinction, M8-L4), the distance and age of the star, and cooling models to be $\sim$ 10-15 M_{Jup}.

CD-35 2722: The first direct imaging AO detections with Gemini-S/NICI of this companion $3.172 \pm 0.005''$ (i.e. 67.56 ± 4.6 AU) off the star CD-35 2722 (50 to 150 Myr as member of the AB Dor moving group, 21.3 ± 1.4 pc, M1) together with common proper motion confirmation and spectral classification as L4 $\pm$ 1 were published by Wahhaj et al. (2011), who estimate the mass of the companion from its luminosity, the age of the star, and cooling models to be 31 ± 8 M_{Jup}, or lower when using the temperature of the companion, so that it could be a planet imaged directly. As shown by Wahhaj et al. (2011), the fact that the position angle of this companion compared to the host star does not change signficantly is inconsistent with a non-moving background object by 3 σ; the fact that at the same time the separation between companion and host star does change signficantly, namely exactly according to what would be expected for a non-moving background object, can either be interpreted as concern (really co-moving ?) or as evidence for orbital motion: No change in position angle, but small change in separation would indicate that the orbit is either seen edge-on and/or strongly eccentric; it is of course also possible that the faint object is a moving background object or another, but independent, young member of the AB Dor group.

HIP 78530: The first direct imaging AO detection with Gemini/NIRI of this companion $4.529 \pm 0.006''$ (i.e. 710 ± 60 AU) off HIP 78530 (1 to 10 Myr as member of the ScoCen OB association, B9, 157 ± 13 pc) together with common proper motion confirmation and spectroscopy (M8 $\pm$ 1) were published by Lafrenière et al. (2011), who determined the mass from formation models to be 19-26 M_{Jup}. HIP 78530 is so far the most massive host star with directly imaged planet (candidate).

WD 0806-661: The first direct (normal IR) imaging detections with the Spitzer Infrared Array Camera at 4.5 μm of this companion $103.2 \pm 0.2''$ (i.e. $\sim$ 2500 AU) off the White Dwarf WD 0806-661 ($\sim$ 1.5 Gyr, 19.2 ± 0.6 pc) together with common proper motion confirmation was published by Luhman et al. (2011). From the companion brightness at 4.5 μm and the rough age of the host star (WD age plus progenitor life time), Luhman et al. (2011) estimate the mass of the companion to be $\sim$ 7 M_{Jup}. Rodriguez et al. (2011) argue that the host star age (WD plus progenitor) can be as large as $\sim$ 2.7 Gyr, so that the companion mass can be as large as $\sim$ 13 M_{Jup}, still in the planetary mass range.

SR 12: Direct imaging detection with Subaru/CIAO of the companion SR 12 C (possibly to be called SR 12 b as planetary companion) $8.661 \pm 0.033''$ (1300 ± 220 AU) off the close binary SR 12 A+B (K4+M2.5, 125 ± 25 pc and 0.3-10 Myr as member of the ρ Oph star forming cloud) together with significant common proper motion confirmation (their figure 4 with five epochs) and a spectrum (M8.5-9.5) were published in Kuzuhara et al. (2011); they derived a mass of SR 12 C to be 6-20 M_{Jup}. Given the large separation inside the ρ Oph cloud, the object C could also be an independant member of the cloud. Orbital motion or curvature in orbital motion were not yet detected.

HR 7329: Direct imaging detection with HST/Nicmos of the companion (one epoch) $4.194 \pm 0.016''$ (200 ± 16 AU) off HR 7329 (η Tel, A0, 47.7 ± 1.5 pc, 8 to 20 Myr as member of the β Pic moving group) and a spectrum (M7-8) were published in Lowrance et al. (2000). Common proper motion between HR 7329 A and B/b was shown convincingly (above 3 σ) only recently with new AO imaging (Neuhäuser et al., 2011).

We notice that the objects GG Tau Bb, TWA 5 B, GJ 417 B & C, GSC 08047 B/b, LP 261-75 B/b, HD 203030 B/b, Wolf 940 B/b, G196-3 B/b, PZ Tel B/b, HR 7329 B/b, were not yet listed in Schneider et al. (2011) nor www.exoplanet.eu. We also note that the object CHXR 73 b (Luhman et al., 2006) listed in Schneider et al. (2011) and www.exoplanet.eu as planet imaged directly, is not included in our listing, because common proper motion has not been shown, yet. A few sub-stellar, possibly planetary mass companions to brown dwarfs, listed as possible planets in Schneider et al. (2011) and www.exoplanet.eu, are also not listed in this paper, because they are probably not bound, namely Oph J1622-2405 (also called Oph 1622 or Oph 11, Jayawardhana & Ivanov (2006), UScoCTIO-108 (Béjar et al., 2008), 2MASS J04414489+2301513 (Todorov et al., 2010), and CFBDIRJ1458+1013 AB (also called CFBDS 1458,

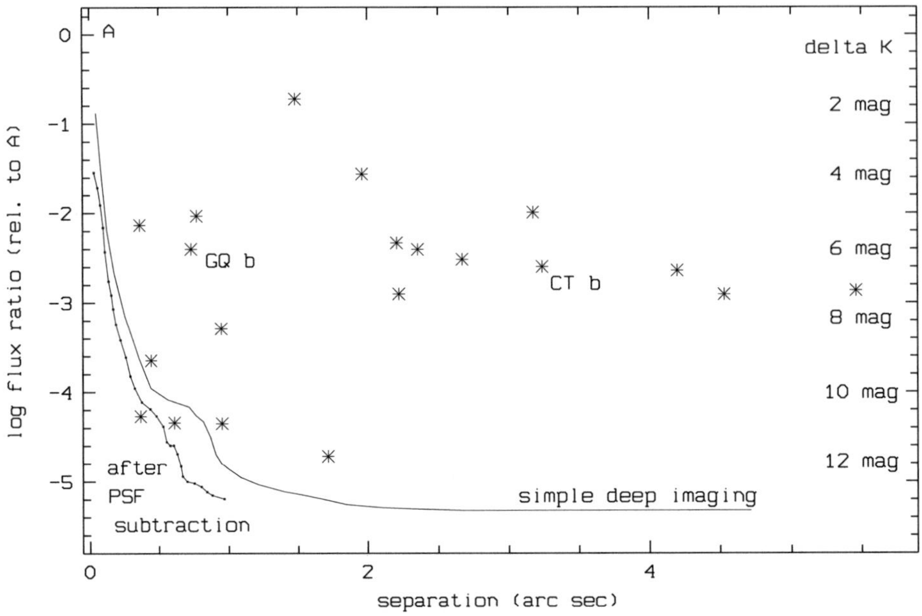

Fig. 4. We plot the (log of the) flux ratio between the noise level (S/N=3) and the primary host star (left y axis) or the K-band magnitude difference (right y axis) versus the separation between companion and host star (in arc sec). The primary host star is indicated in the upper left by the letter A (log of flux ratio and separation being zero). The flux ratios of all 20 companions listed in Tables 1 and 2 with separations up to 6″ and known K-band flux ratio (known for all but Fomalhaut) are plotted as star symbols (references for K-band magnitudes for the stars and their companions and for the separations between them can be found in Sect. 5). The companions discovered by us (GQ Lup b and CT Cha b) are indicated. The curve is the dynamic range achieved in our deep AO imaging on GQ Lup with 102 min total integration time with VLT/NACO; the lower curve with dots is the dynamic range for the same data achieved after PSF subtraction (figure adapted from figure 7 in Neuhäuser et al. (2008)). All companions above the curve(s) can be detected by this (simple AO imaging) method. Companions that are fainter and/or closer, i.e. below the curve, cannot be detected. The PSF subtraction technique can improve the dynamic range and detection capability at 0.5 to 1″ by about one magnitude. The only two companions below the upper dynamic range curve (before PSF subtraction) are HR 8799 d and e, which were detected by the ADI technique (Marois et al., 2008; 2010). Error bars are omitted for clarity.

Liu et al. (2011)). 2M1207 should therefore also not be listed here; however, we do include it for completness and comparison, because it is often included in lists of planets imaged directly.

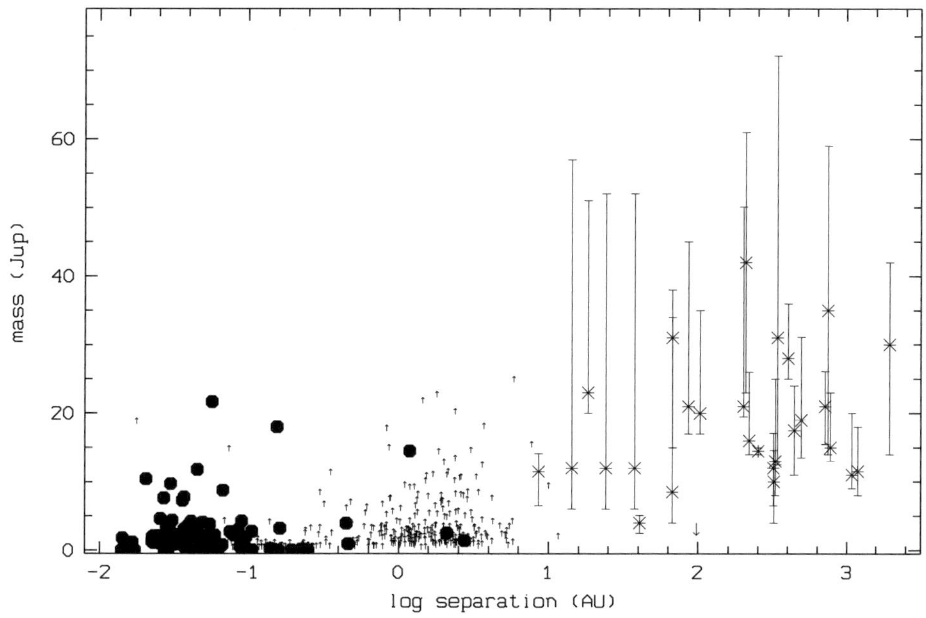

Fig. 5. We plot the mass of the companion (in Jupiter masses) versus the (log of the) separation (in AU) for (i) planet candidates detected by radial velocity only (lower mass limits plotted), (ii) radial velocity planets confirmed by either astrometry or transit (filled symbols), where the true masses are known, and (iii) planets and candidates detected by direct imaging (star symbols) with masses from the Burrows et al. (1997) model as in Tables 2, because only the Burrows et al. model gives a correct mass for the eSB2 brown dwarf 2M0535 B, and projected physical separations (calculated from angular separations and distances as given in Sect. 5); for Fomalhaut b, we plot few M_{Jup} as upper mass limit at $\sim$ 100 AU; only WD 0806-661 is not plotted, because we could not determine its mass in the same (homogeneous) way as for the others, since the WD 0806-661 companion is detected only at 4.5 μm, but not in the near-IR. The four radial velocity planets with the largest separations (8.9 to 11.6 AU) are ν Oph c, HIP 5158 c, HIP 70849 b and 47 UMa d. The directly detected planets (or candidates) with the smallest projected separations (8.6 to 18.3 AU) are β Pic b, HR 8799 e, and PZ Tel b; these systems are all younger than $\sim$ 100 Myr, so that the companions are still bright enough for direct detection. This plot shows that the parameter regimes, in which the radial velocity technique and the direct imaging technique are working, are about to overlap. This is due to longer monitoring periods for the radial velocity technique and due to improved technology and, hence, dynamic range in direct imaging. Hence, it might soon be possible to observe the same planets with both direct imaging and radial velocity, which would be best possible for young nearby systems. The data for radial velocity, transit, and astrometry planets were taken from www.exoplanet.eu (on 28 July 2011), where the references for those planets can also be found. We omitted most error bars for clarity, but we do show the mass errors for the directly imaged planets and candidates.

Object name	Burrows 97 (L, age)	Chabrier 00 (L, M_K, T, t)	Baraffe 03 (L, M_K, T, t)	Marley 07 ($\leq$ 10 Jup)	Baraffe 08 ($\geq$ 10 Myrs)	Wuchterl (Neuh05)
Reference object (eSB2 brown dwarf - brown dwarf binary 2M0335):						
2M0535 A	50 (45-60)	55 (30-60)	50 (45-80)			5-13
B	37 (33-46)	45 (40-50)	43 (40-65)			$\leq$ 13
Directly detected planet candidates:						
GG Tau Bb	42 (23-61)	52 ($\geq$ 35)	56 ($\geq$ 41)			
TWA 5 B	21 (17-45)	23 (20-50)	25 (20-50)			
GJ 417 B & C	30 (14-42)	26 (18-35)	25 (20-35)			
GSC 08047 B/b	16 (14-26)	19 (17-25)	18 (14-25)			
DH Tau B/b	13 (8-25)	20 (6-47)	20 (6-50)	10 ($\geq$ 7)		5
GQ Lup b	20 (17-35)	25 (20-35)	27 (24-37)			1-5
2M1207 b	4 (2.5-5)	5 (2.5-13)	5 (2.5-12)	4 (3-5)	4	
AB Pic B/b	14.5 (14-15)	16 (14-18)	15.5 (11-17)			
LP 261-75 B/b	35 (14-59)	26 (16-30)	28 (16-32)			
HD 203030 B/b	19 (13.5-31)	24 (13-28)	23 (11-26)			
HN Peg B/b	15 (13-23)	21 (14-31)	20 (13-27)	$\geq$ 10		
CT Cha b	17.5 (11-24)	14 (13-19)	16 (13-21)			2-5
Fomalhaut b	$\leq$ 4.25		$\leq$ 2	$\leq$ 3	$\leq$ 2	
HR 8799 b	8.5 (4-38)	13 (3-63)	12 (4-32)	7 ($\geq$ 3)	7 ($\geq$ 3)	
c	12 (6-52)	16 (5-42)	12 (6-42)	10 ($\geq$ 6)	$\geq$ 5	
d	12 (6-52)	16 (5-42)	12 (6-42)	10 ($\geq$ 6)	$\geq$ 5	
e	12 (6-57)	16 (5-42)	12 (6-42)	10 ($\geq$ 6)	$\geq$ 5	
Wolf 940 B/b	28 (25-36)		33 (24-43)			
G 196-3 B/b	31 (12.5-72)	44 (14-60)	43 (11-55)			
β Pic b	11.5 (6.5-14)	10 (6-17)	9.5 (8-11)	10 ($\geq$ 9)	9 ($\geq$ 8)	
RXJ1609 B/b	10 (4-14.5)	8 (4-14)	8 (4-13)	8 ($\geq$ 4)		
PZ Tel B/b	23 (20-51)	28 (21-41)	28 (24-41)			
Ross 458 C	11.5 (8-18)		13 (8-15)			
GSC 06214 B/b	12 (6.5-17)	15 (6-18)	14 (6-23)			
CD-35 2722 B/b	31 (15-34)	35 (16-43)	31 (16-41)			
HIP 78530 B/b	21 (15.5-26)	30 (13-84)	32 (11-93)			
WD 0806-661 B/b			$\leq$ 8.5			
SR 12 C	11 (9-20)	10 (5-25)	10 (6-25)	10 ($\geq$ 8)		2-5
HR 7329 B/b	21 (19.5-50)	26 (21-43)	27 (23-36)			

Table 2. Masses derived from evolutionary hot-start models.

6. Conclusion

We noticed that the Burrows et al. (1997) models give correct masses for 2M0535 B, a young brown dwarf in the eclipsing double-lined spectroscopic brown dwarf - brown dwarf binary system in Orion, where masses have been determined without model assumptions (Stassun et al., 2006; 2007). Hence, we apply this model for best mass estimates for the planets and candidates imaged directly, see Fig. 5.

We conclude that direct imaging detection of planets around other stars is possible since several years with both ground-based AO IR imaging and space-based optical imaging. For most planets and candidates imaged and confirmed as companions by common proper motion so far, the planet status is still dubious. Possibly planetary mass companions apparently co-moving with brown dwarfs, i.e. apparently forming very wide very low-mass binaries, may well be unbound, i.e. currently flying apart (Mugrauer et al. (2005),Close et al. (2007)).

Extra-solar planets or candidates as close to their host star as the Solar System planets (within 30 AU) are still very rare with β Pic b, HR 8799 e, PZ Tel B/b, and HR 8799 d being the only exceptions at 8.5, 14.3, 18.3, and 24.2 AU, respectively, all nearby young stars (19 to 52 pc). As far as angular separation is concerned, the closest planets or candidates imaged directly are PZ Tel B/b, HR 8799 e, β Pic b, HR 8799 d, and GQ Lup b with separations from 0.36 to 0.75 arc sec.

New AO imaging techniques like ADI, SAM, and locally optimized combination of images have improved the ability to detect such planets. Future AO instruments at 8-meter ground-based telescopes will improve the accessible dynamic range even further. Imaging with a new space based telescope like JWST (Beichman et al., 2010) or AO imaging at an extremely large telescope of 30 to 40 meters would improve the situation significantly. Imaging detection of planets with much lower masses, like e.g. Earth-mass planets, might be possible with a space-based interferometer like Darwin or TPF, but also only around very nearby stars.

Acknowledgements. We have used ADS, Simbad, VizieR, WDS, NOMAD, 2MASS, www.exoplanet.eu, and www.exoplanets.org. For the image of ϵ Eri shown in Fig. 1, we would like to thank Matthias Ammler - von Eiff, who took the observation at VLT (ESO program ID 073.C-0225(A), PI Ammler), ESO staff at Paranal and Garching for their help with the observations, and Ronny Errmann, who helped with the data reduction.

7. References

Absil, O. & Mawet, D. (2010). Formation and evolution of planetary systems: the impact of high-angular resolution optical techniques, *A&AR* 18: 317–382.

Baraffe, I., Chabrier, G., Allard, F. & Hauschildt, P. H. (1998). Evolutionary models for solar metallicity low-mass stars: mass-magnitude relationships and color-magnitude diagrams, *A&A* 337: 403–412.

Barman, T. S., Macintosh, B., Konopacky, Q. M. & Marois, C. (2011). Clouds and Chemistry in the Atmosphere of Extrasolar Planet HR8799b, *ApJ* 733: 65–+.

Beichman, C. A., Krist, J., Trauger, J. T., Greene, T., Oppenheimer, B., Sivaramakrishnan, A., Doyon, R., Boccaletti, A., Barman, T. S. & Rieke, M. (2010). Imaging Young Giant Planets From Ground and Space, *PASP* 122: 162–200.

Béjar, V. J. S., Zapatero Osorio, M. R., Pérez-Garrido, A., Álvarez, C., Martín, E. L., Rebolo, R., Villó-Pérez, I. & Díaz-Sánchez, A. (2008). Discovery of a Wide Companion near the Deuterium-burning Mass Limit in the Upper Scorpius Association, *ApJL* 673: L185–L189.

Benedict, G. F., McArthur, B. E., Forveille, T., Delfosse, X., Nelan, E., Butler, R. P., Spiesman, W., Marcy, G., Goldman, B., Perrier, C., Jefferys, W. H. & Mayor, M. (2002). A Mass for the Extrasolar Planet Gliese 876b Determined from Hubble Space Telescope Fine Guidance Sensor 3 Astrometry and High-Precision Radial Velocities, *ApJL* 581: L115–L118.

Benedict, G. F., McArthur, B. E., Gatewood, G., Nelan, E., Cochran, W. D., Hatzes, A., Endl, M., Wittenmyer, R., Baliunas, S. L., Walker, G. A. H., Yang, S., Kürster, M., Els, S. & Paulson, D. B. (2006). The Extrasolar Planet ϵ Eridani b: Orbit and Mass, *AJ* 132: 2206–2218.

Biller, B. A., Liu, M. C., Wahhaj, Z., Nielsen, E. L., Close, L. M., Dupuy, T. J., Hayward, T. L., Burrows, A., Chun, M., Ftaclas, C., Clarke, F., Hartung, M., Males, J., Reid, I. N., Shkolnik, E. L., Skemer, A., Tecza, M., Thatte, N., Alencar, S. H. P., Artymowicz, P., Boss, A., de Gouveia Dal Pino, E., Gregorio-Hetem, J., Ida, S., Kuchner, M. J., Lin, D. & Toomey, D. (2010). The Gemini NICI Planet-finding Campaign: Discovery of a Close Substellar Companion to the Young Debris Disk Star PZ Tel, *ApJL* 720: L82–L87.

Bonnefoy, M., Chauvin, G., Rojo, P., Allard, F., Lagrange, A.-M., Homeier, D., Dumas, C. & Beuzit, J.-L. (2010). Near-infrared integral-field spectra of the planet/brown dwarf companion AB Pictoris b, *A&A* 512: A52+.

Bouy, H., Brandner, W., Martín, E. L., Delfosse, X., Allard, F. & Basri, G. (2003). Multiplicity of Nearby Free-Floating Ultracool Dwarfs: A Hubble Space Telescope WFPC2 Search for Companions, *AJ* 126: 1526–1554.

Bowler, B. P., Liu, M. C., Dupuy, T. J. & Cushing, M. C. (2010). Near-infrared Spectroscopy of the Extrasolar Planet HR 8799 b, *ApJ* 723: 850–868.

Burgasser, A. J., Simcoe, R. A., Bochanski, J. J., Saumon, D., Mamajek, E. E., Cushing, M. C., Marley, M. S., McMurtry, C., Pipher, J. L. & Forrest, W. J. (2010). Clouds in the Coldest Brown Dwarfs: Fire Spectroscopy of Ross 458C, *ApJ* 725: 1405–1420.

Burningham, B., Pinfield, D. J., Leggett, S. K., Tinney, C. G., Liu, M. C., Homeier, D., West, A. A., Day-Jones, A., Huelamo, N., Dupuy, T. J., Zhang, Z., Murray, D. N., Lodieu, N., Barrado Y Navascués, D., Folkes, S., Galvez-Ortiz, M. C., Jones, H. R. A., Lucas, P. W., Calderon, M. M. & Tamura, M. (2009). The discovery of an M4+T8.5 binary system, *MNRAS* 395: 1237–1248.

Burrows, A., Marley, M., Hubbard, W. B., Lunine, J. I., Guillot, T., Saumon, D., Freedman, R., Sudarsky, D. & Sharp, C. (1997). A Nongray Theory of Extrasolar Giant Planets and Brown Dwarfs, *ApJ* 491: 856–+.

Charbonneau, D., Brown, T. M., Latham, D. W. & Mayor, M. (2000). Detection of Planetary Transits Across a Sun-like Star, *ApJL* 529: L45–L48.

Chauvin, G., Lagrange, A.-M., Dumas, C., Zuckerman, B., Mouillet, D., Song, I., Beuzit, J.-L. & Lowrance, P. (2004). A giant planet candidate near a young brown dwarf. Direct VLT/NACO observations using IR wavefront sensing, *A&A* 425: L29–L32.

Chauvin, G., Lagrange, A.-M., Dumas, C., Zuckerman, B., Mouillet, D., Song, I., Beuzit, J.-L. & Lowrance, P. (2005). Giant planet companion to 2MASSW J1207334-393254, *A&A* 438: L25–L28.

Chauvin, G., Lagrange, A.-M., Lacombe, F., Dumas, C., Mouillet, D., Zuckerman, B., Gendron, E., Song, I., Beuzit, J.-L., Lowrance, P. & Fusco, T. (2005). Astrometric and spectroscopic confirmation of a brown dwarf companion to GSC 08047-00232. VLT/NACO deep imaging and spectroscopic observations, *A&A* 430: 1027–1033.

Chauvin, G., Lagrange, A.-M., Zuckerman, B., Dumas, C., Mouillet, D., Song, I., Beuzit, J.-L., Lowrance, P. & Bessell, M. S. (2005). A companion to AB Pic at the planet/brown dwarf boundary, *A&A* 438: L29–L32.

Chauvin, G., Thomson, M., Dumas, C., Beuzit, J.-L., Lowrance, P., Fusco, T., Lagrange, A.-M., Zuckerman, B. & Mouillet, D. (2003). Adaptive optics imaging survey of the Tucana-Horologium association, *A&A* 404: 157–162.

Close, L. M., Zuckerman, B., Song, I., Barman, T., Marois, C., Rice, E. L., Siegler, N., Macintosh, B., Becklin, E. E., Campbell, R., Lyke, J. E., Conrad, A. & Le Mignant, D. (2007). The Wide Brown Dwarf Binary Oph 1622-2405 and Discovery of a Wide, Low-Mass Binary in Ophiuchus (Oph 1623-2402): A New Class of Young Evaporating Wide Binaries?, *ApJ* 660: 1492–1506.

D'Antona, F. & Mazzitelli, I. (1994). New pre-main-sequence tracks for M less than or equal to 2.5 solar mass as tests of opacities and convection model, *ApJS* 90: 467–500.

D'Antona, F. & Mazzitelli, I. (1997). Evolution of low mass stars, *MdSAI* 68: 807–+.

D'Antona, F. & Mazzitelli, I. (1998). A Role for Superadiabatic Convection in Low Mass Structures?, *in* R. Rebolo, E. L. Martin, & M. R. Zapatero Osorio (ed.), *Brown Dwarfs and Extrasolar Planets*, Vol. 134 of *Astronomical Society of the Pacific Conference Series*, pp. 442–+.

Duchêne, G. (2008). High-angular resolution imaging of disks and planets, *NewAR* 52: 117–144.

Dupuy, T. J., Liu, M. C. & Ireland, M. J. (2009). Dynamical Mass of the Substellar Benchmark Binary HD 130948BC, *ApJ* 692: 729–752.

Faherty, J. K., Burgasser, A. J., West, A. A., Bochanski, J. J., Cruz, K. L., Shara, M. M. & Walter, F. M. (2010). The Brown Dwarf Kinematics Project. II. Details on Nine Wide Common Proper Motion Very Low Mass Companions to Nearby Stars, *AJ* 139: 176–194.

Freistetter, F., Krivov, A. V. & Löhne, T. (2007). Planets of β Pictoris revisited, *A&A* 466: 389–393.

Goldman, B., Marsat, S., Henning, T., Clemens, C. & Greiner, J. (2010). A new benchmark T8-9 brown dwarf and a couple of new mid-T dwarfs from the UKIDSS DR5+ LAS, *MNRAS* 405: 1140–1152.

Grether, D. & Lineweaver, C. H. (2006). How Dry is the Brown Dwarf Desert? Quantifying the Relative Number of Planets, Brown Dwarfs, and Stellar Companions around Nearby Sun-like Stars, *ApJ* 640: 1051–1062.

Guenther, E. W., Neuhäuser, R., Huélamo, N., Brandner, W. & Alves, J. (2001). Infrared spectrum and proper motion of the brown dwarf companion of HR 7329 in Tucanae, *A&A* 365: 514–518.

Hatzes, A. P., Cochran, W. D., McArthur, B., Baliunas, S. L., Walker, G. A. H., Campbell, B., Irwin, A. W., Yang, S., Kürster, M., Endl, M., Els, S., Butler, R. P. & Marcy, G. W. (2000). Evidence for a Long-Period Planet Orbiting ϵ Eridani, *ApJL* 544: L145–L148.

Ireland, M. J., Kraus, A., Martinache, F., Law, N. & Hillenbrand, L. A. (2011). Two Wide Planetary-mass Companions to Solar-type Stars in Upper Scorpius, *ApJ* 726: 113–+.

Itoh, Y., Hayashi, M., Tamura, M., Tsuji, T., Oasa, Y., Fukagawa, M., Hayashi, S. S., Naoi, T., Ishii, M., Mayama, S., Morino, J.-i., Yamashita, T., Pyo, T.-S., Nishikawa, T., Usuda, T.,

Murakawa, K., Suto, H., Oya, S., Takato, N., Ando, H., Miyama, S. M., Kobayashi, N. & Kaifu, N. (2005). A Young Brown Dwarf Companion to DH Tauri, *ApJ* 620: 984–993.

Janson, M., Bergfors, C., Goto, M., Brandner, W. & Lafrenière, D. (2010). Spatially Resolved Spectroscopy of the Exoplanet HR 8799 c, *ApJL* 710: L35–L38.

Janson, M., Brandner, W., Henning, T., Lenzen, R., McArthur, B., Benedict, G. F., Reffert, S., Nielsen, E., Close, L., Biller, B., Kellner, S., Günther, E., Hatzes, A., Masciadri, E., Geissler, K. & Hartung, M. (2007). NACO-SDI Direct Imaging Search for the Exoplanet ϵ Eri b, *AJ* 133: 2442–2456.

Janson, M., Brandner, W., Henning, T. & Zinnecker, H. (2006). Early ComeOn+ adaptive optics observation of GQ Lupi and its substellar companion, *A&A* 453: 609–614.

Jayawardhana, R. & Ivanov, V. D. (2006). Discovery of a Young Planetary-Mass Binary, *Science* 313: 1279–1281.

Kalas, P., Graham, J. R., Chiang, E., Fitzgerald, M. P., Clampin, M., Kite, E. S., Stapelfeldt, K., Marois, C. & Krist, J. (2008). Optical Images of an Exosolar Planet 25 Light-Years from Earth, *Science* 322: 1345–.

Konopacky, Q. M., Ghez, A. M., Duchêne, G., McCabe, C. & Macintosh, B. A. (2007). Measuring the Mass of a Pre-Main-Sequence Binary Star through the Orbit of TWA 5A, *AJ* 133: 2008–2014.

Kuzuhara, M., Tamura, M., Ishii, M., Kudo, T., Nishiyama, S. & Kandori, R. (2011). The Widest-separation Substellar Companion Candidate to a Binary T Tauri Star, *AJ* 141: 119–+.

Lacour, S., Tuthill, P., Amico, P., Ireland, M., Ehrenreich, D., Huelamo, N. & Lagrange, A.-M. (2011). Sparse aperture masking at the VLT. I. Faint companion detection limits for the two debris disk stars HD 92945 and HD 141569, *A&A* 532: A72+.

Lafrenière, D., Doyon, R., Nadeau, D., Artigau, É., Marois, C. & Beaulieu, M. (2007). Improving the Speckle Noise Attenuation of Simultaneous Spectral Differential Imaging with a Focal Plane Holographic Diffuser, *ApJ* 661: 1208–1217.

Lafrenière, D., Jayawardhana, R., Janson, M., Helling, C., Witte, S. & Hauschildt, P. (2011). Discovery of an 23 MJup Brown Dwarf Orbiting 700 AU from the Massive Star HIP 78530 in Upper Scorpius, *ApJ* 730: 42–+.

Lafrenière, D., Jayawardhana, R. & van Kerkwijk, M. H. (2008). Direct Imaging and Spectroscopy of a Planetary-Mass Candidate Companion to a Young Solar Analog, *ApJL* 689: L153–L156.

Lafrenière, D., Jayawardhana, R. & van Kerkwijk, M. H. (2010). The Directly Imaged Planet Around the Young Solar Analog 1RXS J160929.1 - 210524: Confirmation of Common Proper Motion, Temperature, and Mass, *ApJ* 719: 497–504.

Lafrenière, D., Marois, C., Doyon, R., Nadeau, D. & Artigau, É. (2007). A New Algorithm for Point-Spread Function Subtraction in High-Contrast Imaging: A Demonstration with Angular Differential Imaging, *ApJ* 660: 770–780.

Lagrange, A.-M., Bonnefoy, M., Chauvin, G., Apai, D., Ehrenreich, D., Boccaletti, A., Gratadour, D., Rouan, D., Mouillet, D., Lacour, S. & Kasper, M. (2010). A Giant Planet Imaged in the Disk of the Young Star β Pictoris, *Science* 329: 57–.

Lagrange, A.-M., Gratadour, D., Chauvin, G., Fusco, T., Ehrenreich, D., Mouillet, D., Rousset, G., Rouan, D., Allard, F., Gendron, É., Charton, J., Mugnier, L., Rabou, P., Montri, J. & Lacombe, F. (2009). A probable giant planet imaged in the β Pictoris disk. VLT/NaCo deep L′-band imaging, *A&A* 493: L21–L25.

Latham, D. W., Stefanik, R. P., Mazeh, T., Mayor, M. & Burki, G. (1989). The unseen companion of HD114762 - A probable brown dwarf, *Nature* 339: 38–40.

Liu, M. C., Delorme, P., Dupuy, T. J., Bowler, B. P., Albert, L., Artigau, E., Reyle, C., Forveille, T. & Delfosse, X. (2011). CFBDSIR J1458+1013B: A Very Cold (>T10) Brown Dwarf in a Binary System, *ArXiv e-prints* .

Lowrance, P. J., McCarthy, C., Becklin, E. E., Zuckerman, B., Schneider, G., Webb, R. A., Hines, D. C., Kirkpatrick, J. D., Koerner, D. W., Low, F., Meier, R., Rieke, M., Smith, B. A., Terrile, R. J. & Thompson, R. I. (1999). A Candidate Substellar Companion to CD -33 deg7795 (TWA 5), *ApJL* 512: L69–L72.

Lowrance, P. J., Schneider, G., Kirkpatrick, J. D., Becklin, E. E., Weinberger, A. J., Zuckerman, B., Plait, P., Malmuth, E. M., Heap, S. R., Schultz, A., Smith, B. A., Terrile, R. J. & Hines, D. C. (2000). A Candidate Substellar Companion to HR 7329, *ApJ* 541: 390–395.

Luhman, K. L., Burgasser, A. J. & Bochanski, J. J. (2011). Discovery of a Candidate for the Coolest Known Brown Dwarf, *ApJL* 730: L9+.

Luhman, K. L., Patten, B. M., Marengo, M., Schuster, M. T., Hora, J. L., Ellis, R. G., Stauffer, J. R., Sonnett, S. M., Winston, E., Gutermuth, R. A., Megeath, S. T., Backman, D. E., Henry, T. J., Werner, M. W. & Fazio, G. G. (2007). Discovery of Two T Dwarf Companions with the Spitzer Space Telescope, *ApJ* 654: 570–579.

Luhman, K. L., Wilson, J. C., Brandner, W., Skrutskie, M. F., Nelson, M. J., Smith, J. D., Peterson, D. E., Cushing, M. C. & Young, E. (2006). Discovery of a Young Substellar Companion in Chamaeleon, *ApJ* 649: 894–899.

Marois, C., Lafrenière, D., Doyon, R., Macintosh, B. & Nadeau, D. (2006). Angular Differential Imaging: A Powerful High-Contrast Imaging Technique, *ApJ* 641: 556–564.

Marois, C., Macintosh, B. & Barman, T. (2007). GQ Lup B Visible and Near-Infrared Photometric Analysis, *ApJL* 654: L151–L154.

Marois, C., Macintosh, B., Barman, T., Zuckerman, B., Song, I., Patience, J., Lafrenière, D. & Doyon, R. (2008). Direct Imaging of Multiple Planets Orbiting the Star HR 8799, *Science* 322: 1348–.

Marois, C., Zuckerman, B., Konopacky, Q. M., Macintosh, B. & Barman, T. (2010). Images of a fourth planet orbiting HR 8799, *Nature* 468: 1080–1083.

Mason, B. D., Wycoff, G. L., Hartkopf, W. I., Douglass, G. G. & Worley, C. E. (2001). The 2001 US Naval Observatory Double Star CD-ROM. I. The Washington Double Star Catalog, *AJ* 122: 3466–3471.

Mayor, M. & Queloz, D. (1995). A Jupiter-mass companion to a solar-type star, *Nature* 378: 355–359.

McElwain, M. W., Metchev, S. A., Larkin, J. E., Barczys, M., Iserlohe, C., Krabbe, A., Quirrenbach, A., Weiss, J. & Wright, S. A. (2007). First High-Contrast Science with an Integral Field Spectrograph: The Substellar Companion to GQ Lupi, *ApJ* 656: 505–514.

Metchev, S. A. & Hillenbrand, L. A. (2006). HD 203030B: An Unusually Cool Young Substellar Companion near the L/T Transition, *ApJ* 651: 1166–1176.

Metchev, S. A. & Hillenbrand, L. A. (2009). The Palomar/Keck Adaptive Optics Survey of Young Solar Analogs: Evidence for a Universal Companion Mass Function, *ApJS* 181: 62–109.

Metchev, S., Marois, C. & Zuckerman, B. (2009). Pre-Discovery 2007 Image of the HR 8799 Planetary System, *ApJL* 705: L204–L207.

Mohanty, S., Jayawardhana, R., Huélamo, N. & Mamajek, E. (2007). The Planetary Mass Companion 2MASS 1207-3932B: Temperature, Mass, and Evidence for an Edge-on Disk, *ApJ* 657: 1064–1091.

Mugrauer, M., Neuhaeuser, R., Guenther, E. & Mazeh, T. (2005). The multiplicity of exoplanet host stars., *Astronomische Nachrichten* 326: 629–630.

Mugrauer, M., Seifahrt, A., Neuhäuser, R. & Mazeh, T. (2006). HD3651B: the first directly imaged brown dwarf companion of an exoplanet host star, *MNRAS* 373: L31–L35.

Mugrauer, M., Vogt, N., Neuhäuser, R. & Schmidt, T. O. B. (2010). Direct detection of a substellar companion to the young nearby star PZ Telescopii, *A&A* 523: L1+.

Nakajima, T., Oppenheimer, B. R., Kulkarni, S. R., Golimowski, D. A., Matthews, K. & Durrance, S. T. (1995). Discovery of a cool brown dwarf, *Nature* 378: 463–465.

Neuhäuser, R. (2008). Homogeneous Comparison of Directly Detected Planet Candidates: GQ Lup, 2M1207, AB Pic, *in* S. Hubrig, M. Petr-Gotzens, & A. Tokovinin (ed.), *Multiple Stars Across the H-R Diagram*, pp. 183–+.

Neuhäuser, R., Brandner, W., Eckart, A., Guenther, E., Alves, J., Ott, T., Huélamo, N. & Fernández, M. (2000). On the possibility of ground-based direct imaging detection of extra-solar planets: the case of TWA-7, *A&A* 354: L9–L12.

Neuhäuser, R., Ginski, C., Schmidt, T. O. B. & Mugrauer, M. (2011). Further deep imaging of HR 7329 A (η Tel A) and its brown dwarf companion B, *MNRAS* pp. 1135–+.

Neuhäuser, R., Guenther, E., Brandner, W., Húelamo, N., Ott, T., Alves, J., Cómeron, F., Cuby, J.-G. & Eckart, A. (2002). Direct Imaging and Spectroscopy of Substellar Companions Next to Young Nearby Stars in TWA, *in* J. F. Alves & M. J. McCaughrean (ed.), *The Origin of Stars and Planets: The VLT View*, pp. 383–+.

Neuhäuser, R. & Guenther, E. W. (2004). Infrared spectroscopy of a brown dwarf companion candidate near the young star GSC 08047-00232 in Horologium, *A&A* 420: 647–653.

Neuhäuser, R., Guenther, E. W., Alves, J., Huélamo, N., Ott, T. & Eckart, A. (2003). An infrared imaging search for low-mass companions to members of the young nearby β Pic and Tucana/Horologium associations, *Astronomische Nachrichten* 324: 535–542.

Neuhäuser, R., Guenther, E. W., Petr, M. G., Brandner, W., Huélamo, N. & Alves, J. (2000). Spectrum and proper motion of a brown dwarf companion of the T Tauri star CoD-33 7795, *A&A* 360: L39–L42.

Neuhäuser, R., Guenther, E. W., Wuchterl, G., Mugrauer, M., Bedalov, A. & Hauschildt, P. H. (2005). Evidence for a co-moving sub-stellar companion of GQ Lup, *A&A* 435: L13–L16.

Neuhäuser, R., Mugrauer, M., Seifahrt, A., Schmidt, T. O. B. & Vogt, N. (2008). Astrometric and photometric monitoring of GQ Lupi and its sub-stellar companion, *A&A* 484: 281–291.

Neuhäuser, R., Schmidt, T. O. B., Hambaryan, V. V. & Vogt, N. (2010). Orbital motion of the young brown dwarf companion TWA 5 B, *A&A* 516: A112+.

Neuhäuser, R., Schmidt, T. O. B., Seifahrt, A., Bedalov, A., Helling, C., Witte, S. & Hauschildt, P. (2009). Medium-resolution infrared integral field spectroscopy of the brown dwarf TWA 5 B, *in* E. Stempels (ed.), *American Institute of Physics Conference Series*, Vol. 1094 of *American Institute of Physics Conference Series*, pp. 844–847.

Oppenheimer, B. R. & Hinkley, S. (2009). High-Contrast Observations in Optical and Infrared Astronomy, *ARofA&A* 47: 253–289.

Oppenheimer, B. R., Kulkarni, S. R., Matthews, K. & Nakajima, T. (1995). Infrared Spectrum of the Cool Brown Dwarf Gl 229B, *Science* 270: 1478–1479.

Quanz, S. P., Meyer, M. R., Kenworthy, M. A., Girard, J. H. V., Kasper, M., Lagrange, A.-M., Apai, D., Boccaletti, A., Bonnefoy, M., Chauvin, G., Hinz, P. M. & Lenzen, R. (2010). First Results from Very Large Telescope NACO Apodizing Phase Plate: 4 μm Images of The Exoplanet β Pictoris b, *ApJL* 722: L49–L53.

Rebolo, R., Zapatero Osorio, M. R., Madruga, S., Bejar, V. J. S., Arribas, S. & Licandro, J. (1998). Discovery of a Low-Mass Brown Dwarf Companion of the Young Nearby Star G 196-3, *Science* 282: 1309–+.

Reid, I. N. & Walkowicz, L. M. (2006). LP 261-75/2MASSW J09510549+3558021: A Young, Wide M4.5/L6 Binary, *PASP* 118: 671–677.

Reidemeister, M., Krivov, A. V., Schmidt, T. O. B., Fiedler, S., Müller, S., Löhne, T. & Neuhäuser, R. (2009). A possible architecture of the planetary system HR 8799, *A&A* 503: 247–258.

Rodriguez, D. R., Zuckerman, B., Melis, C. & Song, I. (2011). The Ultra Cool Brown Dwarf Companion of WD 0806-661B: Age, Mass, and Formation Mechanism, *ApJL* 732: L29+.

Sahlmann, J., Segransan, D., Queloz, D. & Udry, S. (2010). A possible dividing line between massive planets and brown-dwarf companions, *ArXiv e-prints* .

Schmidt, T. O. B., Neuhäuser, R. & Seifahrt, A. (2009). Homogeneous Comparison of Planet Candidates Imaged Directly Until 2008, *in* T. Usuda, M. Tamura, & M. Ishii (ed.), *American Institute of Physics Conference Series*, Vol. 1158 of *American Institute of Physics Conference Series*, pp. 231–234.

Schmidt, T. O. B., Neuhäuser, R., Seifahrt, A., Vogt, N., Bedalov, A., Helling, C., Witte, S. & Hauschildt, P. H. (2008). Direct evidence of a sub-stellar companion around CT Chamaeleontis, *A&A* 491: 311–320.

Schneider, J., Dedieu, C., Le Sidaner, P., Savalle, R. & Zolotukhin, I. (2011). Defining and cataloging exoplanets: the exoplanet.eu database, *A&A* 532: A79+.

Scholz, R.-D. (2010). Hip 63510C, Hip 73786B, and nine new isolated high proper motion T dwarf candidates from UKIDSS DR6 and SDSS DR7, *A&A* 515: A92+.

Seifahrt, A., Neuhäuser, R. & Hauschildt, P. H. (2007). Near-infrared integral-field spectroscopy of the companion to GQ Lupi, *A&A* 463: 309–313.

Stassun, K. G., Mathieu, R. D. & Valenti, J. A. (2006). Discovery of two young brown dwarfs in an eclipsing binary system, *Nature* 440: 311–314.

Stassun, K. G., Mathieu, R. D. & Valenti, J. A. (2007). A Surprising Reversal of Temperatures in the Brown Dwarf Eclipsing Binary 2MASS J05352184-0546085, *ApJ* 664: 1154–1166.

Thalmann, C., Carson, J., Janson, M., Goto, M., McElwain, M., Egner, S., Feldt, M., Hashimoto, J., Hayano, Y., Henning, T., Hodapp, K. W., Kandori, R., Klahr, H., Kudo, T., Kusakabe, N., Mordasini, C., Morino, J.-I., Suto, H., Suzuki, R. & Tamura, M. (2009). Discovery of the Coldest Imaged Companion of a Sun-like Star, *ApJL* 707: L123–L127.

Todorov, K., Luhman, K. L. & McLeod, K. K. (2010). Discovery of a Planetary-mass Companion to a Brown Dwarf in Taurus, *ApJL* 714: L84–L88.

Udry, S. (2010). Detection and characterization of exoplanets: from gaseous giants to super-Earths, *In the Spirit of Lyot 2010*.

Wahhaj, Z., Liu, M. C., Biller, B. A., Clarke, F., Nielsen, E. L., Close, L. M., Hayward, T. L., Mamajek, E. E., Cushing, M., Dupuy, T., Tecza, M., Thatte, N., Chun, M., Ftaclas, C., Hartung, M., Reid, I. N., Shkolnik, E. L., Alencar, S. H. P., Artymowicz, P., Boss, A., de Gouveia Dal Pino, E., Gregorio-Hetem, J., Ida, S., Kuchner, M., Lin, D. N. C. & Toomey, D. W. (2011). The Gemini NICI Planet-finding Campaign: Discovery of a Substellar L Dwarf Companion to the Nearby Young M Dwarf CD-35 2722, *ApJ* 729: 139–+.

Webb, R. A., Zuckerman, B., Platais, I., Patience, J., White, R. J., Schwartz, M. J. & McCarthy, C. (1999). Discovery of Seven T Tauri Stars and a Brown Dwarf Candidatein the Nearby TW Hydrae Association, *ApJL* 512: L63–L67.

White, R. J., Ghez, A. M., Reid, I. N. & Schultz, G. (1999). A Test of Pre-Main-Sequence Evolutionary Models across the Stellar/Substellar Boundary Based on Spectra of the Young Quadruple GG Tauri, *ApJ* 520: 811–821.

Woitas, J., Leinert, C. & Köhler, R. (2001). Mass ratios of the components in T Tauri binary systems and implications for multiple star formation, *A&A* 376: 982–996.

Wolszczan, A. & Frail, D. A. (1992). A planetary system around the millisecond pulsar PSR1257 + 12, *Nature* 355: 145–147.

Wright, J. T., Fakhouri, O., Marcy, G. W., Han, E., Feng, Y., Johnson, J. A., Howard, A. W., Fischer, D. A., Valenti, J. A., Anderson, J. & Piskunov, N. (2011). The Exoplanet Orbit Database, *PASP* 123: 412–422.

Zacharias, N., Monet, D. G., Levine, S. E., Urban, S. E., Gaume, R. & Wycoff, G. L. (2005). NOMAD Catalog (Zacharias+ 2005), *VizieR Online Data Catalog* 1297: 0–+.

Zapatero Osorio, M. R., Rebolo, R., Bihain, G., Béjar, V. J. S., Caballero, J. A. & Álvarez, C. (2010). Infrared and Kinematic Properties of the Substellar Object G 196-3 B, *ApJ* 715: 1408–1418.

4

AO-Based High Resolution Image Post-Processing

Changhui Rao, Yu Tian and Hua Bao
Institute of Optics and Electronics, Chinese Academy of Sciences
China

1. Introduction

In this chapter, we will discuss the AO-based high resolution image post-processing, which is applied in astronomical observation, solar observation, and retinal imaging, etc..

2. The high resolution restoration of adaptive optics image

The adaptive optics (AO) can be used to compensate for wave aberrations of atmosphere to achieve the resolution of diffraction limited in real time [1-5]. However, the correction is generally partial due to the hardware restrictions, such as the finite degree of freedom of the deformable mirror, the latency of the AO loop, the measure error in the wavefront sensor (WFS) and so on. So post-processing is necessary to improve the quality of AO images, which are still corrupted by residual aberrations. Generally, the corruption can be described as a convolving process:

$$g(x,y) = f(x,y) * h(x,y) + n(x,y) \tag{1-1}$$

where "*" denotes the two-dimensional convolution, and g(x,y) represents the observed degraded image; f(x,y), the real object, that is, the object to be restored; h(x,y) is the residual PSF, convolved with the real object of the optical system; n(x,y) stands for the system noise. In frequency space, the Eq.(1-1) is equal to

$$G(u,v) = F(u,v)H(u,v) + N(u,v) \tag{1-2}$$

The primary goal of post-processing is to find the true object f(x,y), and the process can be described as a deconvolution. In AO system, h(x,y) can be detected partially from WFS. If h(x,y) is known, a deconvolution process can be applied to g(x,y) to estimate f(x,y) easily. The method can be generally called deconvolution from wavefront sensing (DWFS), which assumed that the measurement of WFS was accurate. In practical system, however, the WFS is suffered from various kind of noises such as the photon noise and the readout noise of CCD, moreover the situation is more risky in closed-loop [6]. Also, the frame rate and exposure time of the camera taken the corrected images is different from those of WFS camera in practice. So it is difficult to match the short exposure images with the corresponding WFS data.

Therefore, ordinary DWFS is not suitable for post-processing of AO images. In this chapter, three methods of post-processing with different technical routines were described. In section 1, a post-processing method based on self-deconvolution was introduced. The scheme of the algorithm was following the self-deconvolving data reconstruction algorithm (SeDDaRA) which proposed by Caron [7,8]. And the method had been verified in solar images. In section 2, a post-processing method for the partially-corrected AO image was proposed, in which, the WFS data is used and revised while the AO system performs the close-loop correction on the given object. The algorithm has been tested in human eye fundus images. In section 3, a method based on blind deconvolution was introduced. The method combined the frame selection technique to get the most suitable images from AO close-loop series and then implemented the blind deconvolution. Also, the method was verified in celestial images captured by AO system.

2.1 A modified self-deconvolution restoration algorithm

Caron et al. [7,8] proposed a blind deconvolution method SeDDaRA. The method uses a power law relation applied to the Fourier transform of degraded data to extract a filter function and is proved to work well on a wide variety of signal types. The key operation in SeDDaRA is to determine the transfer function (extracted filter function) to deconvolve the images and the most important step is to choose the tuning parameter. In this section, a method based on SeDDaRA is proposed to process deconvolution to solar images corrected by 37-element AO system. The method utilizes the information recorded by WFS in closed-loop status to estimate the deconvolved operator.

From Eq. (1-2), the most important in deconvolution is to find the deconvolved operator H(u,v) .The SeDDaRA gives the following equation to estimate the H(u,v) [8]

$$H_e(u,v)=\left[K_G S\left\{\left|G(u,v)-N(u,v)\right|\right\}\right]^{\alpha(u,v)} \tag{1-3}$$

where α(u,v) is a tuning parameter and K_G is a real positive scalar chosen to ensure |He(u,v)|≤1. S{·} , which denotes a smoothing operation. There are three condition equations, whcih should subject to: (1) 0≤α(u,v) <1, (2) the S{·} is separable and (3) F(u,v) and N(u,v) are uncorrelated[4]. The SeDDaRA chooses α(u,v) by solving

$$\frac{S\left\{\left|G(u,v)\right|\right\}}{S\left\{\left|F(u,v)\right|\right\}}=\left[K_G S\left\{\left|G(u,v)\right|\right\}\right]^{\alpha(u,v)} \tag{1-4}$$

As the α(u,v) is dependent on the spatial frequency and is related to the unknown object F(u,v), it is not easy to determine. The α(u,v) could be set to a constant value, but that would cause distortion in high frequency band [8]. Sudo et al. has applied SeDDaRA to process solar images and the set α=0.5 as a constant according to experience [9]. In this section, in order to choose a suitable tuning parameter which can estimate the extracted filter more precisely, a method that utilizes the long exposure WFS data to determine the α(u,v) was proposed. The method was proven to be effective in AO-corrected solar images post-processing.

In adaptive optics, the WFS provides Zernike coefficients which could generate PSF h(x,y) and corresponding H(u,v), though it is not accurate due to the measurement noise. Increasing exposure time could alleviate the impact of noise. The longer exposure PSF sums multi-frames short exposure PSF simply.

In spite of the fact that the long exposure PSF cannot match the image taken by AO camera as well, it could describe some rules of wavefront aberrations during the time which long exposure PSF measured. That means the longer exposure PSF would be beneficial to provide prior knowledge to the SeDDaRA, i.e., it would be helpful to determine the α(u,v).

In practical system, the frame rate of camera taken corrected AO images is much slower than frame rate of WFS camera (30Hz vs. 1600Hz). Therefore, the latter one's exposure time is much shorter (10ms vs. 0.625ms) and the AO images are longer exposure compared with the WFS data. One AO image can be decomposed as N frames short exposure images, the right side of Eq. (1-4) then can be written as (for simplicity, the noise is ignored)

$$\left[K_G S\{|G(u,v)|\}\right]^{\alpha(u,v)} = \frac{S\{|G_1(u,v)|\}}{S\{|F(u,v)|\}} + \frac{S\{|G_2(u,v)|\}}{S\{|F(u,v)|\}} + \cdots + \frac{S\{|G_N(u,v)|\}}{S\{|F(u,v)|\}} \tag{1-5}$$

Considering the images has been corrected by AO once, the residual aberration is small, so we assume that the Fourier transform of relative longer exposure PSF, i.e. long exposure optical transfer function is real and smooth and $OTF_{LE}(u,v) \approx |OTF_{LE}(u,v)| \leq 1$. Then the right side of Eq, (1-5) is actually the long exposure optical transfer $OTF_{LE}(u,v)$. that is

$$\frac{S\{|G_1(u,v)|\}}{S\{|F(u,v)|\}} + \frac{S\{|G_2(u,v)|\}}{S\{|F(u,v)|\}} + \cdots + \frac{S\{|G_N(u,v)|\}}{S\{|F(u,v)|\}} = |OTF_{LE}(u,v)| \tag{1-6}$$

So there is

$$\left[K_G S\{|G(u,v)|\}\right]^{\alpha(u,v)} = |OTF_{LE}(u,v)| \tag{1-7}$$

which makes

$$\alpha(u,v) = \frac{\ln|OTF_{LE}(u,v)|}{\ln K_G S\{|G(u,v)|\}} \tag{1-8}$$

The α(u,v) is determined and dependented on $|OTF_{LE}(u,v)|$. He(u,v) can be easily got according to Eq. (1-3) and can be implemented in many deconvolution algorithms. The deconvolution scheme we employed is constrained least square filtering [10,11] which is given by

$$F(u,v) = \left[\frac{H_e^*(u,v)}{|H_e(u,v)|^2 + \mu|P(u,v)|^2}\right] G(u,v) \tag{1-9}$$

where μ is a parameter to control the regularization strength and P(u,v) is the Fourier transform of Laplacian operator. Here, the "*" denotes the conjugate part. The regularization P(u,v) can alleviate the noise sensitivity during the deconvolution. Because the AO system has pupil limit, it is necessary to add band-limit constraint to the deconvolution process which also can prevent noise amplification.

A 37-element AO system had been developed and installed on 26cm Solar Structure Telescope located on the Phoenix Hill, Kunming, Yunnan, China. Fig. 1.1 shows the optical layout of the AO system, which works in visible spectral range. The imaging CCD's frequency rate is 30Hz and the WFS CCD's frequency rate is about 1600Hz, more details can be found in [1,2].

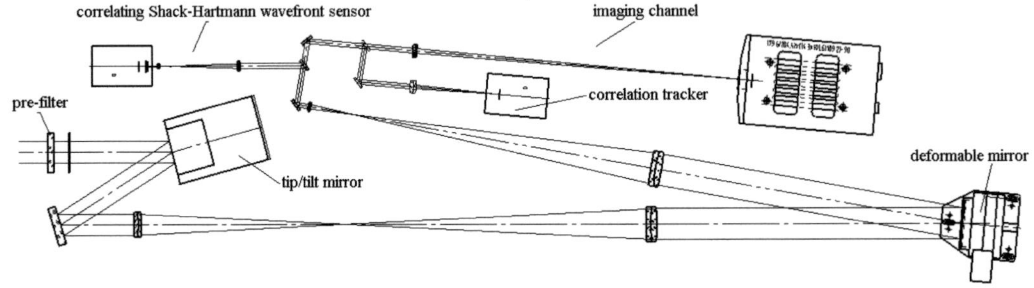

Fig. 1.1. The optical layout of the adaptive optics

Fig. 1.2 shows the image of solar granulation corrected by AO system and the restored images by different algorithms including our method result, DWFS with long exposure PSF and SeDDaRA with constant α=0.45 respectively. The four images have been energy normalized.

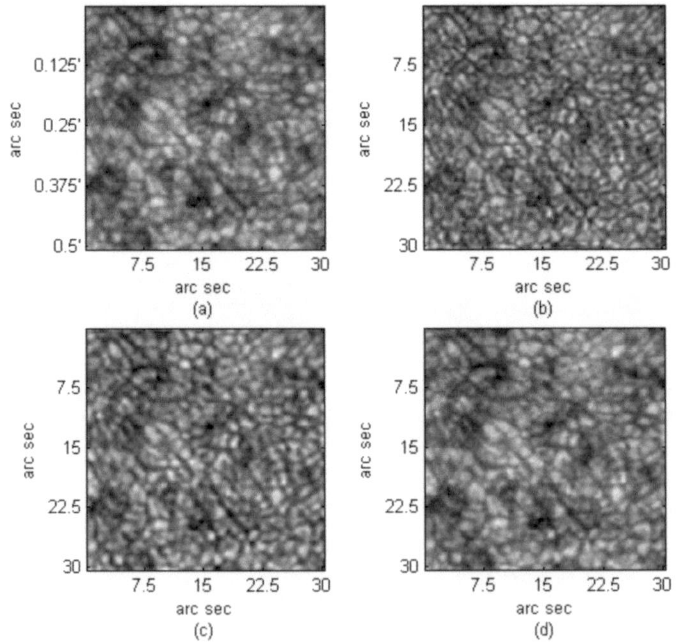

Fig. 1.2. Image of solar granulation. (a) The image corrected by AO system; (b) The image restored by our method; (c) The image restored by long exposure PSF; (d) The image restored by SeDDaRA with a constant α=0.45.

To measure the quality of solar granulation, we employed the following criterion [11]

$$r = \frac{(\Delta I)_{rms}}{\overline{I}} \times 100\% \tag{1-10}$$

where $\overline{I}$ denotes the mean of image I and $(\Delta I)_{rms}$ represents the RMS value of image I. The results of the four images are (a) 2.05%, (b) 4.06%, (c) 3.03% and (d) 2.95%. It shows our method can improve the image quality efficiently.

In order to illustrate the improvement of image quality, the cross sections of the images are shown in Fig. 1.3 It shows that the contrast of the solar granulation image has been improved by deconvolution method we proposed.

Also, the cross section curve of deconvolution result is smooth. The noise was not amplified as the present of Laplacian operator was introduced in the deconvolution scheme.

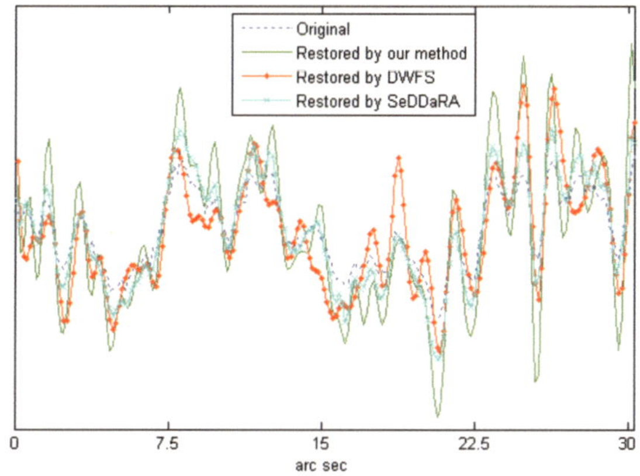

Fig. 1.3. The cross section of Fig.2

Fig. 1.4 gives the image of sunspot and the restored image deconvolved by our method. The images are also energy normalized.

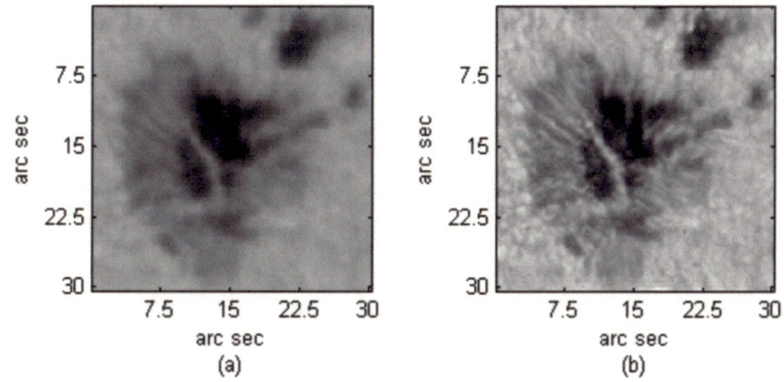

Fig. 1.4. The solar spot images. (a) the original images corrected by AO; (b) the corresponding images restored by our method.

According to the image structure characteristic, a contrast standard to evaluate the improvement was defined as

$$C = \frac{I_{max} - I_{min}}{I_{max} + I_{min}} \tag{1-11}$$

where I_{max} and I_{min} are the maximum and minimum values in the image. The contrast of original image and deconvolution result are 0.4491 and 0.7008, respectively. The outcome proves the validity of our method.

In summary, the method bases on SeDDaRA and costs shorter time compared with those based on blind deconvolution or other iterative methods. As the prior knowledge of long exposure PSF is introduced, it is simple to determine the tuning parameter $\alpha(u,v)$ which is crucial to the quality of restored images. The method is more useful when the target is low contrast and extended widely such as solar images and human eye fundus images which would lead the method based on blind deconvolution more difficult. The method can be combined with AO data process package to offer higher resolution images.

2.2 Hybrid deconvolution

The retina of human eye is a multi-layers structure which consists of photoreceptor cells and capillary blood vessels locate at the back of the human eyeballs. The unique structure of the blood vessels and cells in the retina has been used for biometric identification and disease diagnosis. The AO can compensate the wave aberrations of the human eyes to achieve the resolution of diffraction limited in real-time. However, the correction by AO system is always partial due to the hardware restrictions as stated before. So the post-processing is necessary to improve the quality of the images which is still corrupted by the residual aberrations. In the section, a method of obtaining high-quality retinal images based on image post-processing, which uses wavefront error measured simultaneously by a wavefront sensor, is described.

The post-processing involves the image restoration (deconvolution) technique which needs accurate knowledge of the residual aberration when the AO system is achieving the close-loop status and simultaneously recorded images which is partially corrected by AO system to estimate the undistorted images of the human eyes.

According to Eq. (1-1, 1-2), the deconvolution processing can be described with the expression according to Eq. (1-12)

$$o = FT^{-1}(H^{-1}G) = FT^{-1}(H^{-1}HO + H^{-1}N) \tag{1-12}$$

Where FT denotes Fourier transform; "o" represents the estimation of the undistorted image and "O" is its corresponding Fourier transform; "H", "G" and "N" are the Fourier transforms of point spread function (PSF), degraded image and the Gaussian white noise respectively. The "H" is also called optical transfer function (OTF) which can be computed from the measured aberration by wavefront sensor like Hartmann-Shack wavefront sensor (HSWFS). As the existence of the noise, directly inverse filtering in (1-12) would result in amplification of noise, introducing unwanted variance into the deconvolved images. Also the HSWFS is contaminated by the CCD random noise, so the PSF calculated from the wavefront measurements are not accurate. The key to the problem is to use a joint estimation to estimate the undistorted image and the "real" PSF. The method can be formalized to the minimization of the criterion function

$$J(o,h) = J_1(o,h) + J_2(o) + J_3(h) \tag{1-13}$$

Where $J_1(o,h) = \frac{1}{\sigma^2}\|FT^{-1}(HO - G)\|^2$ is the term of fidelity to image data (considering the noise of Gaussian noise), and $\|\bullet\|^2$ is the norm-squared operator, σ^2 is the noise variance; $J_2(o) = \|\nabla^k(o)\|^2$ is the term of regularization to constrain the amplification of the noise in high frequencies; $\nabla^k(\bullet)$ denotes the differential operator with k_{th} order. Here, k=2 is chosen. So actually $J_2(o)$ is a measure of smoothness by Laplacian operator, and it would constrain the information in high frequency of estimate image in order to alleviate the noise sensitivity problem. $J_3(h) = \|h - h_0\|^2$ is the penalty term to control the estimation of PSF, which should not be too different from h_0 measured by HSWFS. The reason to do so is that the "real" PSF is impossible to be far away from h_0 which is considered to be part of "real" PSF.

The normal blind deconvolution usually gets into local minima and hardly finds a unique solution especially when there is only one single blurred image to be restored. We use a joint estimation to overcome this drawback because the estimation of PSF is related to the PSF measured by HSWFS. There are two advantages: on one hand PSF can be revised by blurred image and estimation of undistorted image; on the other hand the revised PSF can give a better estimation of undistorted image. Also, the existence of h_0 reduces the uncertainty of blind deconvolution hugely.

A table-top small adaptive AO system for vivid human retinal high resolution imaging [5] was built. The system is working for real-time (25Hz) closed-loop measurement and correction of ocular aberration. Fig. 1.5 shows a schematic diagram of the setup. A laser diode with wavelength 780nm was used for illumination. After retina reflection, the light exits from the pupil (6mm) and passes through dichroic mirror, beam expender telescope, deformable mirror (DM) with 37 elements, mirror and beam splitter, projects into HSWFS (97 subapertures in11 by 11 arrays). The slope data of wavefront of HSWFS is acquired by a computer and transferred to control DM.

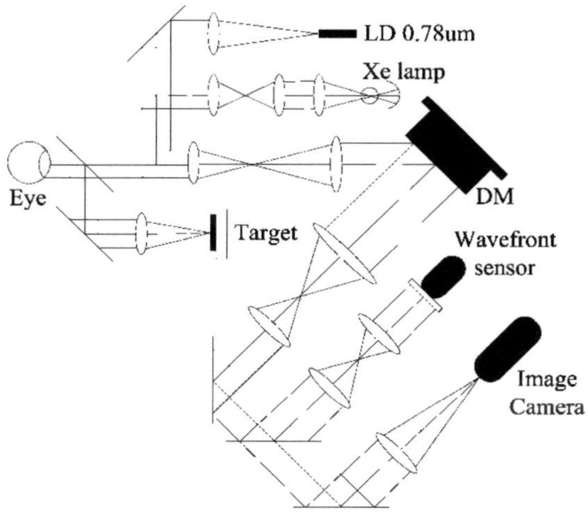

Fig. 1.5. The Schematic diagram of the table-top AO system for human eye retina

The device was tested with a vivid eye of a volunteer. Fig. 1.6 (a) and (b) show that the original blurred image of blood vessels in the retina from a vivid human eye without AO correction and the image corrected by AO. It is obvious that the latter one has better resolution and visual quality than the previous one.

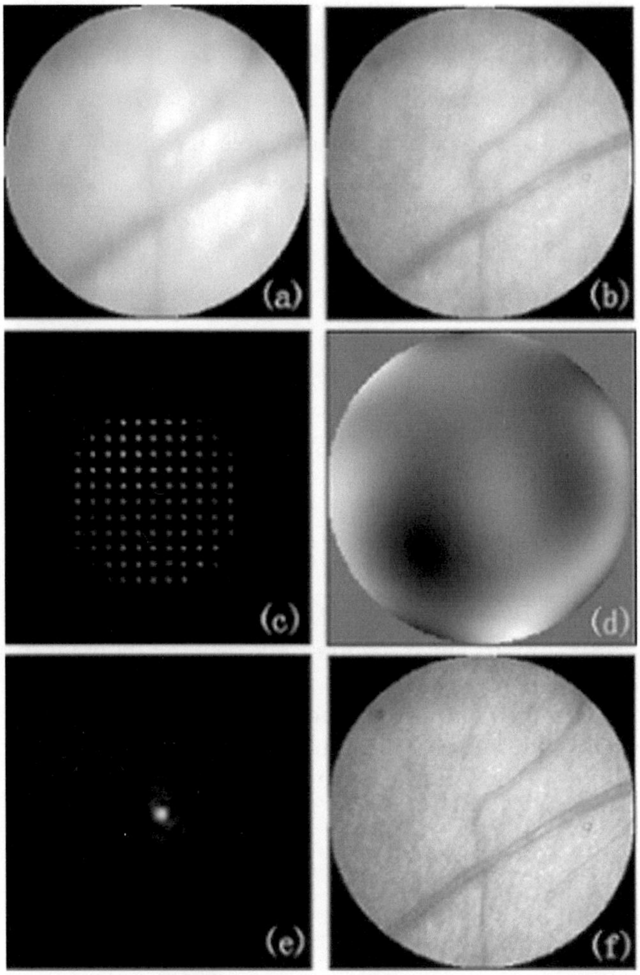

Fig. 1.6. Real eye experiment: (a) Original blood vessels image (400×400 pixels), (b) the image corrected by AO (400×400 pixels), (c) simultaneously recorded HS image (240×240 pixels), (d) reconstructed residual phase (PV=0.606μm), (e) revised PSF, (f) image restored by post-processing(400×400 pixels).

As illustrated before, the AO system cannot achieve the full correction due to the limitation of the hardware, mainly the correction capability of the DM. The Fig. 1.6(c) shows the HS image when AO system is in closed-loop status, and the image in Fig. 1.6 (b) is recorded simultaneously. In the post-processing stage, the residual phase to restore the partially corrected image would be utilized. OTF/PSF which characterizes the residual phase should be estimated firstly. The residual OTF of the ocular can be approximated as the product of the pupil OTF and a function that called AO-OTF

$$H = \mathrm{OTF}_{\mathrm{p}} \times \mathrm{OTF}_{\mathrm{AO}} \tag{1-14}$$

The AO-OTF is related to the residual phase in the pupil plane

$$\mathrm{OTF}_{\mathrm{AO}} = FT\left\{\left|FT\left\{\exp\left[j\phi(\vec{x})\right]\right\}\right|^2\right\} \tag{1-15}$$

Where $\varphi(\overline{X})$ represents the residual phase in the pupil plane which can be computed from measured aberration by HSWFS with modal reconstruction [12]. It is shown in figure 6 (d). And the pupil OTF can be easily calculated from the pupil geometry. Fig. 6 (e) gives the corresponding PSF.

The next step is to estimate the undistorted image from degraded image G and ocular OTF H with the help of Eq. (1-13). Fig. 1.6 (f) is the picture of the restored image by the post-processing.

According to Fig. 1.6 (f), the visual quality is better than the image in Fig. 1.6 (b) which is partially corrected by AO system. The blood vessels are more distinctive. Also, the cells in the fundus of eye are much clearer. There are more structures and small features available.

The contrast is defined as

$$C = \frac{I_{\max} - I_{\min}}{I_{\max} + I_{\min}} \tag{1-16}$$

where I_{max} and I_{min} are the maximum and minimum values in the image. The contrast of the image without AO correction and the image partially corrected by AO is 0.723 and 0.902 respectively. The contrast of the image handled by post-processing achieves 0.982. The contrast of the restored image is higher that would be helpful to find more useful information of the vessels in the retina.

As shown in Fig. 1.7(d), the restored image presents more details. The contrast is improved: 0.9135 versus 0.9797. The performance of the post-processing depends on the accuracy of wavefront sensor and the deconvolution technique. So there are lots of improvement can be done to ameliorate the capability of correction. In this chapter, we have shown that the partially corrected retinal blood vessel as well as cells in fundus of eye images can be restored to get higher visual quality, which would be benefit for the biometric identification and disease diagnosis in the clinical use.

2.3 Multi-frame blind deconvolution

To obtain high-resolution observational results is one of the important goals of astronomical observations, but under the influence of atmospheric turbulence, the ground-based optical telescope can hardly obtain high-quality images. The method of DWFS provides an easy approach for wavefront correction, but as stated before, the DWFS performances is highly restricted by the detection accuracy of the WFS.

In the section, a post-processing method based on frame selection and multi-frames blind deconvolution to improve images partially corrected by AO is proposed. The method tries to give the estimation on the real object by using the multiple short-exposure degraded images captured by AO closed-loop correction. It could not only improve the image quality of the object, but also make corrections on the wavefront detection which would help the engineer to analyze the ability of AO system. This algorithm does not need any a priori knowledge, only a positivity constraint and band-limit constraint are applied to the iterative process, for the convergence of the algorithm. It is benefit for the use of multi-frame images to improve the stability and convergence of the blind deconvolution algorithms. The method had been applied in the image restoration of celestial bodies which were observed by 1.2m telescope equipped with 61-element adaptive optical system at Yunnan Observatory. The results showed that the method can effectively improve the images partially corrected by AO.

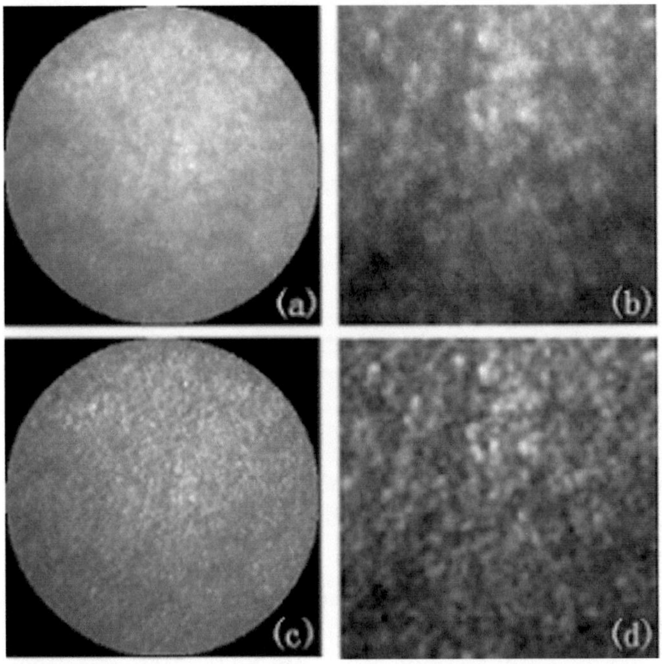

Fig. 1.7. Experiment on cells in fundus of eye: (a) the image corrected by AO (400×400 pixels; (b) corresponding details of (a) (100×100 pixels); (c) and (d) are image restored by post-processing and corresponding details

The multi-frame degradation model for a given object can be written as:

$$g_m(x,y) = f(x,y) \otimes h_m(x,y) + n(x,y), 1 < m \le M \tag{1-17}$$

in which $g_m(x, y)$ and $h_m(x, y)$ represent, respectively, the m-th frame of short-exposure degraded image of the given object f(x, y) and the corresponding PSF. It required $h_m(x, y)$ to be co-prime, and that is generally satisfied in 2-dimensional images. Comparing Eq.(17) with Eq.(1), it was clear that the multi-frame joint estimation transforms the problem of "knowing one deriving two" to the problem of "knowing M deriving M+1". Although the equation was still underdetermined, the degree of uncertainty had been greatly reduced. And the greater M, the greater the possibility that f(x, y) approaches to the real solution.

Atmospheric turbulence is a random process: its intensity varies with time. When the AO system was used for the compensation of atmospheric turbulence, the effect of its correction also changed with time. So, not all frames of the recorded short-exposure degraded images are suitable for participating in the blind deconvolution process, and a selection of the images or a frame selection is necessary.

The frame selection technique introduces a standard to evaluate the quality of partially compensated AO images, and picks up those with good qualities.

Hence, the frame selection is actually to pick up the best part of AO compensations from the series of short-exposure images. The Shannon entropy is taken as the quality evaluation standard, and it is easy to define the Shannon entropy of the i-th frame to be

$$S_i = -\sum_{x,y} \psi_i(x,y) \log \psi_i(x,y) \tag{1-18}$$

in which $\psi_i(x, y)$ represents the probability density function of the i-th frame, namely,

$$\psi_i(x,y) = \frac{g_i(x,y)}{\sum_{x,y} g_i(x,y)} \tag{1-19}$$

The better the AO compensation, the smaller the value of S_i [13]. So, the M frames with the smallest S_i are selected as the degraded images waiting for the blind deconvolution processing (denoted by $g_m(x, y)$ in Eq.(1-17)). To adopt the frame selection as a preprocessing can improve the iterative quality and convergence of the blind deconvolution.

For convenience, by omitting the coordinates the Eq.(1-17) can be rewritten simply as:

$$g_m = f \otimes h_m + n, 1 < m \le M \tag{1-20}$$

Under the condition that n is a Gaussian noise, the maximum likelihood estimate of f is:

$$\tilde{f} = \arg\min E(| | g_m - h_m \otimes \tilde{f} | |^2) \tag{1-21}$$

and similarly, the maximum likelihood estimate of h_m can be written as:

$$\tilde{h}_m = \arg\min E(| | g_m - \tilde{h}_m \otimes f | |^2) \tag{1-22}$$

in which E(•) denotes the mathematical expectation and | | • | |, the norm. According to the maximum likelihood criterion, we can define the cost function:

$$J(f,h_m) = \left\| \tilde{h}_m \otimes \tilde{f} - g_m \right\|^2 \tag{1-23}$$

and by minimizing $J(f, h_m)$, f and h_m could be estimated.

In an incoherent imaging system the positivity constraint, which means both f and h_m are non-negative, could be satisfied. So in the iterative process, to add in the positivity constraint is reasonable:

$$f = p^2 \tag{1-24}$$

$$h_m = q_m^{\ 2} \tag{1-25}$$

Miura[14] has proved that by exchanging the variables in Eqs.(1-24, 1-25), the number of the unknown image elements to be estimated can be effectively reduced. Even for the single frame blind deconvolution of Eq.(1-17), it can change an under-determined equation into an over-determined (i.e., the number of equations is greater than the number of unknowns). On this basis, Miura proposed a single-frame blind deconvolution method under the conditions of band-limitation and canonical constraint [14]. But this method is easy to be trapped into an ineffective solution when the initial guess of the result is improper. To avoid this, multiple degraded images are added to be the first guess of true object in the estimation on the real object and point spread function.

In 2000, we built a set of 61-element AO system on the Yunnan Observatory 1.2m telescope, and used it in the visible-light high-resolution imaging observations of astronomical objects [3, 4]. In May 2004, we completed the update and reform of the system. The AO system is placed on the Coude optical path of the 1.2m telescope. The corrected area of the AO system is the 1.06m full-aperture, the wavefront detection is operated at the 400nm-700nm wavelength, and the imaging observation is conducted at the 700nm-900nm wavelength. In the AO system, the AO correction loop consists of a 9×9 array of high-frame-frequency weak-light Hartmann-Shack wavefront sensors, the 61-element deformable mirror is matched with the former, the real-time wavefront processor is responsible for compensating the high-order aberration caused by atmospheric turbulence, and the image detection system is responsible for the high-resolution imaging observation.

Observations on a FK50 star with the magnitude of $2.^{m}06$ were made. From the 100 frames of degraded images after the close-loop correction of the AO system, 5 frames were picked up by the frame selection as the images ready for the blind deconvolution processing. These are shown in the panels (a)~(e) of Fig. 1.8, while Fig. 1.8(f) is the image after the post-processing. The 6 frame images are normalized to the same energy. 15 iterations were involved in the calculations.

The normalized Strehl ratio is defined as:

$$SR = \frac{i / E}{i_{diff} / E_{diff}} \times 100\% \tag{1-26}$$

In this formula, i and E indicate the maximum value and total energy of the given image respectively; and i_{diff} and E_{diff} are, correspondingly, the maximum value and total energy of the image in the diffraction limit. By calculations the Strehl ratio of Fig. 1.8(f) is 69.9%, which represents improvements on the five degraded images, respectively, by factors of 1.56, 1.45, 1.53. 1.49 and 1.44.

After the blind deconvolution, the full width at half-maximum (FWHM) of the image is improved. Fig. 1.9 compares the cross-sectional diagrams of Fig. 1.8(a) and Fig. 1.8(f). It is clear that after the post-processing, the FWHM of Fig. 1.8(f) is 4.1 pixels (about 0.15arcsec), reaching the diffraction limit of the system, and it is an improvement by 1.4 pixels on Fig. 1.8(a), a partially-corrected image of the AO system.

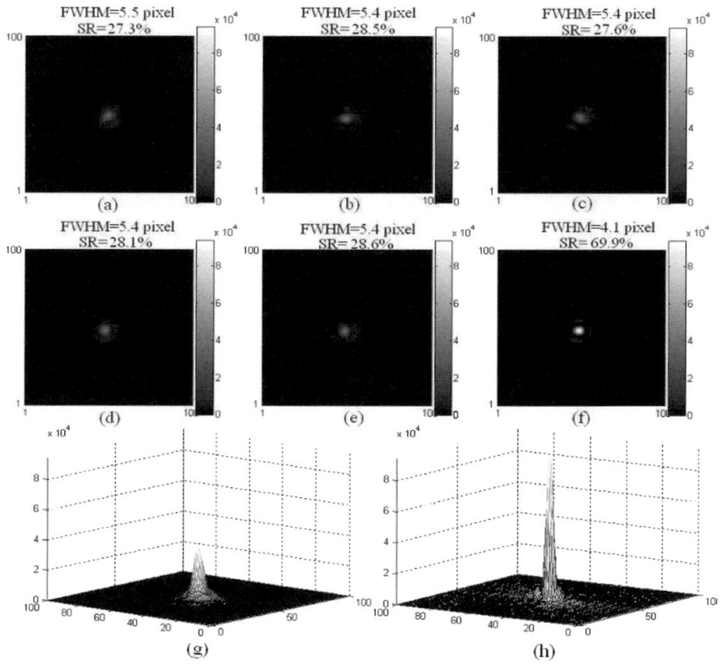

Fig. 1.8. Multi-frame blurred images and the corresponding restored image. (a)~(e) the five blurred images picked up by frame-selection respectively; (f) the restored image and (h) the corresponding 3D view.

In summary, the image quality was fairly improved by the blind deconvolution, the energy became more concentrated, and the resolution was raised.

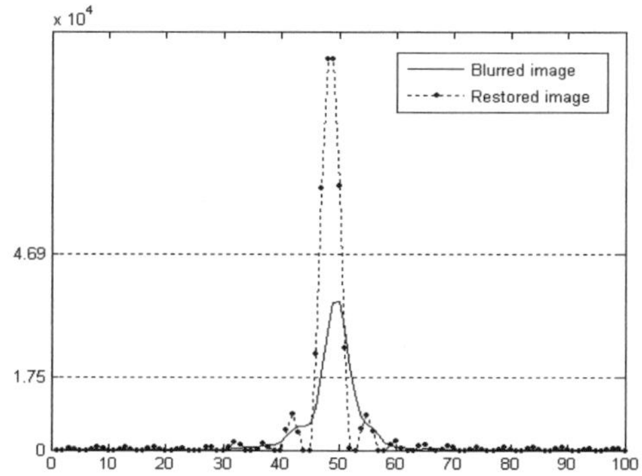

Fig. 1.9. Comparison of FWHM

Observation was also made on the binary star Hei42aa, with the magnitudes of 2. m46 and 3. m76. From 100 frames of short-exposure images after the close-loop correction of the AO system, 5 frames with the smallest Shannon entropies were selected as the images ready for the blind deconvolution processing. Fig. 1.10(a) shows the average of the 5 degraded images, and Fig. 1.10(b) is the restored image after 20 iterations, the two images having the same total energy. The panels (c,d) are the corresponding 3D diagrams of the panels (a,b).

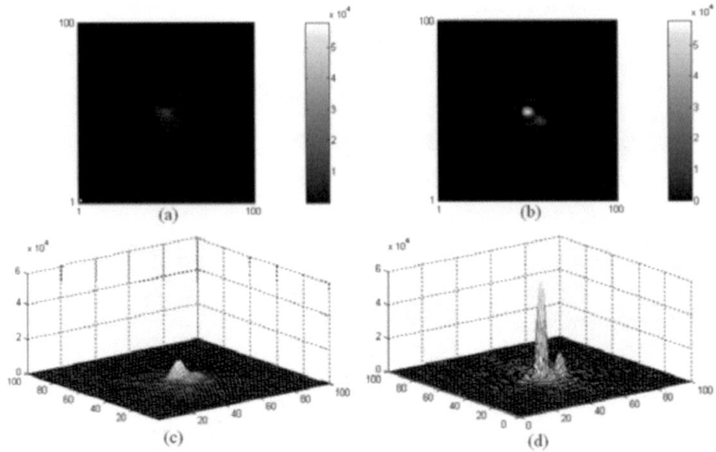

Fig. 1.10. Blurred images of double star and the restored image. (a) the average of blurred image; (b) the restored image; (c) and (d), the 3D view of (a) and (b)

After post-processing, the image energy becomes more concentrated to the two stars, making them more easily distinguishable. Also the separation between the two peaks is 7.9 pixels (0.29arcsec), consistent with the actual spacing of 0.3arcsec in star map.

The experimental results show that, although the AO system can provide restored results close to the diffraction limit in real time, limitations of the hardware means that the correction is generally partial. The post-processing based on frame selection and multi-frame blind deconvolution can effectively compensate for the insufficiency of the AO system and give a corrected image of better quality.

3. The visual enhancement of adaptive optics image

Due to the pollution of various photoelectric noises of an AO imaging system, the degraded images are mostly not very clear and weak edges are difficult to recognize, so the visual enhancement of AO-corrected images is necessary to exhibit abundant details covered by large scale noise and to enhance low-contrast edges [15]. This work is mainly divided into three parts: noise reducing, edge smoothing and contrast enhancement. In order to illustrate the effective of these processing better in our work, we will use AO-corrected Solar and photoreceptor cell images as examples.

3.1 Noise reducing

In this section, we will discuss the method of noise type detection and noise intensity estimation. This part of work is very import in image denoising, because it will direct us to choose filtering algorithms and control the filtering parameters.

In an AO imaging system, there are many noise sources, such as thermal noise, photon noise, dark current noise, CCD readout noise as well as camera background noise and so on. It is very important to choose suitable denoising methods according to noise types. In the following, we will discuss two typical approaches used to process the infrared image and the visible image, respectively.

For the infrared images, the camera background noise is the major noise pollution source, and this kind of noise is usually considered as additive noise, which can be expressed as follows:

$$J(i,j) = I(i,j) + N(i,j) \tag{2-1}$$

where, $J(i,j)$ is the actually captured image; $I(i,j)$ is the unpolluted ideal image; $N(i,j)$ is the additive background noise.

To filter the background noise, the camera dark and flat-field correction is widely used. The basic idea of this method is to get additive noise term $N(i,j)$ in Eq.(2-1) and the flat-field coefficient. The expression can be formulated as follows:

$$g(i,j) = T(i,j) \times |J(i,j) - N(i,j)| \tag{2-2}$$

where, $T(i,j)$ is the flat-field coefficient, $g(i,j)$ is the filtered image, and $|\cdot|$ is the absolute operator.

Fig. 2.1(a) and (b) shows the two solar rice images before and after the dark and flat field correction respectively, which are acquired by the 37-element solar AO imaging system [1, 2], and Fig. 2.1 (c) shows the background noise. According to this experiment, it can be seen that the dark and flat-field correction could suppress the camera background noise effectively.

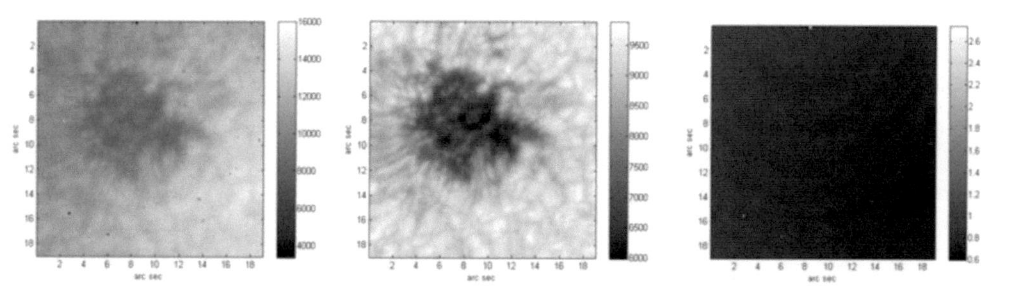

(a) (b) (c)

Fig. 2.1. (a) the solar image before flat-field correction; (b) the solar image after the dark and flat-field correction; (c) the camera background noise image.

Because the camera background noise can be corrected by imaging system, reducing this kind of noise is relatively simple and easy. But unfortunately, the other noise resources are very difficult to detect in AO imaging system, and it seems impossible to distinguish them from each other. Due to the noise model of an AO imaging system is very complex, we suggest considering the other noise pollutions as "combined noise", and using the hybrid filtering methods to suppress noise. Moreover we assume the combined noise as the additive noise, because if the combined noise in real application is more in line with the characteristics of multiplicative noise, the logarithms can be simply used to change it into the additive form as follows:

$$\begin{aligned} \because \quad & J(i,j) = I(i,j) \times N(i,j) \\ \therefore \quad & \log J(i,j) = \log I(i,j) + \log N(i,j) \end{aligned} \tag{2-3}$$

To illustrate how to reduce the combined noise in AO images, we use the human photoreceptor cell image as example, which are acquired by the table-top 37-elements AO flood illumination retinal imaging system [5].

Because we have assumed the combined noise as the additive noise, the neighborhood average algorithm can be used to suppress the undesirable pixel variance in flat areas. However, it is well known that the simply neighborhood average processing blurs image edges violently. To solve this problem, the bilateral filtering proposed by Tomasi [16], is used to reduce large scale noise in photoreceptor cell image. The basic idea of this method is that, relevant pixels are not only geometric closeness but also photometric similarity, and the combination between pixels prefers near values to distant values in both domain and range. This method can be expressed as follows:

$$B_{x,y} = k(x,y) \sum_{i,j \in G} I(i,j) \times e^{\frac{(x-i)^2+(y-j)^2}{-2\sigma^2} + \frac{\left(I_{x,y} - I(i,j)\right)^2}{-2\delta^2}} \tag{2-4}$$

where G is the filter window located at (x,y); $I(i,j)$ presents pixels in filter window; σ is the pixel geometric closeness; δ is the RMS of image noise, standing for noise intensity; $K(x,y)$ is used for normalization.

In the real applications, the results of bilateral filter are highly depends on the parameter of noise intensity: if δ is too small, the bilateral filter degrades as the Gaussian filter and edges will be blurred; if δ is too large, the bilateral filter can not suppress the noise sufficiently. In order to solve this difficulty, we select a noise estimation area in the background. Fig. 2.2 shows a real human photoreceptor cell image, in which a 128×128 pixels estimation area is selected at left-bottom corner.

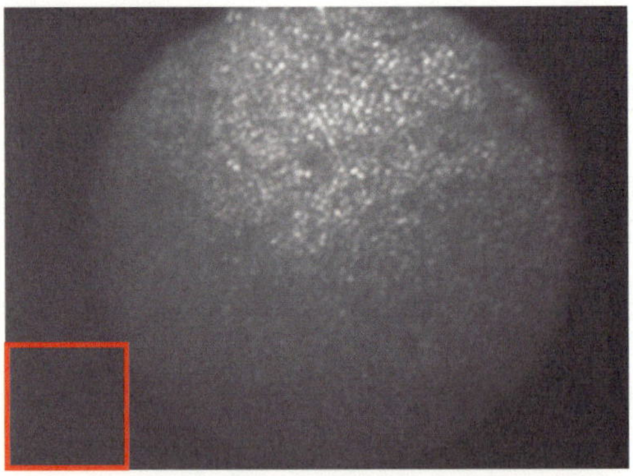

Fig. 2.2. The photoreceptor cell image with noise estimation area

Fig. 2.3(a) shows the noise area in background, Fig. 2.3(b) displays the gray histogram of this area, according to which we could compute the noise intensity, namely parameter δ used in the bilateral filtering, which takes the form as follows:

$$\delta = \sqrt{\frac{\sum_{i=0}^{m-1}\sum_{j=0}^{n-1}(g(i,j)-\overline{g})^2}{m \times n}} \tag{2-5}$$

where m and n are the size of estimating zone; $g(i,j)$ represents pixels in the zone and $\overline{g}$ is the corresponding mean value of these pixels.

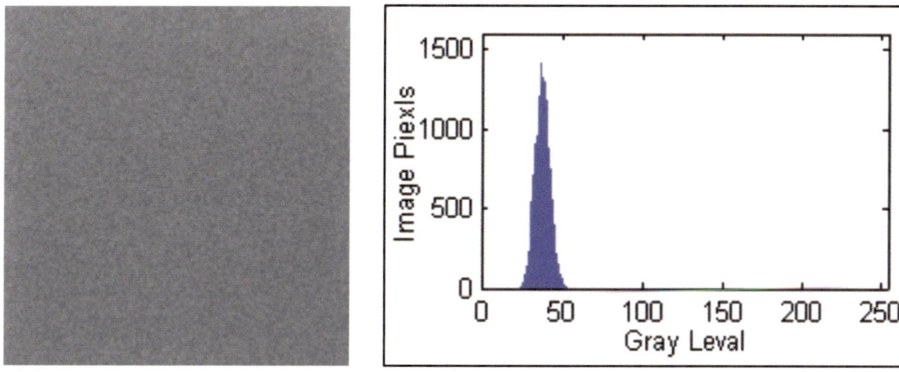

(a) (b)

Fig. 2.3. (a) the noise in the background; (b) the gray histogram in noise estimation area

Fig. 2.4(a) is a part of the photoreceptor cell image; Fig. 2.4(b) and Fig. 2.4(c) shows the filtered results by the bilateral filter and the Gaussian filter, respectively. It can be seen that, through the noise intensity estimation, the bilateral filter suppressed the large scale noise effectively; but the traditional Gaussian filter blurred image seriously.

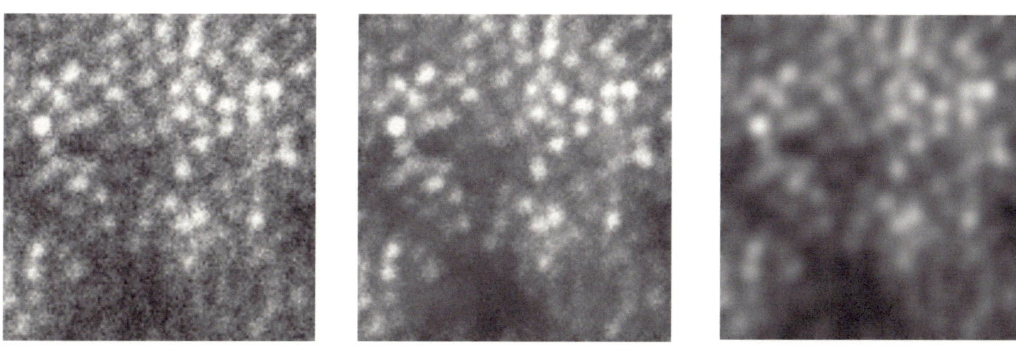

(a) (b) (c)

Fig. 2.4. (a) the photoreceptor cell image; (b) the result of the bilateral filter controlled by noise intensity estimation; (c) the result of Gaussian filter

3.2 Edge smoothing

In this section, we will discuss a difficult task in low level image processing, that how to smooth image edges polluted by noise, but not blur them. In our experiments, we find the coherent diffusion can be used to solve this problem under delicate control.

The partial derivative equation (PDE) based the anisotropic diffusion introduced by Perona and Malik [17] shows good capabilities of noise reducing and maintaining edges, by using an edge stopping function to carry out thermal diffusion between adjacent pixels, where diffusion are implemented within homogeneous areas and stopped near edge, the discrete P-M function can be expressed as follows:

$$I_{i,j}^{t+1} = I_{i,j}^{t} + \lambda\left[c_N \cdot \nabla_N I + c_S \cdot \nabla_S I + c_E \cdot \nabla_E I + c_W \cdot \nabla_W I\right] \tag{2-6}$$

where $I_{i,j}^{t}$ and $I_{i,j}^{t+1}$ are image pixels at time t and $t+1$; c_α is conduction coefficient, where α denotes four diffusion directions; λ is used to control diffusion intensity; ∇ is the gradient operator.

Generally, the PDE based algorithms can be roughly categorized into two classes: one is based on local pixel statistics, to balance the extent of noise smoothing and edge preserving, e.g. P-M anisotropic diffusion and speckle reducing anisotropic diffusion [18-19]; the other is based on local geometries, to control the smoothing behaviors both in intensity and direction, e.g. coherence-enhancing diffusion filtering [20] and nonlinear coherent diffusion [21].

In our study, we find that if the image contains the structural features, the geometric anisotropic diffusion performances much better than the local statistics anisotropic diffusion. The geometric anisotropic diffusion is not only protect the blurring of edges, but also coherent enhance jaggy edges. But the use of this kind of methods should be careful, that large scale noise must be suppressed sufficiently before this processing, so as to avoid the appearance of false edges. In our filtering framework, this part of work has been finished by bilateral filtering in previous section. In order to smooth the photoreceptor cell edges and filter the residual noise, the nonlinear coherent diffusion is utilized in our method, which can be expressed as follows:

$$\frac{\partial I(i,j,t)}{\partial t} = div(D \cdot \nabla I_{i,j}) \tag{2-7}$$

where, parameter D is diffusion tensor, which is usually constructed by structure tensor C with the following description:

$$C = \begin{bmatrix} \sum_w I_x^2 & \sum_w I_x I_y \\ \sum_w I_x I_y & \sum_w I_y^2 \end{bmatrix} \tag{2-8}$$

in which, W is the detecting window, used to capture image local patterns. After eigenvalue decomposition, structure tensor C can be written as follows:

$$C = \begin{bmatrix} U & V \end{bmatrix} \begin{bmatrix} \mu_1 & 0 \\ 0 & \mu_2 \end{bmatrix} \begin{bmatrix} U^T \\ V^T \end{bmatrix} \tag{2-9}$$

Where the eigenvectors U, V correspond to the directions of edge tangent direction and edge normal direction; the eigenvalues μ_1 and μ_2 stand for the strengths of maximum and minimum brightness variations.

As we have known, the two-dimensional structure tensor corresponds to an ellipse function (see Fig. 2.5), whose geometric properties are summarized as follows [22]:

1. if μ_1 and μ_2 are both small, the windowed image region is approximately located in flat areas;
2. if μ_1 is large and μ_2 is small, the estimation window is mostly located nearby boundaries;
3. if μ_1 and μ_2 are both large, this indicates that a corner probably locates in this region.

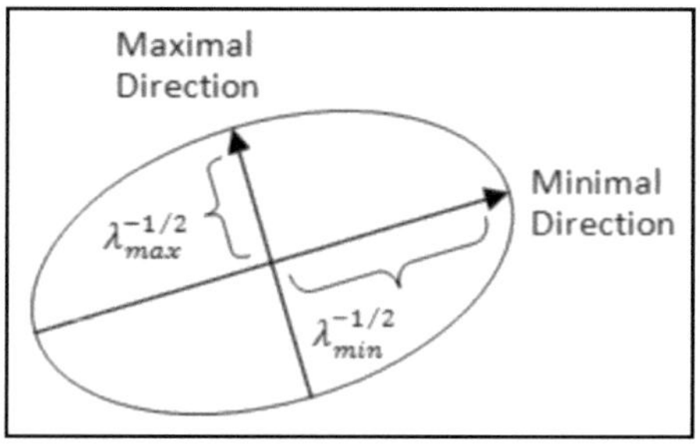

Fig. 2.5. Geometric expression of structure tensor

Within the framework of structure tensor driven diffusion, structure tensor C can be designed to control diffusion behaviors, based on the principle that diffusion tensor D should has the same principal directions with structure tensor C, but with different diffusion intensities. We use this conclusion to design a new diffusion tensor with the following description:

$$D(x,y) = \begin{bmatrix} U & V \end{bmatrix} \begin{bmatrix} \beta_{\min} & 0 \\ 0 & \beta_{\max} \end{bmatrix} \begin{bmatrix} U^T \\ V^T \end{bmatrix} \tag{2-10}$$

where $\beta_{\min}$ and $\beta_{\max}$ are the diffusion intensities of the tangent and normal directions, which take the form:

$$\begin{aligned} \beta_{\min} &= \begin{cases} 0, & if \quad \lambda_{\max} - \lambda_{\min} < \delta_{\min},\ \lambda_{\min} > \delta_{\max} \\ \alpha, & else \end{cases} \\ \beta_{\max} &= \begin{cases} \alpha \times \left(1 - (\dfrac{\lambda_{\max} - \lambda_{\min}}{\lambda_{\max}})^2\right), & if\ \beta_{\min} \neq 0 \\ 0, & else \end{cases} \end{aligned} \tag{2-11}$$

In our method, the coherence diffusion will take four different diffusion behaviors: if the detection window locates at corners, then $\lambda_{\max} \approx \lambda_{\min} > \delta_{\max}$ and $\beta_{\max} \approx \beta_{\min} \approx 0$, the diffusion will be stop there, see Fig. 2.6(b-a); if the detection window locates at edges, then $\lambda_{\max} >> \lambda_{\min} \approx 0$ and $\beta_{\min} \approx \alpha$, $\beta_{\max} \approx 0$, and the coherence diffusion will diffuse along the tangent direction, see Fig. 2.6(b-b) and (b-c); finally, if the detection window locates in the flat region, then $\lambda_{\max} \approx \lambda_{\min} \approx 0$ and $\beta_{\max} \approx \beta_{\min} \approx 0$, the isotropic diffusion will be carried out, see Fig. 2.6(b-d).

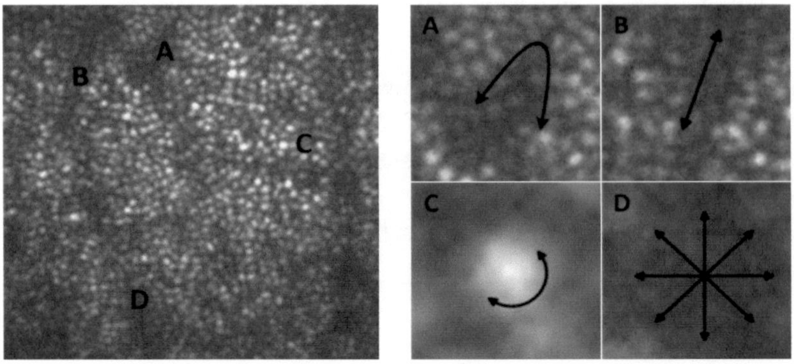

(a) (b)

Fig. 2.6. (a) the photoreceptor image with four structure features; (b) four diffusion behaviors used in our method

In the real applications, the parameters $\delta_{\min}$ and $\delta_{\max}$ used in the coherence diffusing were set 10 and 200, and the iteration was stopped when mean square error (MSE) is smaller than 0.001.

$$MSE = \frac{1}{M \times N} \sqrt{\sum_{i,j \in M.N} (I^t(i,j) - I^{t-1}(i,j))^2} \tag{2-12}$$

where M and N are the image size; $I^t(i,j)$ and $I^{t-1}(i,j)$ are the filtered images at time t and $t-1$.

Fig. 2.7(a) shows a part of photoreceptor cell image; Fig. 2.7(b) is the result of coherent diffusion controlled by our diffusion tensor; Fig. 2.7(c) is the result simply processed by the traditional nonlinear coherent diffusion. It can be seen that, our method filtered the residual noise containing in the bilateral filtering result, smoothed the irregular photoreceptor cell edges, and most important is that the combination of bilateral filtering and coherent diffusion effectively avoid the appearance of false edges, which are often produced by the traditional coherent diffusion methods.

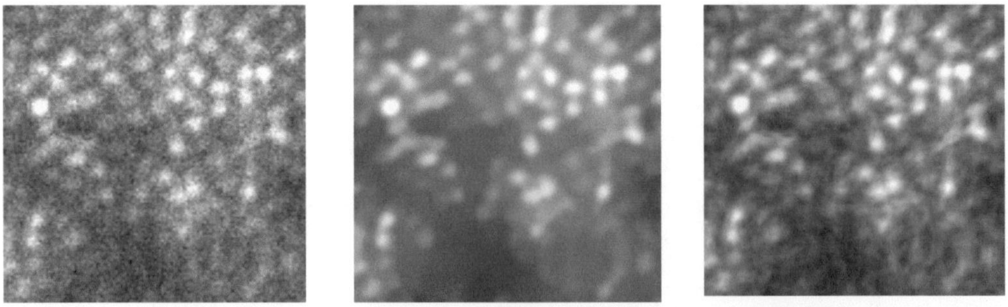

(a) (b) (c)

Fig. 2.7. (a) the photoreceptor image; (b) the result of our method; (c) the result of the traditional nonlinear coherent diffusion.

3.3 Contrast enhancement

In this section, we will discuss two methods of enhancing image contrast based on spatial and frequency domain respectively, and we will show that if combined with priori knowledge, these methods are more effective.

There are many contrast enhancing methods in low level image processing, such as gray level based transformation, histogram equalization and frequency domain high pass filtering and so on. However, these common methods usually partially improve image contrast in real application. Following, we will introduce a special method for photoreceptor cell images, and a band pass filtering method used by other articles [23].

In our method, image edges and brightness are extracted, and both of them are merged into the coherent diffused result. The proposed algorithm takes the form:

$$R = a \times E(I) + b \times B(I) + c \times I \tag{2-13}$$

where a, b and c are the weight coefficients; I is the coherent diffused image; $E(\cdot)$ is the edge image, detected by Sobel operator; $B(\cdot)$ is the brightness image, extracted by two overlapped windows, see Fig. 8, which is defined as follows:

$$B(x,y) = \frac{\sum_{i,j \in M_1} I_{i,j} \times g(i,j)}{M_1} - \frac{\sum_{\substack{i,j \in M_2 \\ i,j \notin M_1}} I_{i,j} \times g(i,j)}{M_2 - M_1} \tag{2-14}$$

where $g(i,j)$ is the normalized Gauss function; M_1 and M_2 are the detecting windows.

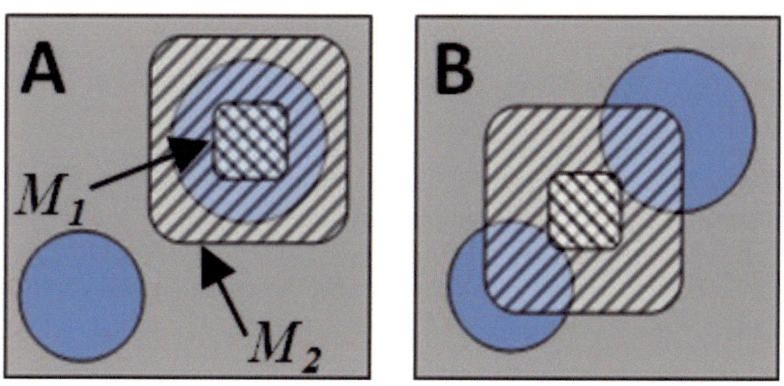

Fig. 2.8. Retinal cell image brightness detecting.

In our study, we have the priori knowledge that, the size of human photoreceptor cells is in the region of $(3\mu m, 6\mu m)$, the magnification of AO optical system is 11.7, and camera pixel size is $8\mu m$. With this information, we can determine the cells size in image within the region of $(5 \times 5, 9 \times 9)$ pixels, so that the size of the two overlapped windows is 3×3 and 11×11 pixels, respectively.

There are three relationships between detecting windows and photoreceptor cells to be consider: if M_1 locates in bright region and M_2 covers dark region, the output of $B(x,y)$ is positive, see Fig. 2.8(a); reversely, if M_2 locates in dark region and M_1 covers bright region, the output of $B(x,y)$ is negative, see Fig. 2.8 (b); finally, if both M_1 and M_2 locate in the flat region, the output of $B(x,y)$ is close to zero.

To illustrate our contrast enhancement processing more clearly, the interim edge image and the brightness image are displayed in Fig. 2.9(a) and (b), and Fig. 2. 9(c) is the merged image.

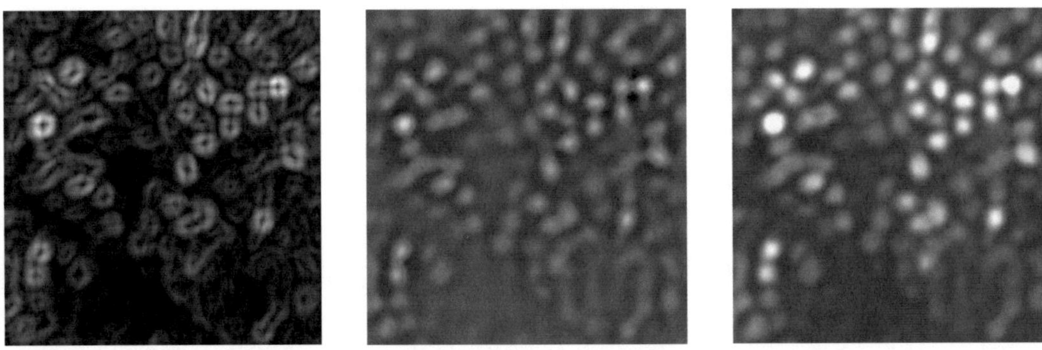

(a) (b) (c)

Fig. 2.9. (a) edge image; (b) brightness image; (c) merged result of edge, brightness and coherent diffused image.

In real applications, our method is applied in two group objects. Group 1 is the snapshot in the condition of dark-adapted pupil, and group 2 is the snapshot with dilated pupil (with Tropicamide 1%). Fig. 2.10(a) and (c) present two examples of the original AO photoreceptor cell image of group 1 and group 2 respectively, and the corresponding results are shown in Fig. 2.10(b) and (d).

To provide a quality measurement for our results, contrast criterion is used:

$$r = \frac{(\Delta I)_{rms}}{\bar{I}} \times 100\% \qquad (2\text{-}15)$$

The quality indexes defined by Eq. (2-15) before and after processing are 4.84% and 8.88% for Fig. 10(a) and (b), and 3.24% and 5.61% for Fig. 2.10(c) and (d). In order to illustrate the validity of our method further, Fig. 2.11(a) and (b) present the contrast indexes of 15 images of both groups respectively, which shows that the contrast of AO photoreceptor cell images are effectively improved by our method.

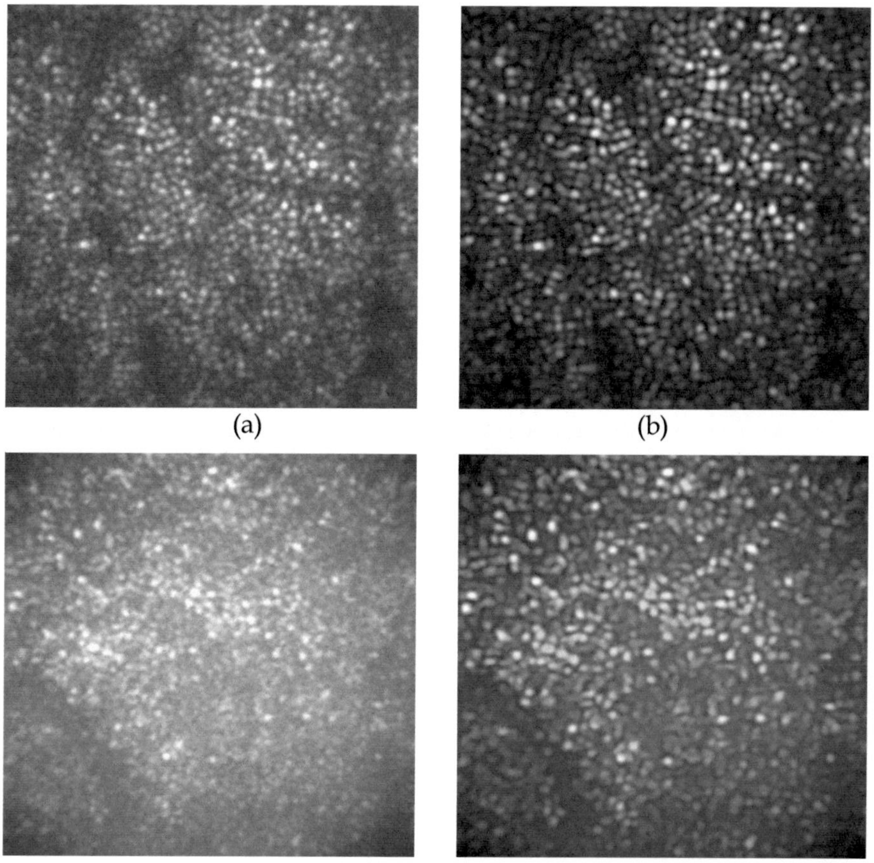

(a) (b) (c) (d)

Fig. 2.10. Results of hybrid filtering and enhancing (a) image captured with dilated pupil; (b) the processed result of (a); (c) image captured with dark-adaptive pupil; (d) the processed result of (c).

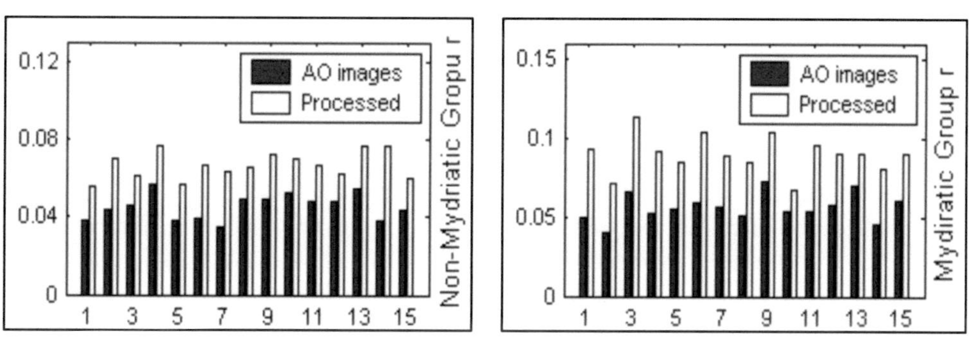

(a) (b)

Fig. 2.11. Statistical contrast improvements of both groups: (a) improvement of contrast criterions of group 1; (b) improvement of contrast criterions of group 2.

In contrast with our hybrid filtering method, the frequency domain algorithms could also be used in AO image processing. Here, we introduce a frequency enhancement method used for the photoreceptor cell images. As we have known, frequency processing tends to enhance image details and amplify noise at the same time, so it is important to estimate the spatial frequency of imaging objects. For the human photoreceptor cells, the spatial frequency can be estimation by following method:

$$f = \frac{2 \times f_{eye} \times tg0.5}{d_{cone}} \text{ (cycles/degree)} \qquad (2\text{-}16)$$

where f_{eye} is the focal length of human eye to be tested, with the experience value $17mm$; d_{cone} is the photoreceptor cell diameter, within the range $3 \sim 6um$. So the reasonable photoreceptor cell spatial frequency is in the range of $50 \sim 100$ cycles/degree. Fig. 2.12 is an illustration, in which Fig. 2.12 (a) shows the power spectrum of a photoreceptor cell image; Fig. 2.12(b) is the corresponding spectrum curve, from which the photoreceptor cell start-stop frequency is displayed.

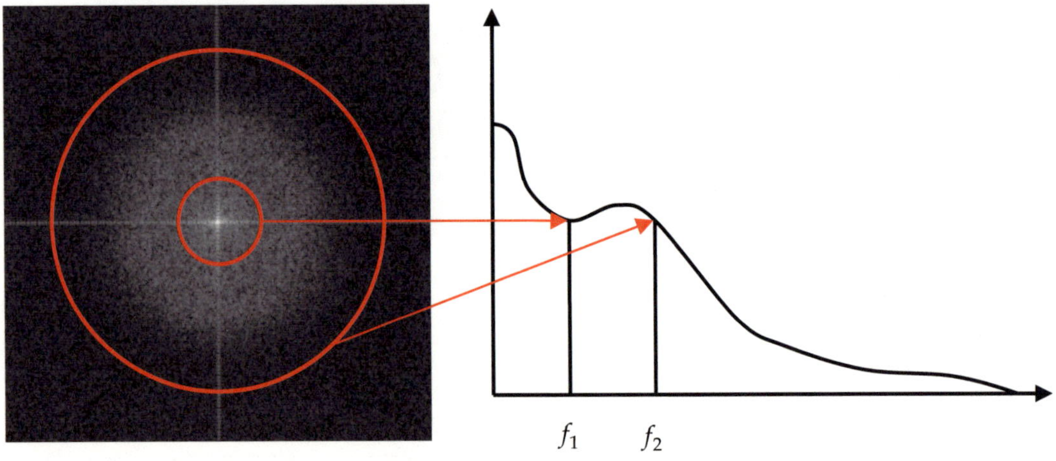

(a) (b)

Fig. 2.12. Illustration of spatial frequency estimation: (a) power spectrum of photoreceptor cell image; (b) corresponding spectrum curve

With this information, Butterworth band pass filtering can be used to suppress the high frequency noise, above cell stop frequency; and enhance the image energy, within the start-stop frequency. Fig. 2.13(a) is a typical photoreceptor cell image captured at 1.0 degree central eccentricity; Fig. 2.13(b) shows the result of frequency domain processing; Fig. 2.13(c) is the result of our hybrid filtering and enhancing method.

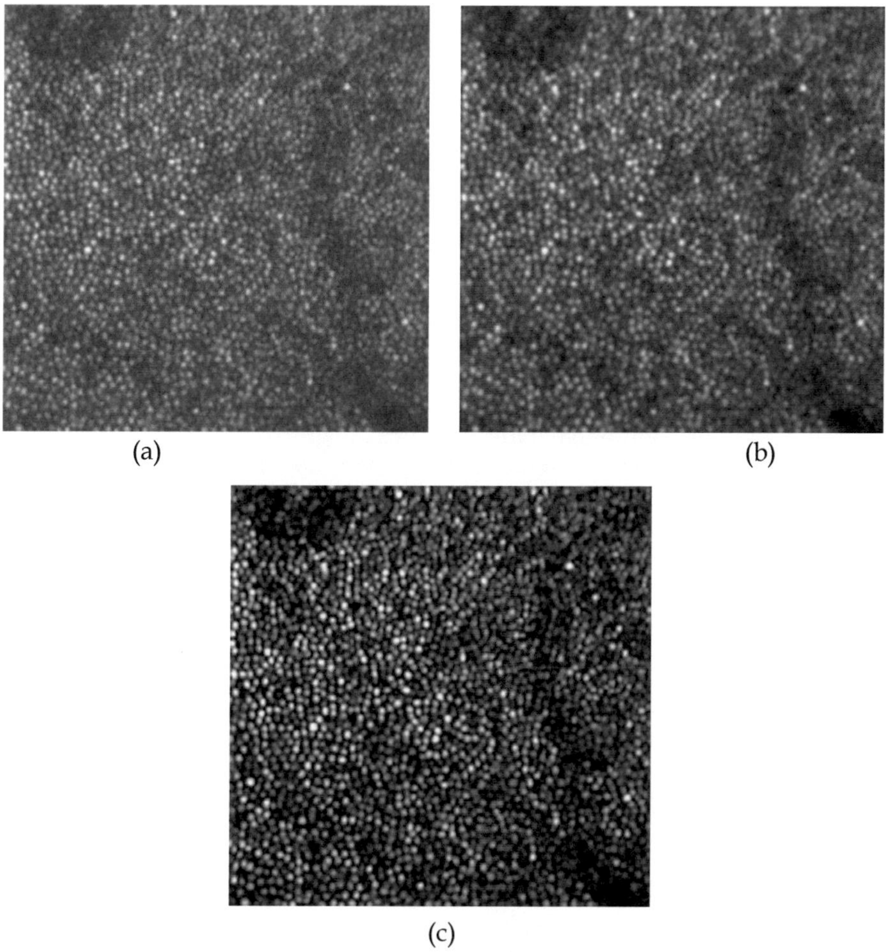

Fig. 2.13. The comparison of frequency enhancement with the hybrid filtering and enhancing method. (a) the photoreceptor cell image; (b) result of frequency processing; (c) result of our hybrid filtering and enhancing method.

It can be concluded from our experiments that the hybrid filtering and enhancing methods can effectively improve the photoreceptor cell image quality; due to the complex of the noise model in an AO imaging system, the combined denoising method performs much better than the traditional single filtering.

4. References

[1] Changhui Rao, Lei Zhu, Xuejun Rao, etc., "Performance of the 37-element solar adaptive optics for the 26cm solar fine structure telescope at Yunnan Astronomical Observatory", Applied Optics, Vol.49, No. 31, G129-135 (2010)
[2] Changhui Rao, Lei Zhu, Xuejun Rao, etc., "37-Element Solar Adaptive Optics for 26cm Solar Fine Structure Telescope at Yunnan Astronomical Observatory," Chinese Optics Letters 10, 966-968 (2010)

[3] Changhui Rao, Wenhan Jiang, Yudong Zhang and etc., "Upgrade on 61-element adaptive optical system for 1.2m telescope of Yunnan Observatory", Proc. SPIE. 5490, 943 (2004).

[4] Changhui Rao, Wenhan Jiang, Yudong Zhang and etc., "Performance on the 61-element upgraded adaptive optical system for 1.2-m telescope of the Yunnan Observatory", Proc. SPIE 5639, 11 (2004)

[5] Ning Ling, Yudong Zhang, Xuejun Rao and etc., "Small Table-Top Adaptive Optical Systems for Human Retina Imaging," Proc. SPIE, 4825, 99~108(2002)

[6] J.-P. Veran and F. Rigaut and H. Maitre and D. Rouan, "Estimation of the Adaptive Optics Long-exposure Point Spread Function using Control Loop Data," J.Opt.Soc.Am.A 14, 3057–3069(1997)

[7] J. N. Caron, N. M. Namazi, R. L. Lucke, C. J. Rollins and P. R. Lynn Jr.,"Blind data restoration with an extracted filter function," Optics Letters 26, 1164 (2001)

[8] J. N. Caron, N. M. Namazi, R. L. Lucke, C. J. Rollins,and P. R. Lynn Jr., "Noniterative blind data restoration by use of an extracted filter function," Applied Optics 32, 6884-6889 (2002)

[9] Yoshifumi Sudo, Naoshi Baba, Noriaki Miura, Satoru Ueno, and Reizaburo Kitai, "Application of self-deconvolution method to shift-and-add solar imaging", Applied Optics 45, 2707-2710 (2006)

[10] Rafael C. Gonzales and Richard E. Woods, "Digital Image Processing (2nd edition)," Pearson Education North Asia Limited, (2002)

[11] Tian Yu, Rao Chang-Hui and etc., "Hybrid Deconvolution of Adaptive Optics Retinal Images from Wavefront Sensing," Chinese Physics Letters, 25,105-107(2008)

[12] Rao Changhui, Zhang Xuejun, Jiang Wenhan et al.. Image deconvolution from Hartman-Shack wavefront sensing: indoors experimental result [J]. Acta Optica Sincia, 22(7), 789～793 (2002)

[13] J. J. Green and B. R. Hunt, "Improved restoration of space object imagery", J. Opt. Soc.Am. A, 12, 2859-2865 (1999)

[14] Noriaki Miura, "Blind deconvolution under band limitation", Optics Letters 23, 2312-2314, (2003)

[15] Hua Bao, Changhui Rao, Yudong Zhang, Yun Dai, Xuejun Rao and Yubo Fan. "Hybrid filtering and enhancement of high-resolution adaptive-optics retinal images", Opt. Letters 34(22): 3484~3486 (2009)

[16] C. Tomasi and R. Manduchi, Bilateral filtering for gray and color images, Proc. IEEE Int. Conf. On Computer Vision, Bombay, India, 836-846 (1998)

[17] P. Perona and J. Malik, "Scale-space and edge detection using anisotropic diffusion," IEEE Trans. Pattern Anal Machine Intell., vol. 12, no. 7, pp. 629-639, Jul. (1990)

[18] Y. Yu and S. T. Acton, "Speckle reducing anisotropic diffusion," IEEE Trans. Image Process., vol. 11, no. 11, pp. 1260-1270, Nov (2002)

[19] Y. Yu and S. T. Acton, "Edge detection in ultrasound imagery using the instantaneous coefficient of variation," IEEE Trans. Image Process., vol. 7, no. 3, pp. 1640-1655, Dec. (2004)

[20] J. Weickert, "Coherence-enhancing diffusion filtering," Computer Vision, vol. 31, no. 2/3, pp. 111-127 (1999)

[21] K. Z. Abd-Elmoniem, A. B. M. Youssef, and Y. M. Kadah, "Real-time speckle reduction and coherence enhancement in ultrasound imaging via nonlinear anisotropic diffusion," IEEE Trans. Biomedical Eng., vol. 49, no. 9, pp.997-1014, Sep. (2002)

[22] C. Harris and M. Stephens, A combined corner and edge detector, Proc. Vision Conference, Manchester, UK, 147-151 (1988)

[23] J. Z. Liang, D. R. Williams. Aberrations and retinal image quality of the normal human eye [J]. Opt. Soc. Am. A, 14(11):2873~2883 (1997)

5

Adaptive Optics for High-Peak-Power Lasers – An Optical Adaptive Closed-Loop Used for High-Energy Short-Pulse Laser Facilities: Laser Wave-Front Correction and Focal-Spot Shaping

Ji-Ping Zou[1] and Benoit Wattellier[2]
[1]*LULI, Ecole Polytechnique, CNRS, CEA, UPMC; Palaiseau,*
[2]*PHASICS, Ecole Polytechnique, Palaiseau,*
France

1. Introduction

Since its invention in 1960, the laser has become a powerful tool not only for researches in laboratory, but also in our everyday life. Numerous discoveries in many fields have been led by laser developments. Tremendous progress has been achieved for high-peak-power laser systems during the last decades thank to the use of the Chirped Pulse Amplification (CPA) technique (Strickland & Mourou, 1985) and due to the advance in laser materials and technologies .

Two families of high-peak-power lasers are distinguished:

- laser facilities working in nanosecond (10^{-9} s) regime but producing pulse energy as high as 1 MJ (1MJ=10^{6} J);
- laser facilities generating ultra-short femtosecond (10^{-15} s), ultra-intense (~10^{24} W/cm^{2}) pulses.

With lasers of the first group, such as National Ignition Facility (NIF) (Haynam et al., 2006) and Le Laser Mégajoule (LMJ) (Cavaillier et al., 2003), research can be performed in the field of laser induced dense and hot plasma physics, thermonuclear fusion, astrophysics in laboratory, etc. With lasers of the second group, such as the laser facilities proposed by the European project "Extreme Light Infrastructure" (ELI) (Korn & Antici, 2009), laser power of 200 PW will make possible to enter the ultra-relativistic regime, where protons become relativistic. Fundamental physics can also be investigated, including particle physics, vacuum structure, attosecond science, as well as photonuclear physics. In practice, the lasers of the first group can be used as pump laser for the second group. Also, the combination of nanosecond pulses together with picosecond pulses, or even femtosecond pulses, allows exploiting new research domains and enhancing time-resolved diagnostics in dense and hot plasma experiments.

The way for enhancing such laser system performance consists in increasing its delivered energy, reducing its pulse duration and improving its focal spot quality. In this chapter, we will discuss how focal spot quality can be improved by adaptive optics.

It is well-known that the focal spot quality directly depends on whether the laser wave front is perfect or deformed. Laser beam wave front control is of crucial importance for both of these two laser families. However, for large-scale high-energy laser facilities which belong to the first family, wave front control is particularly requested. Such laser systems are generally composed of a large number of optical components. At the end of the laser chains, the optical components are very large (their diameter exceeds 200 mm) and a good optical quality is hard to obtain at reasonable cost. Static aberrations may be generated by large size optics and their misalignment. In addition, large-scale high-energy laser facilities use flash-lamp-pumped rod and disk amplifiers for laser amplification. Once laser shots are performed, heat is accumulated in the amplifier medium due to the very low conversion efficiency from pump to laser energy. This heat generates a temperature rise in the materials and then an optical index change. These thermal effects deform the wave front of the next laser shots. However the heat is evacuated by thermal diffusion soon after the shot. Due to the poor thermal conductivity of the used laser glass, heat evacuation takes typically more than one hour for 200-mm-diameter class lasers. Thermally induced aberrations are the main origin for dramatic laser wave front distortion. As a result, high-peak-power laser systems suffer three consequences: 1) the focal spot quality degradation induces an important decrease of the effective laser intensity on target; 2) the focal spot shape will not be reproducible from shot to shot; so do experimental conditions on target; 3) the requested long thermal recovery time slows down the shot repetition rate so that the laser operation effectiveness of such facilities is limited to a few shots per day. Within the last decade we have continuously investigated on dynamic wave front control for the LULI (Laboratoire pour l'Utilisation des Lasers Intenses) laser facilities.

In this chapter, an overview will be given of our investigations and the wave front control results for the LULI laser facilities: an optimized procedure of the laser wave front control has been performed by an Adaptive Optics (AO) closed-loop combined with a semi-automatic alignment system. This procedure enables us to obtain a shot-to-shot reproducible focal spot on target with a near-diffraction-limited spot size. Also, the shot repetition rate is improved by almost a factor of 2.

As an application of laser wave front control and for the study of the parametric instabilities driven by laser-plasma interaction, we have successfully demonstrated the feasibility of the generation of a bi-speckle far-field pattern by using the same AO system.

In Section 2, we give a complete wave front analysis based on Zernike polynomial decomposition. Section 3 is dedicated to a description of the LULI AO system. Wave front correction, presented in Section 4, aims at improving the deformed laser wave front to a flat one. Its quality is characterized by the Strehl-ratio of the focal spot obtained for both low energy pulsed beam and full-energy shots. In Section 5, the bi-speckle generation procedure and its experimental results will be shown. The advantages and the limitations of this method will also be analyzed. Finally, in Section 6, we will conclude and discuss on future wave front control developments.

2. Wave front deformation measurement and analysis

The LULI laser facilities consist of two large-scale flash-lamp pumped laser systems. The first one, ELFIE, is based on the chirped pulse amplification (CPA) technique and is an upgrade from the former existing 100TW system (Zou, 1998, 2007). This laser starts with a diode-pumped Ti:sapphire oscillator, which generates a train of 100 fs optical pulses at 1057 nm. These pulses are stretched to nanosecond duration by an Offner stretcher and pre-amplified to the milli-Joule level by a Ti:sapphire regenerative amplifier. The main amplification chains are composed of four stages of flash-lamp pumped rod and disk amplifiers. The laser media is a mix of Nd^{3+}:Silicate and Phosphate glasses to ensure a large spectral bandwidth while optimizing the CPA gain. The beam diameter is as large as 100 mm. This system is capable of producing simultaneously twice 30 J/300 fs pulses, resulting in a peak power as high as 2x100 TW.

The second laser facility, LULI2000 as illustrated in Figure 1, uses large aperture rod and disk amplifiers for kilojoule/nanosecond pulse amplification at 1053 nm (Sauteret, 2000). Its amplification section is composed of two independent optical chains with flash-lamp pumped Nd^{3+}:phosphate glass amplifiers. Each is composed of four sequential amplification stages with different aperture diameters: 50, 94, 150, and 208 mm respectively. Chirped pulse amplification in this system allows to perform plasma experiments by combining the nanosecond pulses with high-energy sub-picosecond ones.

On the amplification chains of the both LULI laser facilities, rod amplifiers are cooled by continuous water circulation. Air and nitrogen gas flows are used to cool down disk amplifiers after each laser shot.

Fig. 1. LULI2000 laser facility: top: amplification chains; button left: command control room; button right: one of the interaction chambers "MILKA".

2.1 Laser beam wave front and far-field patterns

2.1.1 Aberrations and Zernike polynomials

One common way to represent the measured wave front is to use the Zernike polynomial base (Noll et al., 1976). This is a set of polynomials which are orthogonal within the unit disk. Examples of such polynomials are given in Fig. 2.

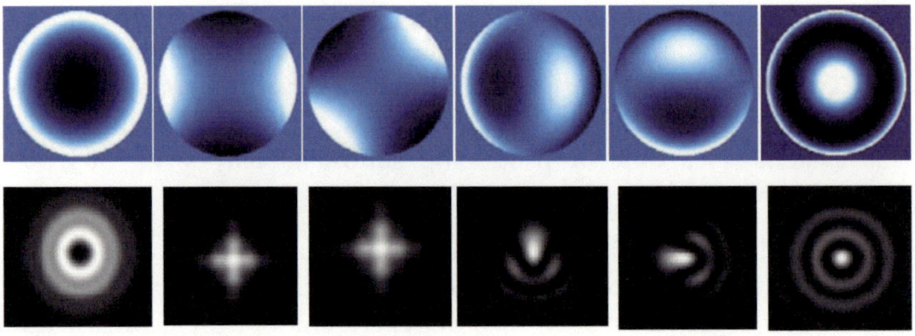

Fig. 2. Examples of Zernike polynomials(top raw) and their corresponding far-field pattern (bottom raw). From left to right, you see defocus, 2 astigmatisms, 2 comas and spherical aberration.

If the orthonormal set of Zernike is used, the Zernike representation leads quickly to the wave front RMS, by the formula:

$$\sigma^2 = \sum_{m,n} Z_{mn}^2 \tag{1}$$

2.1.2 Far-field patterns characterization

What interests mostly the laser physicist is to estimate the intensity where the beam is focused: the highest the intensity, the strongest the interaction of light with matter. In this section, we will present observables that characterize a far-field pattern.

The simplest observables are scalar values. However they lack information and are often only indicative.

- **Peak intensity (W/cm^2):** the maximum laser intensity. This is compared to intensity levels for matter phase changes. Its absolute value is very hard to measure because it is hard to measure how much energy is really in the focus. Indirect measurements are possible, such as evaluating the plasma ionization state, which is related to the peak intensity in the laser that generated the plasma.
- **FWHM (Full Width at Half Maximum) (µm)**: this is the diameter of the focal spot where the intensity is half the peak intensity. This FWHM is usually compared to the FWHM of a diffraction limited spot.
- **Focal spot contrast**: this is the ratio between the absolute peak intensity and the intensity of the first ring.
- **M^2:** this is the ratio between the focal spot 2nd moment and the 2nd moment of the focal spot of a perfect Gaussian beam. This is the most thorough scalar observable, since it is a propagation invariant.

- Strehl ratio: this is the ratio between the beam peak intensity and the peak intensity of a beam with the same near field intensity profile but with a flat wave front. Strehl ratio is often used to quantify how much intensity is lost due to the aberrations in the beam.

Beside these scalar values, the focal spot shape can be described using a radial averaging or the encircled energy. This is also called "Energy in the bucket". This represents for a given distance from the focal spot centroid the fraction of the total energy which lies within this distance.

$$E_e(r) = \frac{\iint_0^r I(x,y)\,dxdy}{\iint_0^\infty I(x,y)\,dxdy} \qquad (2)$$

An example of an encircled energy function is illustrated in Fig. 3 for the Airy pattern (focal spot corresponding to a diffraction limited top hat circular beam).

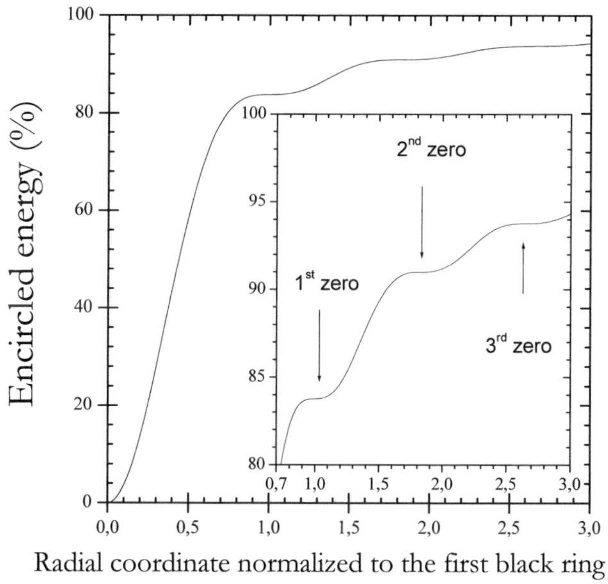

Fig. 3. Encircled energy graph for an Airy pattern. Horizontal tangents correspond to the zeroes of the Airy pattern.

Encircled energy quantifies the concentration of light energy in the focal spot. Nowadays most of the commercial high power lasers are specified using encircled energy. The points where the encircled energy derivative vanishes correspond to local extrema in the intensity pattern. From the above graph, we can deduce that the energy contained in the first Airy disk represents 83.7% of the total energy. Moreover it remains more than 5% of the energy after the third ring.

Directly measuring the encircled energy is very tricky since at the focal spot edge part the intensity goes very close to the noise level of a sensor.

2.1.3 Far-field pattern simulation from wave front measurement

The focal spot is difficult to measure and characterize directly. This is because low noise cameras are required to do so and relay optics that alter the focal spot shape are necessary to increase the focal spot size, so that it fits the CCD sensor aperture.

In the frame of the Fraunhofer approximation, the intensity in the focal plane is equal to the Fourier transform of the electric field in the near-field, or in the wave front measurement plane. Since a wave front sensor gives the intensity and the phase of the beam, the far-field pattern is easily deduced from a wave front measurement by a simple FFT (Fast Fourier Transform) algorithm. Wave front measurement is then an alternative to focal spot imaging.

2.2 Wave front sensors

2.2.1 Wave front sensors for laser beam characterization

There are several techniques to characterize the wave front of laser beams. Among the most widely used, we can cite:

- **Shack-Hartmann**: this is based on the displacement of focal spots of a microlens array, due to the phase gradients.
- **Propagation-based or curvature sensor**: this is based on the fact that phase information is transferred to intensity information during the propagation. Using an iterative algorithm, the wave front is recovered from at least 3 intensity measurements.
- **Quadri-Wave Lateral Shearing Interferometry**: the phase gradients are recovered from the deformation of an interferogram generated by a diffraction grating.

These three kinds of instruments form the family of the so-called wave front sensors (WFS).

2.2.2 Quadri-Wave Lateral Shearing Interferometer

Lateral shearing interferometry is a well-known technique to measure the phase gradients in one direction. The incident wave front is replicated into two identical but tilted wave fronts. After propagation, their mutual interference pattern is recorded with a CCD camera. The phase gradients are recovered from the fringe deformation, by means of a Fourier deconvolution around the interferogram fringe frequency. However, at this point, it lacks some gradient information to recover a full two dimensional phase-field. Multiwave interferometry (Primot & Sogno, 1995) extends this principle to more than one gradient direction. In the case of QWLSI, four replicas are created by a specific 2D diffraction grating. In this case, two gradients along two perpendicular directions are measured and then integrated to determine the field intensity and phase (Primot & Guérineau, 2000).

The interferogram deformation can be interpreted using either the wave or geometrical optics. Wave optics is more rigorous and underlies the interferogram numerical processing.

Geometrical optics is more intuitive to understand physical effects, such as the influence of spatial coherence on interferogram contrast.

2.2.2.1 Interferogram formation (Bon et al., 2009)

The incident light field complex amplitude is given in the frame of the slowly varying envelope by:

$$A(\mathbf{r}) = \sqrt{I(\mathbf{r})}\exp(i[\mathbf{k}\cdot\mathbf{r} - \varphi(\mathbf{r})]) \tag{3}$$

where $\mathbf{r}$ is the position vector, $\mathbf{k}$ the mean wave vector, I the field intensity, and φ the field phase. The wave front sensor measures the field phase.

The beam to be characterized enters the interferometer and is replicated by a diffraction grating. In the case of our QWLS Interferometer, the so-called Modified Hartmann Mask (MHM) is used (Primot & Guérineau, 2000). It is made of the superposition of a Hartmann mask (amplitude grating of period $\mathbf{p}$) and a π-shift checker board (phase grating of period $2\mathbf{p}$) as presented in Fig. 4. This has been optimized to diffract more than 90% of the light energy into the 4 first orders carried by 4 wave vectors (see Fig. 4).

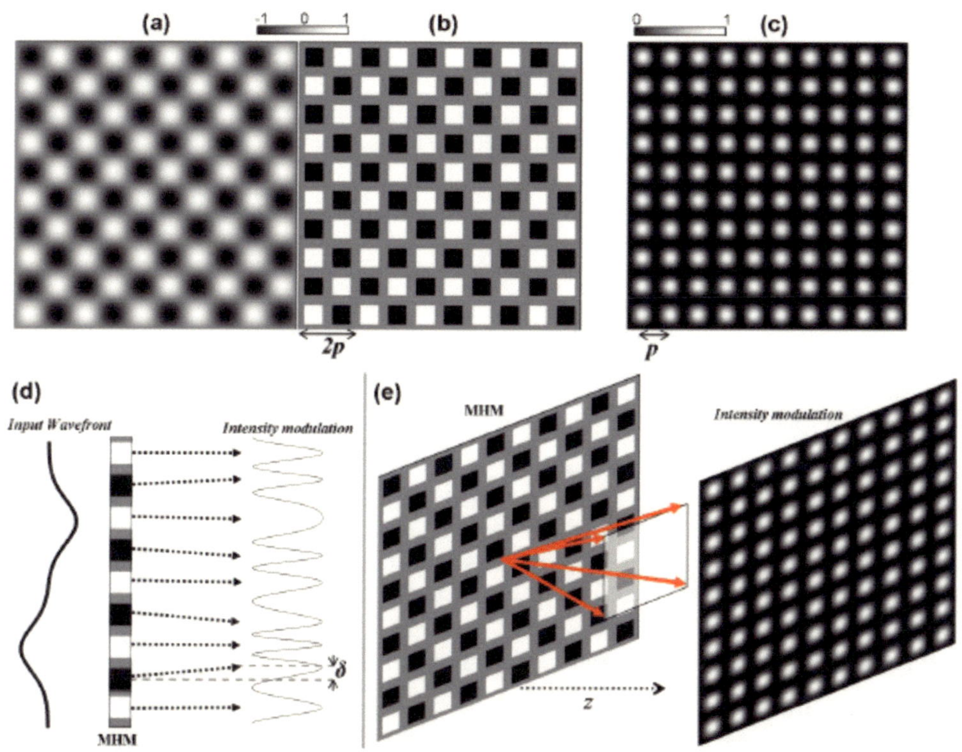

Fig. 4. (a) Ideal transparency for a 4-wave-only 2D diffraction grating. (b) MHM transparency, approximation of (a). (c) Intensity transmission of (a). (d) 1D interferogram formation by a MHM in the case of a disturbed input wave front; geometrical approach. (e) 2D interferogram formation by a MHM in the case of a plane input wave front; visualization of the 4 diffracted waves (arrows).

Each diffracted order propagates along its own wave vector direction. After a propagation length z along the z-axis, in the scope of paraxial propagation and if we neglect free space diffraction, the electromagnetic field is the coherent addition of all the replicas, which are mutually displaced due to their deviation. It has already been shown, in a somewhat different form, that the intensity is given by:

$$I(\mathbf{r},z) = I_0\left\{1+\left[\cos\left(\frac{2\pi}{p}x+\frac{2\pi}{p}z\frac{\partial OPD}{\partial x}\right)+\cos\left(\frac{2\pi}{p}y+\frac{2\pi}{p}z\frac{\partial OPD}{\partial y}\right)\right]+ +\frac{1}{2}\left[\cos\left(\frac{2\pi}{p}(x+y)+\frac{2\pi}{p}z\frac{\partial OPD}{\partial(x+y)}\right)+\cos\left(\frac{2\pi}{p}(x-y)+\frac{2\pi}{p}z\frac{\partial OPD}{\partial(x-y)}\right)\right]\right\} \tag{4}$$

where I_0 is the interferogram maximum of intensity in z = 0.

The interferometer sensitivity is determined by the ratio p=z. This means that the interferometer sensitivity is easily tunable by moving the grating back or forth in front of the CCD sensor.

2.2.2.2 Interferogram analysis

The interferogram is interpreted as a non-uniform OPD (Optical Path Difference) function, frequency-modulating a perfect sinusoidal pattern. The OPD gradients are then recovered by demodulating the recorded interferogram around its specific carrier frequencies. The result is two gradient fields, which are then integrated:

$$\begin{pmatrix} H_x \\ H_y \end{pmatrix} = \frac{2\pi}{p}z\begin{pmatrix} \dfrac{\partial OPD}{\partial x} \\ \dfrac{\partial OPD}{\partial y} \end{pmatrix} \tag{5}$$

Because the phase information is coded in the interferogram by a frequency modulation, the phase and the intensity are independently determined. The intensity image is recovered by applying a low-pass filter on the interferogram.

From Equation(5), both the interferogram period and the modulation due to the OPD are wavelength independent. Under the hypothesis of a wavelength independent beam OPD, the measured OPD with the wave front sensor is the same using either a monochromatic source or a polychromatic one. This makes this technique best suited for short pulses characterization where the spectral bandwidth can be as large as 100 nm.

2.3 Wave front deformation measurement and analysis

A flash-lamp pumped high-peak-power laser is a complex system. It is characterized not only by the use of a large number of large-scale optical and laser elements, set up in a chain extended on an important length, but also by a long thermal recovery cycle between two laser shots. Most of the broadband pumping energy from flash-lamps is actually not transferred to the laser pulses: it is dissipated as heat in the amplifiers host. The consequent temperature rise deforms the amplifying materials, even though the cooling system for each amplifier runs during or immediately after the shot. As a result, thermally induced aberrations appear and considerably evolve.

We measure and analyze laser wave front distortions in three different situations:

1. When the chain is "cold" (the laser had not been shot yet), we investigate the static aberrations induced from the imperfection in optics and its misalignment.
2. After a single full-energy shot and during several hours, we carry out the measurement of the thermal-effect-induced wave front distortions. It leads us to study the nature, the

magnitude and the temporal evolution of the main thermal aberrations in our laser systems.

3. The wave front distortions measured during a shot sequence help us to understand how the cumulative thermal effect degrades further the beam wave front and hence the focal spot quality on target.

2.3.1 Wave front sensor implementation on LULI laser beams

The wave front sensors used for the LULI laser facilities are based on quadri-wave lateral shearing interferometry. This technique is commercialized by Phasics under the name SID4®.

To evaluate the wave front distortions in the LULI laser systems, we implemented a wave front sensor SID4 at the end of each amplification chain. In parallel, a far-field device composed of an afocal system and a 12-bit CCD camera was set up to measure the corresponding focal spot.

2.3.2 Static aberrations

The static aberrations can be characterized when the chain is "cold", i.e. the chain is not subject to any pump generated thermal effect before the first shot of the day. According to their origin, we can distinguish the static aberrations in two categories.

The first category is determined by optical quality of each element, therefore its amount remains constant for laser everyday operation. For instance, the LULI2000 amplification chain is built up of more than 100 optical components. Before implementation, optical quality of each element was tested separately by a ZYGO interferometer to check if its optical quality is as good as given by the specifications. For each, the measured transmitted or reflected wave front deformation (depending on if it is a transmissive or reflective optical component) is found to be less than or equal to $\lambda/4$ peak-to-valley (PtV) and about 0.025 λ rms at λ_0= 1053 nm.

The second category is generated by laser axis misalignment. Its amount may change from laser day-to-day operation. The chain is "misaligned" when the beam axis does not perfectly coincide with the well-defined axes of the optical components. For both of the LULI laser facilities, laser beam alignment consists of centering the laser beam on alignment targets (crosshairs) for each amplification stage. The accuracy is typically 2% of the corresponding beam diameter. In parallel, beam pointing of each section is controlled by a far-field camera. Depending on pupil conjugations, the accuracy is different along the chain. The final beam pointing accuracy should be determined by the focal spot size and its alignment tolerance on target in the center of the interaction chamber. It is about 3 μrad in the case of LULI2000. When beam misalignment occurs, tilt-induced coma is increased. Normally coma remains negligible in our chain, because the laser beam axis should coincide with the axes of optical components, especially the lenses used for spatial filters. The amplitude of the coma, measured when the chain is "cold", can indicate us whether laser alignment is accurate or not.

At the end of the aligned chain the total measured wave front deformation (given in PtV) resulting from the cumulated wave front distortion of categories 1 and 2, is approximately 0.5 λ and the total rms value is 8 to 10 times smaller. It can be decomposed in low order Zernike aberrations with a similar amplitude among them. The corresponding Strehl ratio *Rs* is typically about 0.9.

2.3.3 Wave front deformation evolution after a full-energy shot

As we have shown in the former paragraph, Xe-flash-lamps have a broad spectrum extended from UV to IR (Brown, 1981), whereas the absorption line of Nd^{3+} for laser transition is very narrow. The thermal load originates therefore from the flash-lamp light absorbed within the Nd:glass, the cladding at the laser glass boundary and the mechanical mounts. This induces local dilatation and mechanical stress in the glass. Therefore effects like thermal lens, birefringence, etc. occur. Investigations have been done in order to measure and compensate these effects (Jasbir et al. 1986; Gopi et al. 1990; Liu e al. 2010; Kuzmin et al. 2011). The wave front aberrations induced by thermal gradients can be distinguished in two classes: the so-called pump-shot aberrations, appearing instantaneously during a shot, and the aberrations induced by thermal relaxation during the following hours. In this section we are focusing on the second class.

After a full-energy shot, a low-energy 10 Hz probe beam is immediately injected into the chain and propagates through all the chain towards the wave front sensor. Fig. 5 illustrates the typical temporal evolution of the wave front deformation at the LULI laser facilities. Wave front measurement has been performed once every two minutes during all the thermal recovery time: 40 minutes for ELFIE (Fig. 5.a) and more than 3 hours for LULI2000 (Fig. 5.b). Composed of larger optics and producing higher amplification gain, LULI2000 suffers more significantly from wave front deterioration than ELFIE. Three zones are clearly distinguished from the wave front deformation curve measured at LULI2000: 1) the rapid increases of the distortion during about 10 minutes; 2) the decrease during 50 minutes owing to the cooling performed by water, air and nitrogen flows; 3) the phase of slow evolution from one hour to more than 3 hours after the shot.

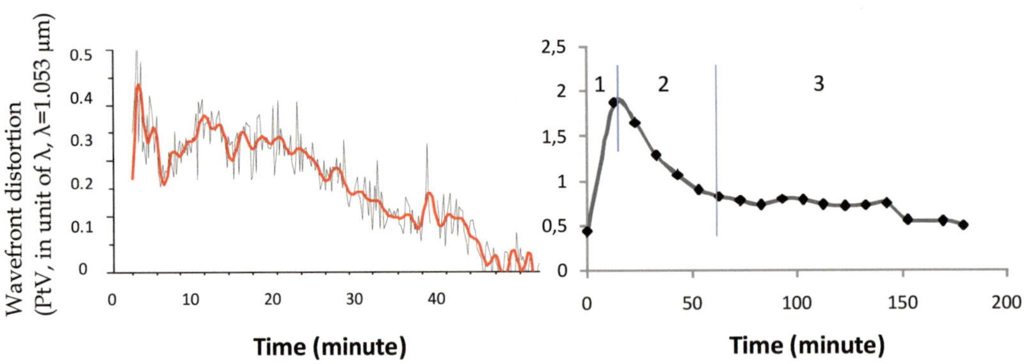

(a) (b)

Fig. 5. Temporal evolution of the wave front distortion measured after a full-energy shot at the end of (a) ELFIE and (b) LULI2000.

At the LULI laser facilities, the main thermal aberrations induced by a full-energy shot are defocus and astigmatism 0°. Fig. 6 presents their temporal evolution during the recovery time, decomposed from Fig. 5.b in the case of LULI2000.

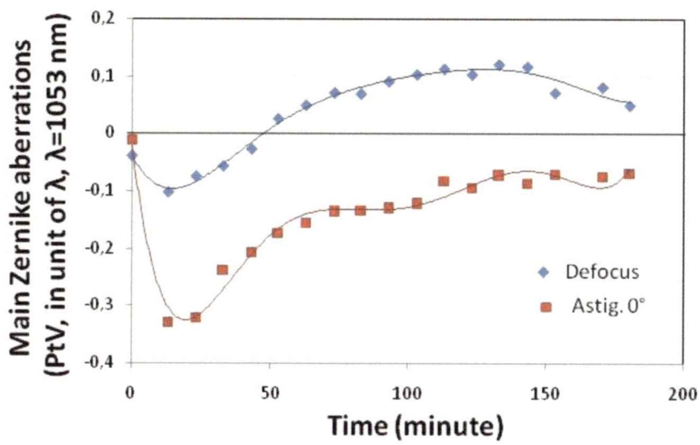

Fig. 6. Temporal evolution of defocus and astigmatism 0° after a full-energy shot (LULI2000).

The defocus aberration results from the thermal-lens effect induced due to the change of the temperature-dependent refractive index $\Delta n(r)_T$ (Koechner, 1999). Its value is determined by the temperature transverse-gradient from the center to the edge of the laser media and thermo-optic coefficient dn/dT:

$$\Delta n(r)_T=[T(r)-T(0)](dn/dT) \tag{6}$$

The same phenomenon has been observed in both laser systems (Figure 6): The defocus changes its sign from negative to positive values as time increases. This is easily understood if we consider that two effects compete: heat diffusion inside the glass and cooling at the edges. Heat diffusion happens when the temperature is not homogeneous in a material : heat flow travel from hot to cold areas. Edge cooling evacuates heats at the material boundaries. Since, in a rod amplifier, the flashlamp energy is more absorbed at the edges, the temperature is higher at the edges. Then the edges are cooled down until the temperature is the same at the center as at the edges: the curvature vanishes because the temperature is uniform. But the cooling continues to evacuate the heat and the edges are then cooler than the center: a new thermal lens appears with an opposite sign curvature. Finally the defocus disappears more than three hours after the shot. This is illustrated in Table 1.

Time (minute)	Operations & effects	Temperature evolution	Beam curvature evolution
0	Laser shot	$T_{edge} = T_{center}$	Beam curvature => 0
0 - 45	Thermal load > cooling	$T_{edge}>T_{center}$	Divergent curvature
45		$T_{edge} = T_{center}$	Zero crossing
45 – 180	Thermal load < cooling	$T_{edge}<T_{center}$	Convergent curvature
> 180	Thermal recovery	$T_{edge} = T_{center}$	Beam curvature => 0

Table 1. Synthesis on time evolution of the defocus, generated by thermal lens effect after a full-energy shot.

As for astigmatism, it can be produced in rod amplifiers by thermal-stress-induced birefringence. However, a strong astigmatism is observed and is identified mainly in our disk amplifier stages. The disks in a disk-amplifier have oval shape and their longer side is set up under Brewster's angle with respect to the beam propagation direction. The astigmatism can be generated under 2 situations: a) a spherical wave front, induced by thermal lens effect in the former amplification stages, propagates in the disks set up at Brewster's angle; b) thermal gradient is more important along the shorter side of the gain medium compared to the longer one, considering the flash-lamp geometry in the disk amplifiers. The second situation is dominant in our case and leads to a thermal cylindrical lens. In LULI2000, to generate a vertical rectilinear polarization (90°) at the end of the chain, the last disk amplifier is oriented with the shorter side of its disks in horizontal direction (0°). As a result, the wave front sensor inspects an important astigmatism 0° and its maximum value is measured 20 minutes after the shot.

2.3.4 Wave front deformation induced from cumulated thermal effect

The time lapse between two successive shots should be well matched with the thermal recovery time. This is necessary for the complete heat dissipation in all amplifiers of the chain. If the next shot takes place before the system is completely thermalized, cumulative thermal effects further degrade the laser wave front. Before the AO system was used, laser shot repetition rate of our chains resulted from a compromised value. For this, we defined the shot cycle by taking into account the facility use effectiveness and the focal spot quality on target. The shot repetition rate is one shot every 20 minutes for ELFIE and one shot every 2 hours for LULI2000. In this case, the complete recovery time is longer than the time delay between two shots for both of these two systems (Fig. 5a and Fig. 5b). Therefore, cumulative thermal effects appear when the laser chains operates in the desired shot sequence.

Fig. 7 presents the temporal evolution of the wave front distortion measured at LULI2000 during a shot sequence of 6 kJ-shots. The repetition rate of this shot sequence is one shot every hour. Two important features can be observed: 1) A complete thermalization after a full-energy shot is characterized by three distinguished phases: a rapid increase, a rapid decrease and a slow evolution of the plateau. However, for a "kJ" shot sequence with one-hour delay between shots, the next shot takes place before the end of the second phase. As a consequence, cumulative thermal effects impact the wave front deformation amplitude more and more important. 2) If we are only interested in the wave front deformation amplitude measured immediately before the shot (see Fig. 7, and Fig. 8 solid marks), we can find the wave front deformation amplitude increases according to an logarithmic tendency.

To characterize the influence of the shot repetition rate on the wave front quality, we compare the wave front distortions induced during a sequence of five high-energy shots with two different shot cycles: for ELFIE, in 20 and 40 minutes cycles; and for LULI2000, in one and two-hour cycles. Immediately before a next shot, the wave front and the associated far-field pattern were systematically recorded. The black curve in Fig. 8 illustrates the measured wave front distortion with one-hour recovery time in the case of LULI2000. We see that the wave front error increases considerably, resulting from accumulated heat during the shot sequence, and that a distortion of more than 3λ is measured at the fifth shot. The wave front degradation is still visible under the cumulated thermal effect even though the shot cycle is one shot every two hours (red curve in Fig. 8). Once decomposed on the set of

Zernike polynomials, we note that under the cumulative thermal effect, the main wave front deformations are always defocus and astigmatism 0° and their amplitude increases continuously along the shot sequence.

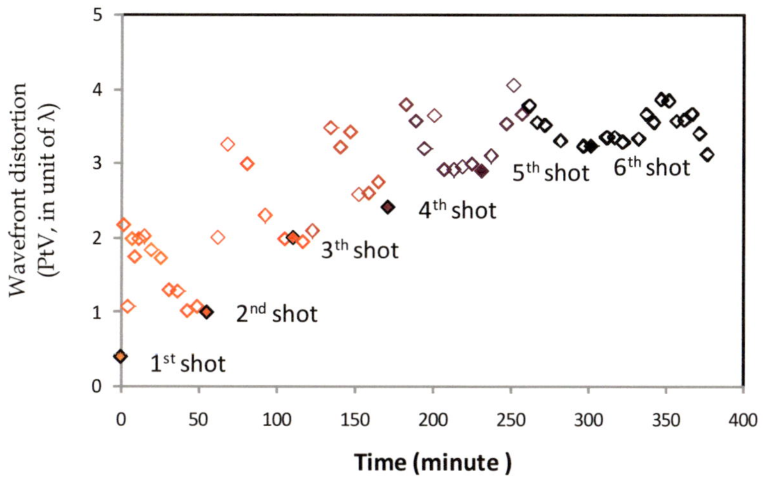

Fig. 7. Wave front distortion temporal evolution measured with a probe beam during a sequence of 6 shots with one shot every hour in LULI2000.

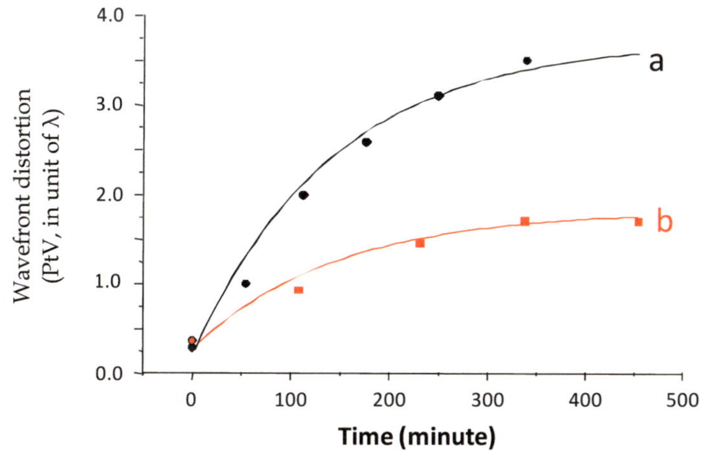

Fig. 8. Cumulated thermal effect comparison: the wave front deformation measured before each kJ shot of LULI2000 during 2 shot sequences: one-hour delay between shots (black curve); two-hour delay between shots (red curve).

Fig. 9 illustrates the corresponding focal spot degradation for a shot sequence of one hour cycle at LULI2000. The initial near-diffraction-limited single spot is enlarged and transformed in a dissymmetrical pattern surrounded by several side lobes. The effective laser intensity decreases dramatically, characterized by a deduced Strehl ratio about 0.2 at the end of the shot sequence.

Shot N°	1st	2nd	3rd	4th	5th
Focal spots	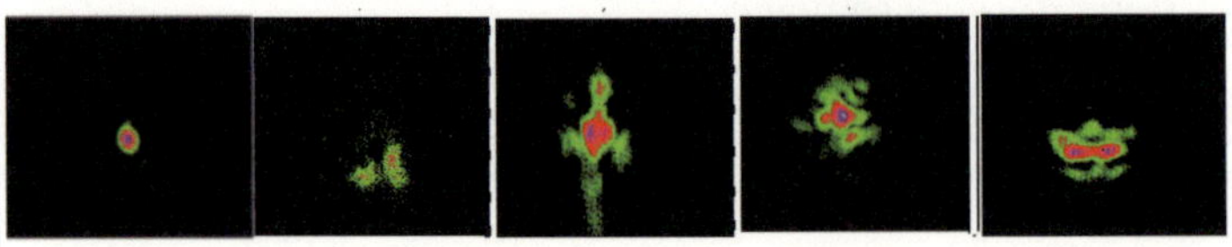				

Fig. 9. Focal spots degradation under the cumulated thermal effects, measured immediately before each shot for a shot sequence of one-hour cycle.

3. Adaptive Optics (AO) systems for the LULI laser facilities

We have previously identified that thermal aberrations are the principal source of the wave front deterioration for the LULI laser facilities. To compensate these wave front errors and thus improve the laser focusability on target, the use of an adequate adaptive optics (AO) system is indispensable. In the following paragraph we describe at first the AO system we designed and implemented into the LULI laser systems, then we present the optimized wave front control procedure and finally we show how the AO system performance can be improved with a suitable high-order-mode filtering.

3.1 LULI Adaptive Optics closed-loop

The LULI AO system is composed of three key elements : a wave front sensor, a wave front corrector and a convergence loop. All these should be well studied to enable a rapid and efficient wave front error compensation. The corrector we use is a bimorph deformable mirror (DM) with 32 actuators distributed on three rings (Kudryashov, 1996). It has a large aperture (diameter of 100 mm) and a high damage threshold guaranteed by its multi-dielectric coating. The applied voltages on the actuators range from -300 to +300 V, corresponding to a dynamics for wave front correction of more than 6λ.

The results of AO wave front control for 100TW/ELFIE are described by the referred article (Wattellier et al. 2004). In this paragraph, only the AO system in LULI2000 is described. The DM was placed between the second and the third amplification stages and was associated with the SID-4 wave front sensor located at the chain output, as shown in Fig. 10. The measurement plane and the correction plane have to be optically conjugated to each other in order to have a linear response between them. The correction loop was applied using a low energy 10 Hz pulsed beam (λ =1053 nm).

To calibrate the mirror, we measure the wave front deformation induced by the DM actuators. Thirty-two phase-map influence functions are therefore obtained by applying a voltage of 150 V to each of them one after another. These 32 wave front maps are the columns of a matrix that links the voltage to the phase maps (Dalimier & Dainty, 2005). By inverting this matrix, we can deduce the voltages to apply in order to generate any phase map.

For each loop iteration, the wave front error is calculated as the difference between the measured aberrated wave front and the reference defined in our case as a flat spatial phase. The adjusted voltage is thus determined and applied for each actuator to minimize this error. In parallel, a far-field measurement characterizes the wave front quality improvement performed by the loop: during several seconds/iterations, the degraded focal spot is transformed to a single spot close to a diffraction-limited Airy pattern. A similar focal spot can also be computed using the measured laser intensity distribution and the phase.

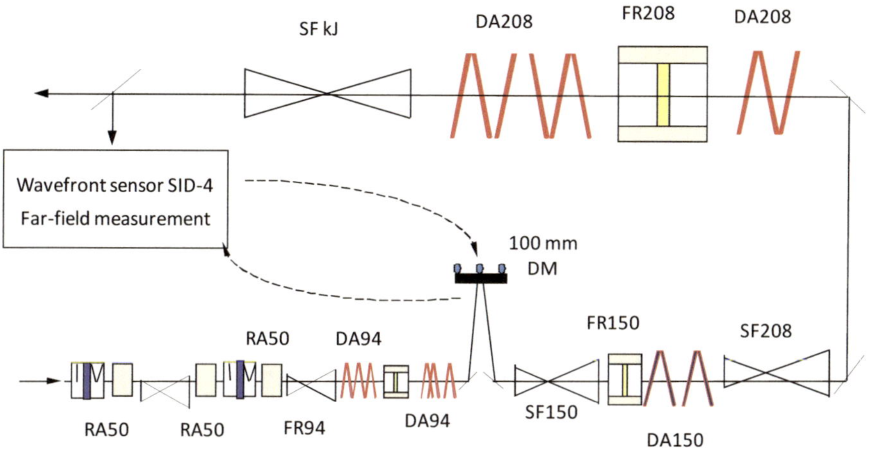

Fig. 10. A 100 mm diameter bimorph DM, implemented at the output of the second amplification stage of LULI2000, is related to a wave front sensor (SID-4) and a far-field measurement module placed at the chain output. RA, rod amplifier; DA, disk amplifier; FR, Faraday rotator; SF, spatial filter.

3.2 Optimized wave front control

We have shown before that the laser misalignment may generate coma. Nevertheless, coma can also be induced by thermal tilt. For the LULI laser facilities, we deal differently with the aberrations due to the thermal tilt and all the remaining thermally induced aberrations. An optimized procedure for our wave front control system has been elaborated and characterized by a precise beam realignment between two successive shots and an AO closed-loop operation. Its main features are as follows. (1) We perform a beam alignment before the first shot, so that the wave front errors, corresponding to the second category of the static aberrations, are reduced to a minimum. The AO system is also activated at this stage. This way, the residual wave front error of the static aberrations is reduced from 0.5λ to 0.2λ PtV. (2) To correct the aberrations induced by thermal tilt, which appear between two successive shots, beam pointing and centering are checked and readjusted systematically with the semi-automatic alignment system. This way, thermally induced tilt and coma are minimized. Also, such an alignment allows a very accurate beam centering and pointing on the DM so that its response matrix is always the same from shot to shot. (3) The correction of the remaining thermal aberrations is made by the AO system.

For an optimized closed-loop performance, a fitted higher-order-mode filtering after a calibration procedure is necessary to define the closed-loop configuration. To understand its importance, we carried out the measurement of the residual rms phase distortion with respect to the mode number of the DM used for correction. As shown in Figure 11a, as the mode number increases, the residual phase distortion decreases continuously from more than 0.2 rad until its minimum value of 0.03 rad with 24 modes. With different mode filtering in the loop configuration, we compare the residual phase maps and their corresponding focal spot patterns. It is obvious that an adequate higher-order mode filtering (8 modes filtered in our case) enables an excellent wave front correction and so the focal spot quality improvement (Fig. 11a and Fig. 11b). If the mode filtering is too important (e.g. 22 modes filtered, 10 modes kept) or not enough (e.g. only 2 modes filtered, 30 modes kept), the wave front correction is not complete; it results in a degraded peak intensity in the focal spot with side-lobes. In the 24-mode loop configuration, the maximum applied voltages are approximately 50 V. However, the voltages applied to the outer ring actuators are much higher and with strong oscillation when higher-order modes are activated. In fact, the higher-order modes have small eigenvalues, so the loop is more sensitive to noise perturbation and is difficult to converge. An adequate higher-order mode filtering for our AO system ensures, two at a time, high correction dynamics and a more stable loop convergence.

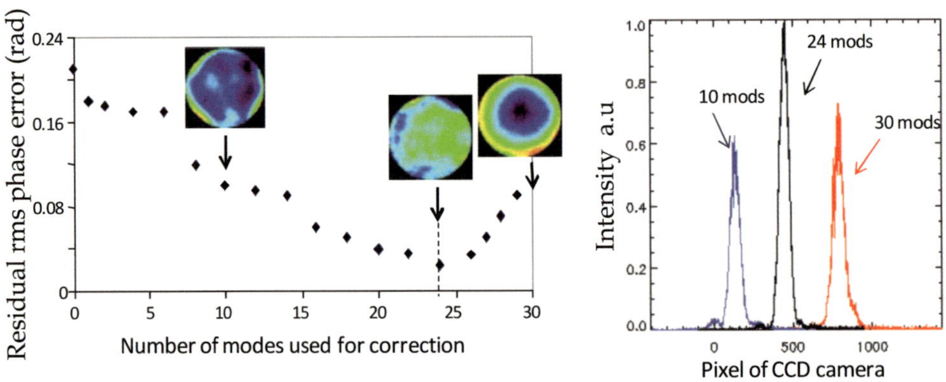

(a) (b)

Fig. 11. Residual rms phase error vs the mode number used for the correction loop. The residual phase maps(a) and the focal spot profiles(b) corresponding to 10, 24 and 30-mode loop configurations.

4. Wave front correction results

The dynamic wave front correction at the LULI laser facilities is performed only after the fast-evolution phase: 20 minutes after the shot for ELFIE and one hour after the shot for LULI2000. The wave front correction results presented in this paragraph have been obtained from LULI2000 facility during a shot sequence of 5 full-energy shots (one shot per hour). Before each shot, we first measure phase distortions and the corresponding focal spots. We perform a semi-automatic alignment to correct the residual thermal tilt before starting the convergence loop with the help of a low energy pulsed-probe beam.

Fig. 12 shows the residual phase measured before the 5th kJ-shot. The phase distortion, composed of essentially defocus and astigmatism 0°, is reduced significantly after using AO closed-loop: from more than 3λ to about 0.3 λ (PtV). In the meantime, a far-field device measures the associated focal spot evolution: the focal spot, strongly degraded by the cumulative thermal effects is restored after the loop convergence. A nearly Airy-disk like focal spot is obtained with a deduced Strehl ratio greater than 0.9. The AO loop is locked when we fire the next shot.

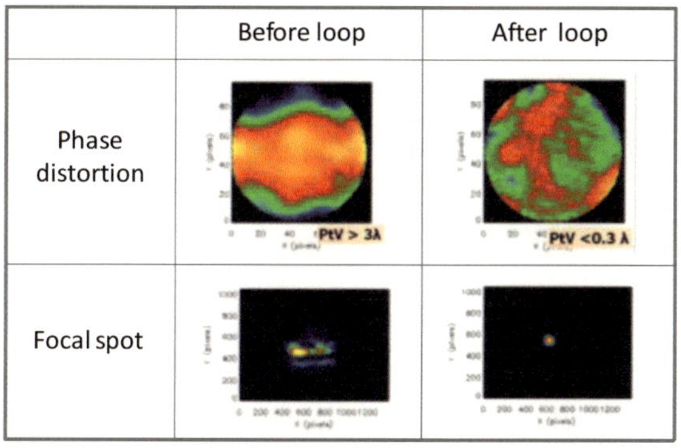

Fig. 12. Phase distortions and their corresponding focal spots measured before and after the loop performed with a 10 Hz pulsed laser for the 5th kJ shot of the shot sequence (one-hour delay between shots).

The focal-spot quality illustrated by Fig. 13 is evaluated during the 5th shot with AO wave front control. Compared with the case without wave front correction, the encircled energy of the focal spot has been considerably improved and is similar to that of the corresponding Airy spot. The Strehl ratio deduced from the measured focal spot is increased from 0.2 (without loop, Figure 10a) to more than 0.7. The difference of the Strehl ratio between 0.9 (probe beam) and 0.7 (kJ beam) comes essentially from the pump-shot induced aberrations.

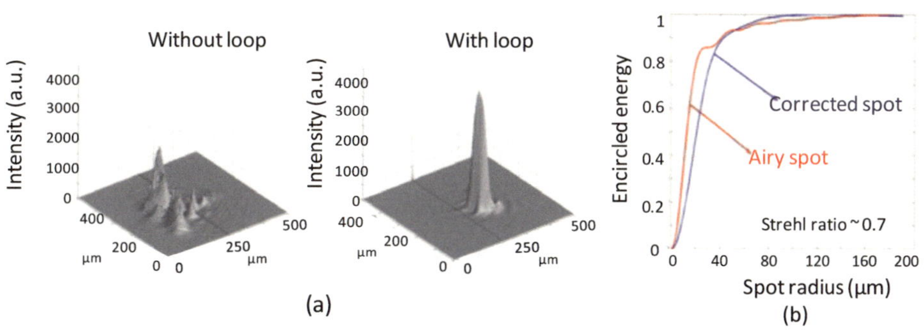

Fig. 13. (a) Focal-spot patterns measured during the fifth kilojoule shot, without and with the optimized AO wave front control. (b) Deduced Strehl ratio and encircled energy of the focal spot with loop convergence.

5. Application to focal-spot shaping: Bi-speckle generation

High-peak-power, short-pulse lasers are well suited for the study of the parametric instabilities driven by laser-plasma interaction, because these instabilities have a fast growth rate in the sub-picosecond range (Rousseaux, 2009). Using an AO closed-loop on one of the LULI facilities, previous experiments have provided measurements to study the parametric instability saturation generated by a mono-speckle in small size plasmas. Another interesting topic consists of investigating the interplay of the instability evolution generated by two hot spots on target.

It is well known that two symmetrical spots can be generated in focal plan if a linear phase φ_{bs} with 2 opposite slopes in one direction is added to the spatial phase of a laser beam. The amplitude of the laser beam becomes:

$$E(x,y)=\exp\left[-(x^2+y^2)^n\right]\exp\left[-i(\varphi+\varphi_{bs})\right] \quad (7)$$

with $\varphi_{bs}=k_x x$ and $-\frac{D}{2}\leq x\leq\frac{D}{2}$ (D : beam aperture). The two spots are separated by a distance of 2a (Fig. 14), determined as following:

$$2a=2k_x f=\frac{4N\lambda f}{D} \quad (8)$$

with N: integer number; f: focal distance. In the case of the LULI laser facility, an off-axis parabola with f = 75 cm is used.

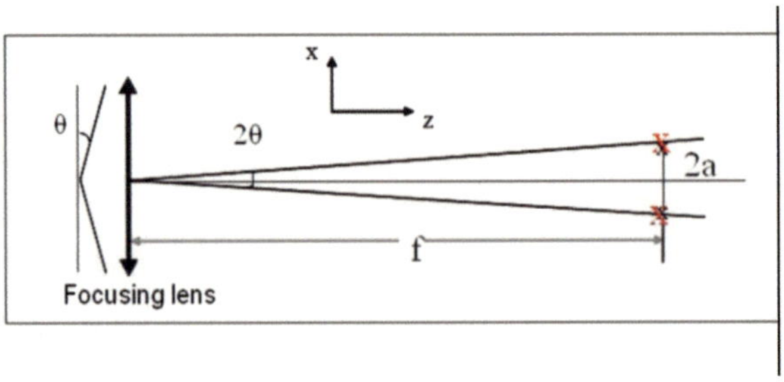

Fig. 14. Double focal spots generation by use of an additional linear phase with two opposite slopes.

Several conventional methods exist for bi-speckles generation. For example, an accurately tilted mirror in bi-sections, a well-designed phase plate or a bi-prism, placed just before the focusing system, can produce double focal spots on target. Nevertheless, we choose the programmable AO closed-loop already integrated in our chain to produce them.

It is important to make sure that all laser spatial features are well studied when the bi-speckles are generated in the interaction chamber centre. In fact, the DM is set up in the middle of the chain, the introduced spatial phase for the bi-speckles generation may induce the modulations on the laser energy profile when laser beam propagates from DM towards the interaction chamber. We performed beam near-field and far-field investigations to determine the maximum phase amplitude for the bi-speckles generation while the beam energy-distribution remains correct.

We have previously computed a 64x64 pixel phase map in a conventional space xyz. This phase has two opposite linear slopes in the plane of x0z and with their interception on x=0. The phase amplitude in z is defined as the phase difference between x=0 and x=D/2. Nevertheless, the phase value should be kept constant in the direction of y. This phase map is then used as the phase target for our AO closed-loop.

The distance between the two focal spots is linear with respect to the additional phase amplitude. When the phase amplitude increases from 0 to 2.5 λ, the bi-speckles are separated from 0 to 100 μm if the focusing distance is 75 cm. The maximum phase amplitude is determined to ≤ 2.5 λ to get rid to the degradation of the laser near-field pattern while the beam continues to propagate in the chain. Fig. 15 illustrates the intensity and the phase measured after the closed-loop convergence with a phase amplitude of 2 λ.

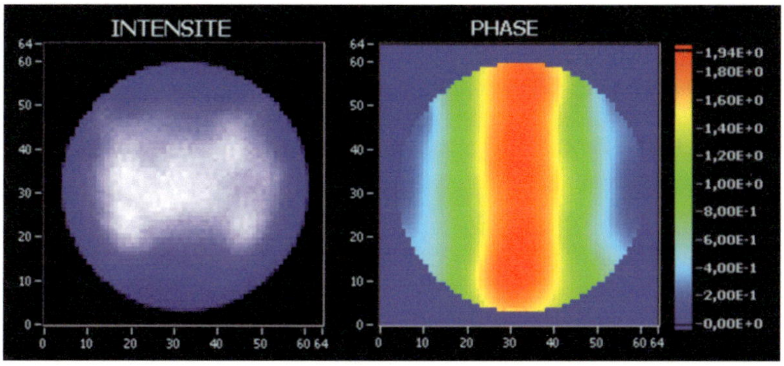

Fig. 15. Laser intensity distribution and spatial phase with a phase amplitude of 2λ between the center and the edge, measured by the SID4 after the AO closed-loop convergence.

Fig. 16a presents the simulated far-field result using a super-Gaussian beam profile and different phase maps: 1) the computed phase target with a linear profile; 2) the measured phase after the AO loop. We note that using the measured phase, an important amount of parasitic light appears around the two main focal spots. It results from a phase profile curved in x=0 instead of a linear one. Generally, for producing a phase close to the phase target, a large number of actuators in the deformable mirror is necessary and also the actuator distribution should be matched with the phase target. In our deformable mirror, the actuator number is quite limited. In addition, they are placed in rings and are not suitable to generate the phase with linear structure. To remove the parasitic light on target, a mask is placed just before the beam focusing system. This way, two spots separated to 85μm (Fig. 16b) have been obtained experimentally for the study of the parametric instabilities.

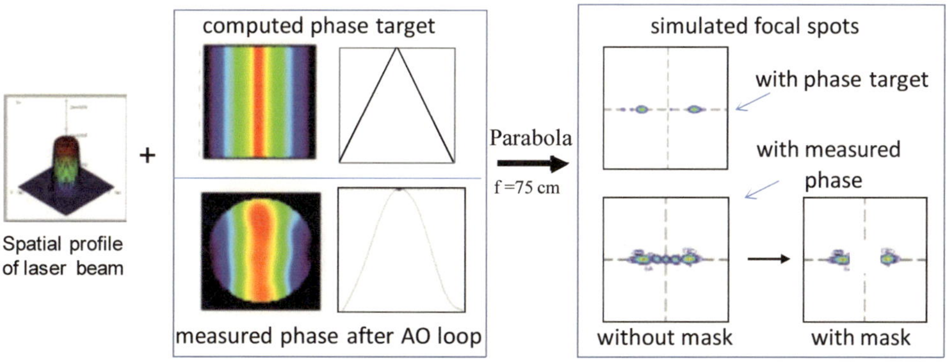

(a)

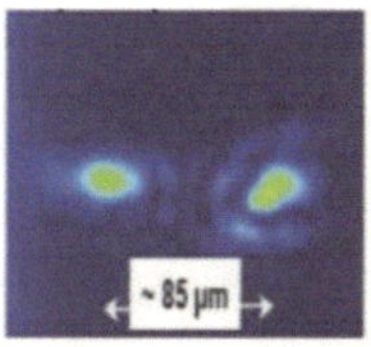

(b)

Fig. 16. (a) Focal spot simulation using (top) the computed phase target with a linear profile and (button) the measured phase after the AO loop. (b) Bi-speckles measured in the interaction chamber centre and realized with a phase amplitude of ~2λ.

Bi-speckles generation by use of a programmable AO closed-loop enables us to control easily the energy contained in each of the two focal spots. By moving the phase curve in the direction of x, it could correct energy non-equilibrium between two spots induced by a near-field distribution defect. Also, it may change the ratio of the energy contained inside each of these two focal spots according to plasma experiment request. However, to minimize the parasitic light around the two focal spots, a corrector with more suitable geometry and much more actuators would be requested.

6. Conclusions and perspectives

We have demonstrated efficient wave front control and important focal spot quality improvement by the use of an AO setup on the LULI laser facilities. Following an optimized operation procedure, we perform the dynamic wave front control associated with an elaborate alignment system. It enables an effective correction of all static and thermal lower-order aberrations generated in the amplification chains. The shot-to-shot reproducibility of the focal spot on target has been obtained with a quality close to the diffraction limit. Also, the shot repetition rate of the LULI laser facilities was increased. For instance, with an AO wave front control, the LULI2000 shot cycle can speed up by a factor of 2. Laser intensity on target is significantly increased for laser-induced high-field physics research.

In the scope of Inertial Confinement Fusion and study of the laser parametric instabilities, we have proposed and validated the bi-speckles generation procedure using the same AO closed-loop as for the wave front control of the laser chain. With an adequate AO system design, it will be feasible to realize more complicated high-quality focal spot shaping.

In addition to the present work, further investigations include the study and compensation of pump-shot induced aberrations, the analysis and correction of the wave front distortion generated by the pulse compressor using large-scale optics and gratings. Additional investigations aim at improving the existing AO system with a more dynamic corrector and its more suitable location in our chains. All these points are key factors to improve the beam quality of the whole laser chain and so the laser intensity on target for laser-induced plasma physics research.

7. References

Bon, P.; Maucort, G.; Wattellier, B. & Monneret, S. (2009). Quadri-wave lateral shearing interferometry for quantitative phase microscopy of living cells, *Opt. Express* Vol. 17, 13080-13094

Brown, D. C. (1981). High-Peak-Power Nd:Glass Laser Systems, *Springer-Verlag*

Cavaillier, C.; Camarcat, N.; Kovacs, F. & André, M. (2003). Status of the LMJ Program, *Proceedings of the Third International Conference on Inertial Fusion Sciences,* American Nuclear Society, pp. 523–528.

Dalimier, E. & Dainty, C. (2005). Comparative analysis of deformable mirrors for ocular adaptive optics, *Opt. Express,*Vol. 13, 4275–4285

Gopi, N.; Nathan, T. P. S. & Sinha, B. K. (1990). Experimental studies of transient, thermal depolarization in a Nd:glass laser rod, *Appl. Opt.* Vol. 29, 2259-2265

Haynam, C. A.; Wegner, P. J.; Auerbach, J. M.; Bowers, M. W.; Dixit, S. N.; Erbert, G. V.; Heestand, G. M.; Henesian, M. A.; Hermann, M. R.; Jancaitis, K. S.; Manes, K. R.; Marshall, C. D.; Mehta, N. C.; Menapace, J.; Moses, E.; Murray, J. R.; Nostrand, M. C.; Orth, C. D.; Patterson, R.; Sacks, R. A.; Shaw, M. J.; Spaeth, M.; Sutton, S. B.; Williams, W. H.; Widmayer, C. C.; White, R. K.; Yang, S. T. & Van Wonterghem, B. M. (2006). National Ignition Facility laser performance status, *Appl. Opt.* Vol. 46, pp. 3276–3303

Korn, G. & Antici, P. (2009). Report on the results of the Grand Challenges Meeting, Extreme Light Infrastructure, 27-28 April 2009, Paris

Koechner, W. (1999). *Solid-State Laser Engineering,* 5th ed., Vol. 1 of Springer Series in Optical Sciences, Springer

Kudryashov, A. V. & Shmalhausen, V. I. (1996). Semipassive bimorph flexible mirrors for atmospheric adaptive optics applications, *Opt. Eng.*Vol. 35, pp. 3064-3073

Kuzmin, A. A.; Khazanov, E. A. & Shaykin, A. A. (2011). Large-aperture Nd:glass laser amplifiers with high pulse repetition rate, *Opt. Express,* Vol. 19, No. 15, pp. 14224-14232

Liu, L.; Wang, X.; Guo, S.; Xu, X. & Lu, Q. (2010). Model of thermally induced wave front distortion and birefringence in side-pumped Nd-doped YAG and phosphate glass heat capacity rod lasers, *Appl. Opt.* Vol. 49, No. 28, pp. 5245-5253

Noll, R.J. (1976). Zernike polynomials and atmospheric turbulence, *J. Opt. Soc. Am.* 66(3), 207-11.

Primot, J. & Sogno, L. (1995). Achromatic three-wave (or more) lateral shearing interferometer, *J. Opt. Soc. Am. A,* 12(12), pp. 2679.

Primot, J. & Guérineau, N. (2000). Extended Hartmann test based on the pseudoguiding property of a Hartmann mask completed by a phase chessboard, *Appl. Opt.* 39(31), pp. 5715–5720.

Rousseaux, C.; Baton, S.; Bénisti, D.; Gremillet, L.; Adam, J. & Héron, A. (2009). Experimental evidence of predominantly transverse electron plasma waves driven by stimulated Raman scattering of picosecond laser pulses, *Phys. Rev. Lett.* Vol. 102, Issue 18

Sauteret, C.; Sautivet, A. M.; Zou, J. P. & Maignan.J. (2000). Architecture of the LULI2000 facility, *LULI Annual Scientific Report,* Available from www.luli.polytechnique.fr

Strickland, D. & Mourou, G. (1985). Compression of amplified chirped optical pulses, *Opt. Commun.* Vol. 56, pp. 219–221

Uppal, J. S. & Monga, J. C. (1985). Contribution of stress-dependent variation of refractive index to thermal lensing in Nd:glass laser rods. *Appl. Opt.* Vol. 24, No. 22, pp. 3690-3692

Velghe, S.; Primot, J.; Guérineau, N.; Cohen, M. & Wattellier, B. (2005). Wave-front reconstruction from multidirectional phase derivatives generated by multilateral shearing interferometers, *Opt. Lett.* Vol. 30, pp. 245–247

Wattellier, B.; Fuchs, J.; Zou, J. P.; Abdeli, K.; Pépin, H. & Haefner, C. (2004). Repetition rate increase and diffraction-limited focal spots for a nonthermal-equilibrium 100-TW Nd:glass laser chain by use of adaptive optics, *Opt. Lett.* Vol. 29, No. 21, pp. 2494-2496

Zou, J. P.; Descamps, D.; Audebert, P.; Baton, S.; Paillard, J. L.; Pesme, D.; Michard, A.; Sautivet, A .M.; Timsit, H. & Migus. A. (1998). The LULI 100-TW TI :sappire/Nd : glass laser : a first step towards a high performance petawatt facility. *Proceedings of the Third International Conference on Solid State Lasers for Application to Inertial Confinement Fusion,* Vol. 3492, pp. 94-97

Zou, J. P.; Le Blanc, C.; Audebert, P.; Janicot, S.; Sautivet, A. M.; Martin, L.; Sauteret, C.; Paillard, J. L.; Jacquemot, S. & Amiranoff, F. (2008). Recent progress on LULI high power laser facilities, *Journal of Physics,* Conf. Series. 112, 032021

Part 3

Adaptive Optics and the Human Eye

6

The Human Eye and Adaptive Optics

Fuensanta A. Vera-Díaz and Nathan Doble
The New England College of Optometry, Boston MA, USA

1. Introduction

Scientists have rapidly taken advantage of adaptive optics (AO) technology for the study of the human visual system. Vision, the primary human sense, begins with light entering the eye and the formation of an image on the retina (Fig 1), where light is transformed into electro-chemical impulses that travel towards the brain. The eye provides the only direct view of the central nervous system and is, therefore, the subject of intense interest as a means for the early detection of a host of retinal and possibly systemic diseases. However, ocular aberrations limit the optical quality of the human eye, thus reducing image contrast and resolution. With the use of AO it is now routinely possible to compensate for these ocular aberrations and image cellular level structures such as retinal cone and rod photoreceptors (Liang et al, 1997; Doble et al, 2011), the smaller foveal cones (Putnam et al, 2010), retinal pigment epithelium (RPE) cells (Roorda et al, 2007), leukocyte blood cells (Martin & Roorda, 2005) and the smallest retinal blood vessels (Tam et al, 2010; Wang et al, 2011), *in vivo* and without the aid of contrast enhancing agents.

The chapter begins with a review of the structure of the human eye before describing the challenges and approaches in using AO to study the visual system.

1.1 The human eye and visual system

The human eye behaves as a complex optical structure sensitive to wavelengths between 380 and 760 nm. Light entering the eye is refracted as it passes from air through the tear film-cornea interface. It then travels through the aqueous humor and the pupil (a diaphragm controlled by the iris) and is further refracted by the crystalline lens before passing through the vitreous humor and impinging on the retina (Fig 1). The tear film-cornea interface and the crystalline lens are the major refractive components in the eye and act together as a compound lens to project an inverted image onto the light sensitive retina. From the retina, the electrical signals are transmitted to the visual cortex via the optic nerve (Fig 1). A summary of this path is presented in this section, for detailed information on the anatomy and physiology of the eye the reader is directed to the references (Snell & Lemp, 1998; Kaufman & Alm, 2002; Netter, 2006).

1.1.1 Tear film-cornea interface

The tear film–cornea interface (Fig 1) is the most anterior refractive surface of the eye as well as the most powerful due to the difference between its refractive index and that of air. The anterior radius of the tear film-cornea interface is approximately 7.80 mm and the refractive index of the tear film is 1.336, which give a dioptric power of approximately 43.00 diopters. Therefore, small variations in its curvature can cause significant changes in the power of the eye.

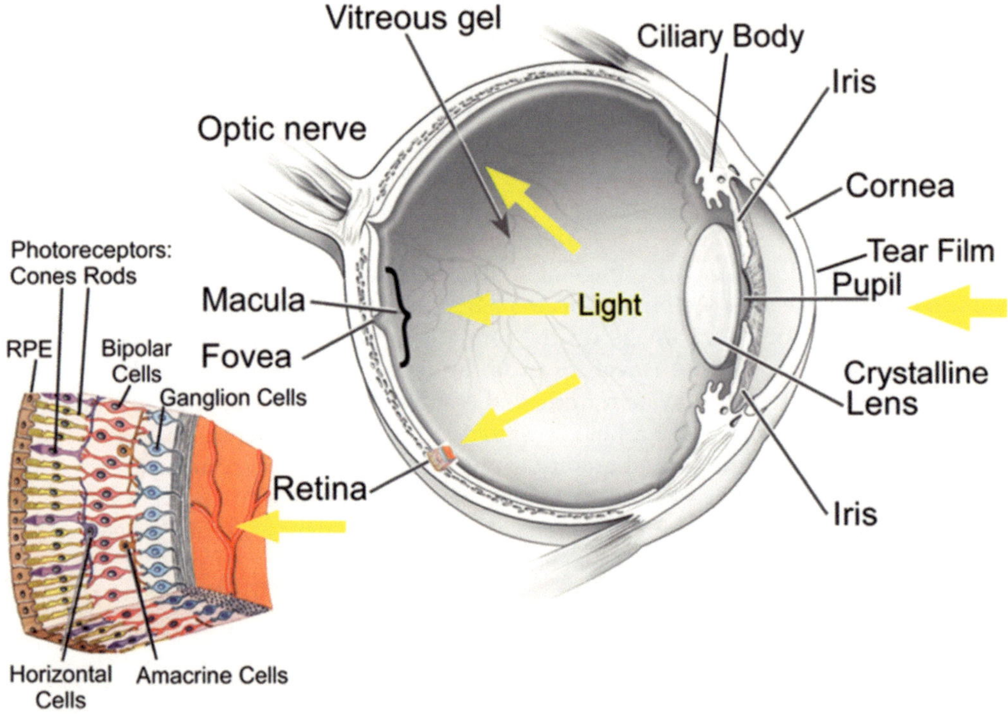

Fig. 1. Gross anatomy of the human eye and detail of the retina.
The major refractive elements and hence primary sources of aberration are the tear film–cornea interface and the crystalline lens. The incident light on the retina is absorbed by the cone and rod photoreceptors after traversing several retinal layers. Image modified from the National Eye Institute, National Institutes of Health.

The cornea is a transparent tissue, achieved by its regular composition of collagen fibers, avascularity and an effective endothelial pump. The cornea in the adult typically measures 10.5 mm vertically and 11.5 mm horizontally and its thickness increases from the center (about 530 μm) to the periphery (about 650 μm). The cornea is more curved than the eyeball and hence protrudes anteriorly. Behind the cornea, the aqueous humor has the same refractive index as the vitreous humor (1.336), whereas the refractive index of the cornea is 1.376. Because the change in refractive index between cornea and aqueous humor is relatively small compared to the change at the air–cornea interface, it has a negligible refractive effect.

1.1.2 Crystalline lens. Accommodation

The crystalline lens is held behind the iris by thin yet strong ligaments, zonules of Zinn, attached to the ciliary processes in the ciliary body (Fig 1). The crystalline lens is flexible and may change its shape using the mechanism of accommodation, by adjusting the ciliary muscle so that the images may be more accurately focused on the retina. It has an ellipsoidal, biconvex shape with the posterior surface being more curved than the anterior. The crystalline lens is typically 10 mm in diameter and has a thickness of approximately 4mm, although its size and shape changes during accommodation, and it continues to grow throughout a person's lifetime. The crystalline lens achieves transparency due to its composition, as 90 % of it is formed by tightly packed proteins and there is an absence of organelles such as a nucleus, endoplasmic reticulum and mitochondria within the mature lens fibers.

The intensity of the light reaching the retina is regulated by the diaphragm formed by the iris: the pupil. The pupil is therefore important in regulating the aberrations of the eye, the magnitude of the aberrations increase with larger pupil diameters - section 1.2.1.

1.1.3 Retina

Upon reaching the retinal surface, the light traverses its many layers (Fig 1) before reaching the photoreceptor cells, where the photons are absorbed and transformed into electro-chemical impulses. The gross anatomy of the retina is composed of a macula or central region, with the fovea as the very center. At the fovea the cone photoreceptors have the smallest diameter (1.9-3.4 μm), the highest average density (199,000 cones per mm^2) (Curcio & Allen, 1990) and the eye has the highest resolution (visual acuity, VA). The signals from these photoreceptors are then processed by the many intervening cell types in the retina before exiting towards the brain via the ganglion cells and the optic nerve.

i. Physiology of the Photoreceptors: Rods and Cones

The photoreceptors are photosensitive cells located in the outermost layer of the retina that are responsible for the phototransduction, i.e. they convert photons into electro-chemical signals that can stimulate biological processes. The proteins (opsins) in the outer segments of these photoreceptors absorb photons and trigger a cascade of changes in the membrane potential; this mechanism is called the signal transduction pathway. In brief, the photoreceptors signal their absorption of photons via a decrease in the release of the neurotransmitter glutamate to the bipolar cells. The photoreceptors are depolarized in the dark, when a high amount of glutamate is being released, and after absorption of a photon they hyperpolarize so less glutamate is released to the presynaptic terminal of the bipolar cells.

The effect of glutamate in the bipolar cells varies depending on the type of receptor imbedded in the bipolar cell's membrane; it may depolarize or hyperpolarize the bipolar cell. This allows one population of bipolar cells to get excited by light whereas another population is inhibited by it, even though all photoreceptors show the same response to light. This complexity is necessary for various visual functions such as detection of colour, contrast or edges. The complexity increases as there are interconnections among bipolar cells, horizontal cells and amacrine cells in the retina. The final result of this complex net is several populations of different classes of ganglion cells that have specific functions in the retina and exit the eye through the optic nerve.

The photoreceptor cells are the rods and cones (Fig 1, 2), named as consequence of their anatomy. Rods are narrower and distributed mostly in the peripheral retina. A third class of light cells are the photosensitive ganglion cells, discovered in the 1990s (for a review see (Do & Yau, 2010)), which use the photopigment melanopsin and are believed to support circadian rhythm but do not contribute significantly to vision. The human retina contains approximately 120 million rods and 5 million cones, although this amount varies with age and certain retinal diseases. There are also major functional differences between the rods and cones. Rods are extremely sensitive, have more pigment and can be triggered by a very small number of photons. Therefore, at very low light levels (scotopic vision), the visual signal is coming solely from rods. Rods are almost absent in the fovea, and only a small amount are present in the macular area. Cones, on the other hand, are only sensitive to direct and large amounts of photons; and are used for photopic vision. In humans there are three different types of cone cells that respond approximately to short (S), medium (M) and long (L) wavelengths.

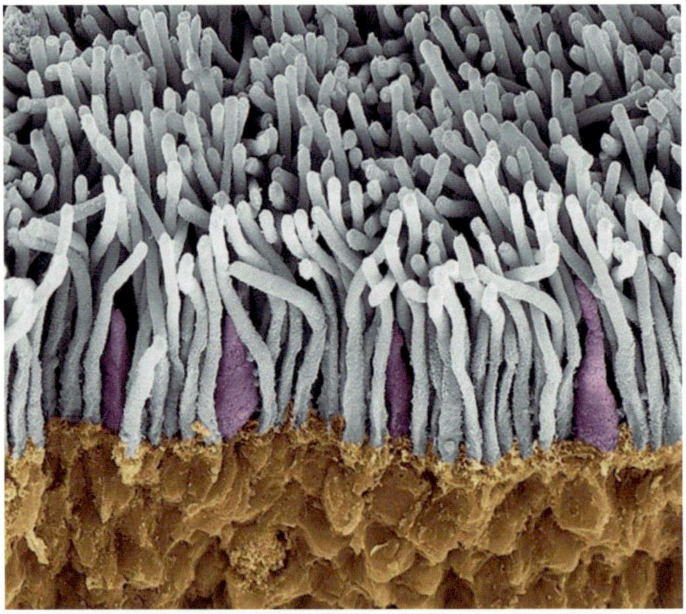

Fig. 2. Colored scanning electron micrograph (SEM) of rods (blue) and cones (purple) in the retina of the eye. The outer nuclear layer is brown. Magnification x1800 when printed at 10 centimetres wide. By Steve Gschmeissner. Reproduced with permission from Science Photo Library.

There is a dependence on photoreceptor arrangement with retinal eccentricity, decreasing in regularity and density from the fovea toward the periphery, although the smallest cones are not always located in the center of the fovea (Chui et al, 2008a). At a given retinal location, there is considerable individual variation in cone photoreceptor packing density, although more than 20 % of the variance could be accounted for by differences in axial length (Chui et al, 2008b).

ii. Waveguide Properties of the Photoreceptors: The Stiles-Crawford Effect

As mentioned above, cones are sensitive to large amounts of light and only if it is directly incident on them. There is, therefore, a reduction in light sensitivity when its entry point is shifted from the center to the edge of the pupil. This phenomenon, called the Stiles-

Crawford Effect (SCE) (Stiles & Crawford, 1933; Westheimer, 2008), plays an important role in vision because unwanted scattered light is rejected. Individual cones have specific waveguiding properties (Enoch, 1963), cone disarray is very small in healthy eyes and ensembles of cones have essentially the same directionality properties as a single cone (Roorda & Williams, 2002). This property of the photoreceptors shows small variations across the retinal field (Westheimer, 1967; Burns et al, 1997). It has been suggested (Vohnsen, 2007) that the photoreceptors may be at least partially adapted to match the average ocular aberrations in order to maximize their light-capturing capabilities.

The study of SCE may provide useful information about subtle structural changes in retinal disease, changes that may not be detected with conventional clinical tests. It has been shown that this property of the photoreceptors is altered in central serous chorioretinopathy (Kanis & van Norren, 2008). Delayed recovery of photoreceptor directionality was found when measuring SCE at a stage of the disease when no abnormalities were found using other common diagnostic techniques such as VA and optical coherence tomography (OCT). Transient changes of the SCE have also been found in the near periphery of myopic eyes with elongated axial lengths (Choi et al, 2004) and in eyes with permanent visual field loss and damage of the inner retinal layers secondary to optic neuropathies (Choi et al, 2008).

iii. Temporal Properties of the Photoreceptors

The photoreceptors outer segments contain discs studded with opsins that capture photons to initiate the phototransduction process. Throughout the day, new discs are added, dozens of discs are shed and phagocytosis occurs at the RPE. *In vivo* detection of disc renewal has only been possible recently with the use of AO. Using an AO flood-illuminated camera, Pallikaris et al (2003) observed changes in cone reflectance over a 24 hour period using non-coherent illumination. These changes were incoherent, not sinusoidal, with both rapid, over minutes, and slow, over hours, changes. They also found these changes to be independent from cone to cone. Hence, they concluded that the changes are not caused by spatiotemporal variation in the optical axes of the cones but were likely caused by changes in the composition of the outer segment-RPE interface due to the migration of melanosomes during disc shedding, or a change in refractive index in the outer segment interface during shedding. If the reflectance changes are related to the renewal process of the receptors, it will be possible to study disruptions in the disc shedding process that occur in diseases such as retinitis pigmentosa.

Other authors have shown faster cone changes. Jonnal et al (2007) showed rapid changes in reflectance in response to visible stimulation of individual photoreceptors. These changes are initiated 5 to 10 msec after the onset of the stimulus flash and last 300 to 400 msec and are believed to be linked to the process of cone phototransduction. Possible mechanisms for this phenomenon are processes taking place within the cone immediately following stimulation, such as changes in the concentration of G-proteins, hyperpolarization or other changes in the properties of the outer segment membrane, or changes in the physical size of the outer segment secondary to swelling.

Jonnal et al (2010) reported the period for cone reflectance oscillation when using long coherent illumination to range between 2.5 and 3 hours, with sinusoidal oscillations occurring during a 24 hour period. The power spectra of most cones peaked at a frequency between 0.3 and 0.4 cycles/hour, although this peak varied within a 24 hour period. They hypothesized that these oscillations are due to elongation of the cones outer segments (OS) (Jonnal et al, 2010). These rates agree with post-mortem studies in mammals on OS renewal rates on rods (~2μm / day) and cones (~ 1-3μm/ day).

1.1.4 The visual system

In brief, the electrical signals at the retina exit each eye via ganglion cells axons through the optic nerve, following a path that crosses at the optic chiasma to later reach the lateral geniculate nucleus (LGN) and from there continue to the primary visual cortex (V1, or striate cortex) first, and to further cortical areas later. The optic chiasm is the point for crossover of information of right and left eyes. The LGN, located at the thalamus, appears to be the first location of feed-forward input from higher levels in the brain to the visual input from the eye before most of the visual input travels to the visual cortex. Note that there is a lateral pathway, that of the superior colliculi, important for eye movement control.

At the visual cortex the signals are processed in V1 and communicated via multiple pathways to numerous visually responsive cortical areas. The visual system comprises a complex network where a cascade of action potentials stream from neuron to neuron forwards, laterally and backwards again. These signals are responsible for our visual perception of the external world, but we are far from understanding how perception of the real world's complex patterns occurs. Visual scientists typically consider that an image can be broken into its components, such as edges, textures, colors, shares, motion, etc. and specialized neurons detect a subset of these components. For a review on receptive field properties of these neurons, retinotopic maps in LGN and V1, orientation and direction selectivity, binocularity and binocular disparity, response timing and other properties of the visual system see online text books (Neuroscience Online, 1997; Webvision, 2011).

1.2 Optical aberrations of the eye

In addition to being the main refractive components, the cornea and the crystalline lens are the main sources of aberrations in the human eye. The relative contribution of each of these components can be deduced from total ocular and corneal aberrometry data. The magnitude of the aberration is strongly dependent on individual factors such as age, the state of accommodation or the particular direction through the ocular media. The human eye has monochromatic, longitudinal (up to 2 diopters across the visible spectrum) and transverse chromatic aberrations, the former being significant when using wide bandwidth imaging light sources.

1.2.1 Describing human ocular aberrations

The standard representation of ocular aberrations is in terms of Zernike polynomials (American National Standards Institute (ANSI) - 2010). Zernike polynomials are a mathematical series expansion that are orthogonal over a unit circle. Any wavefront profile can be decomposed into a weighted sum of these polynomials. The low order terms can be translated into the common sphere and cylinder notations used in optometric fields (Porter et al, 2006) and are easily corrected using, for example, spectacles or contact lenses. The higher order Zernike polynomials are traditionally not correctable by such methods, although recently attempts are being made, and require advanced technologies such as AO. Figure 3 shows the first 15 Zernike terms and their corresponding far field point spread functions (PSF).

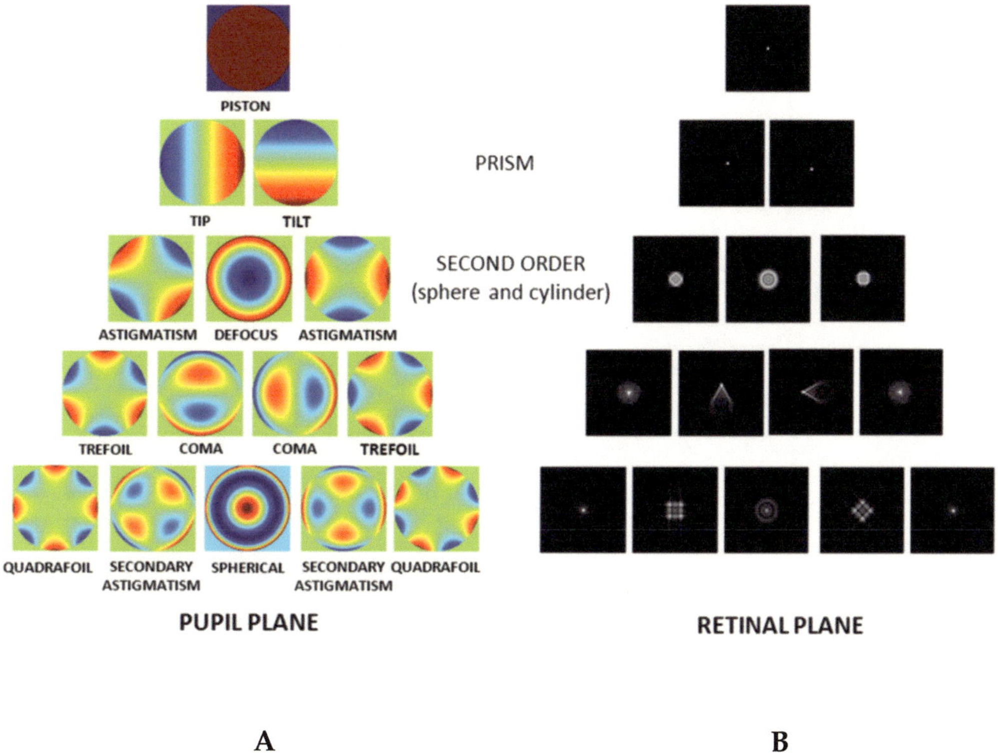

Fig. 3.
(A) The ocular aberrations can be represented as a weighted sum of Zernike polynomials, each representing a specific aberration.
(B) By Fourier transforming and multiplying by the complex conjugate the PSF for each mode can be calculated. Defocus and astigmatism are termed low order modes and are corrected by conventional refractive methods. The higher order modes generally have lower amplitudes but require more elaborate correction technologies.

Porter et al (2001) and Thibos et al (2002) independently measured the wavefront aberration in large human population samples using Shack Hartmann aberrometry. Figure 4 shows measured aberrations coefficients from Porter et al (2001); they measured 109 individuals through a 5.7 mm pupil. The majority of the power lies within the low order modes, i.e. defocus (Z_2^0) and astigmatism (Z_2^{-2}) and (Z_2^2), with these modes accounting for over 92 % of the total wavefront aberration variance. Note that for this particular study the average defocus coefficient was higher than the general population as they were subjects recruited from a clinic at Bausch & Lomb who were mostly myopic. That said, for high resolution imaging applications where even larger pupil sizes are used, any residual power in the higher order modes can become particularly detrimental.

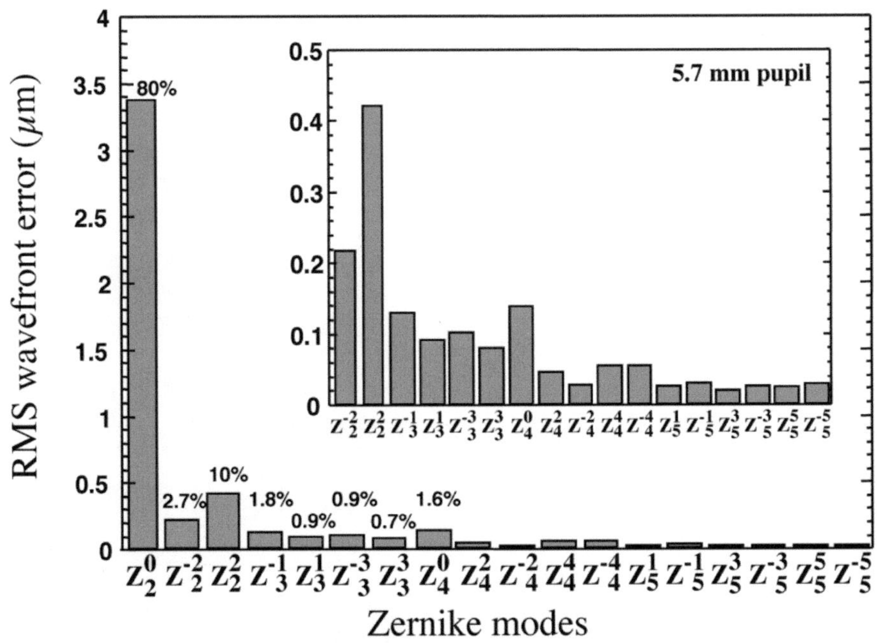

Fig. 4. The wavefront aberration decomposed into Zernike polynomials for a large human population (Porter et al, 2001) over a 5.7 mm pupil. The majority of the aberration power is found in the low order modes, i.e. defocus (Z_2^0) and astigmatism (Z_2^{-2} and Z_2^2).
The percentages above the first eight modes indicate the percentage of the total wavefront variance. Note: the Zernike order follows that of Noll (1976). Reproduced with permission from the Optical Society of America (OSA).

Doble et al (2007) showed the peak to valley (P-V) wavefront error dependence on pupil size (Fig 5) using aberration data from two human population studies; one comprising of 70 healthy eyes based at the University of Rochester/Bausch & Lomb, and the other consisting of 100 healthy eyes measured at the University of Indiana. Figure 5 shows the wavefront values for each of these populations using different corrective states. Data show that to correct 95 % of the normal human population over a 7.5 mm pupil upwards of 20 µm wavefront correction is required even with the benefit of a second order correction (Fig 5B).

Ocular aberrations also vary with time, mostly due to changes in accommodation (He et al, 2000), although there are other significant contributors such as eye movements. Even when paralyzing accommodation with anticholonergic drugs, the microfluctuations of accommodation can cause significant refractive power changes, up to 0.25 diopters. Hofer et al (2001a; 2001b) and Diaz-Santana et al (2003) have performed detailed measurements on wavefront dynamics and their effect on AO system performance. With a static correction of the higher order aberrations, these dynamic changes can reduce the retinal image contrast by 33 % and the Strehl ratio (SR) by a factor of 3 highlighting the need for real time aberration correction (Hofer et al, 2001b). The SR is defined as the ratio of the peak intensity in the aberrated PSF to that of the unaberrated case; an SR greater than 0.8 is considered to be diffraction limited.

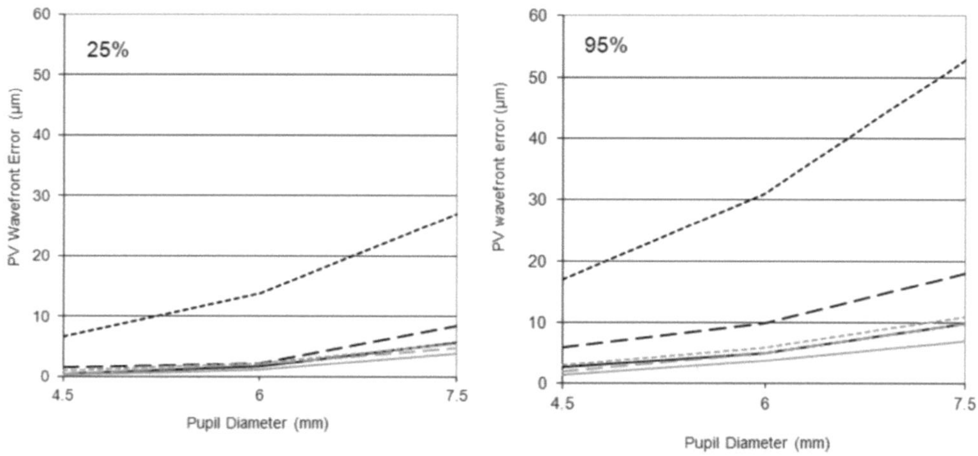

Fig. 5. Peak to valley wavefront error that encompasses 25 (left) and 95 % (right) of the population in the Rochester (black lines) and Indiana (gray lines) populations. For the Rochester data, three correction states are given: (i) all aberrations present (short dashed lines), (ii) all aberrations present with zeroed Zernike defocus (long dashed lines) and (iii) all aberrations present with zeroed defocus and astigmatism (solid lines). For the Indiana data, the three cases are: (i) residual aberrations after a conventional refraction using trial lenses (short dashed lines), (ii) all aberrations present with zeroed Zernike defocus (long dashed lines) and (iii) all aberrations present with zeroed defocus and astigmatism (solid lines) (Doble et al, 2007). Reproduced with permission from the Optical Society of America (OSA).

1.2.2 The resolution of the human eye

The lateral (transverse) resolution of the eye is given by Eq. 1:

$$r = 1.22\, f\, \lambda \,/\, n\, D \tag{1}$$

where r is the distance from the center of the Airy disk to the first minima, f is the focal length (of the reduced eye), λ is the wavelength, n is the refractive index and D is the pupil diameter. The maximum resolution would be achieved using the shortest wavelength, λ and the largest possible aperture, D (f being fixed). As an example, for a human eye, with $D = 8$ mm, $f = 22.2$ mm, n ~1.33 and imaging at $\lambda = 550$ nm, the lateral resolution r is 1.4 µm. In practice, however, ocular aberrations limit this resolution to about 10 µm.

In theory, a lateral resolution of 1.4 µm is sufficient to see the smallest retinal cells. For example, foveal cones have a center to center spacing of 1.9-3.4 µm, and for the rods, the range is 2.2-3.0 µm (Curcio et al, 1990; Jonas et al, 1992). To obtain retinal images with the highest resolution and contrast it is therefore necessary to correct both the low and high order aberrations over a large pupil and moreover track and correct for any associated temporal changes, i.e. we need to employ AO. The ability of AO to dynamically correct (or even induce) higher order spatial modes is becoming increasingly important in the study of the human visual system. The next sections describe how AO is applied to various retinal imaging modalities.

2. The application of adaptive optics to the human eye

The concept of AO was first proposed by the astronomer Horace Babcock in 1953 (Babcock, 1953). However, it was not until the late 1960s/early 70s that the first system was implemented, first by the military followed subsequently by the astronomy community. The first step towards the application of AO to the human eye was the work of Dreher et al (1989) who employed a deformable mirror (DM) to give a static correction of astigmatism in a scanning laser ophthalmoscope (SLO). Later work by Liang et al (1994) saw the first use of a Hartmann-Shack wavefront sensor (HS-WFS) for measurement of the human wavefront aberration who then used a HS-WFS in conjunction with a DM (Liang et al, 1997) to produce some of the first *in vivo* images of the cone photoreceptors. Today, AO has been successfully applied to several retinal imaging modalities employing a variety of DM and WFS technologies.

A detailed discussion of AO is beyond the scope of this chapter and the interested reader is referred to the available reference texts (Hardy, 1998; Porter, 2006; Tyson, 2010).

2.1 Key AO components

Similar to the AO systems used for other applications such as astronomy and communications, a vision science AO system comprises three main parts:

i. The Wavefront Sensor (WFS): Most vision science AO systems employ a Hartmann-Shack WFS (Shack & Platt, 1971), although curvature (Roddier, 1988) and pyramid sensing (Ragazzoni, 1996) have also been employed successfully to the eye (pyramid sensing: Iglesias et al, 2002; curvature sensing: Gruppett et al, 2005). Typically, the ocular wavefront is sampled at 10-20 Hz with closed loop bandwidths of 1-3 Hz which is sufficient to correct most of the ocular dynamics (Hofer et al, 2001a). The basic operating principle and design considerations of a WFS are the focus of other chapters in this book and will not be discussed here.
ii. The Wavefront Corrector: These are typically DMs although liquid crystal spatial light modulators (LC-SLMs) have been used in several systems (Thibos & Bradley, 1997; Vargas-Martin et al, 1998; Prieto et al, 2004). Early vision AO systems used large, expensive DMs that were originally designed for military, astronomy or laser applications. These DMs had apertures that were several centimeters in diameter requiring long optical paths to magnify the 6-7 mm pupil diameter of the human eye. Today, many systems employ microelectromechanical systems (MEMS) (Fernandez et al, 2001; Bartsch et al, 2002; Doble et al, 2002), electromagnetic (Fernandez et al, 2006) or bimorph type mirrors (Glanc et al, 2004), all of which have much smaller active apertures and a lower cost.
iii. The Control Computer: These take the output of the WFS and converts it to voltage commands that are sent to the wavefront corrector.

There are three main ophthalmic imaging modalities that have successfully employed AO (i) flood illuminated fundus cameras that take a short exposure image of the retina, (ii) confocal laser scanning ophthalmoscopes (cSLOs) that acquire the image by rapidly scanning a point source across the retinal surface and (iii) optical coherence tomography (OCT) which again scans a point source but uses low coherence interferometry to form the image. Each of these modalities are discussed in subsequent sections; however, as the flood illuminated technique is conceptually the simplest it is used here to introduce the application of AO to the human eye.

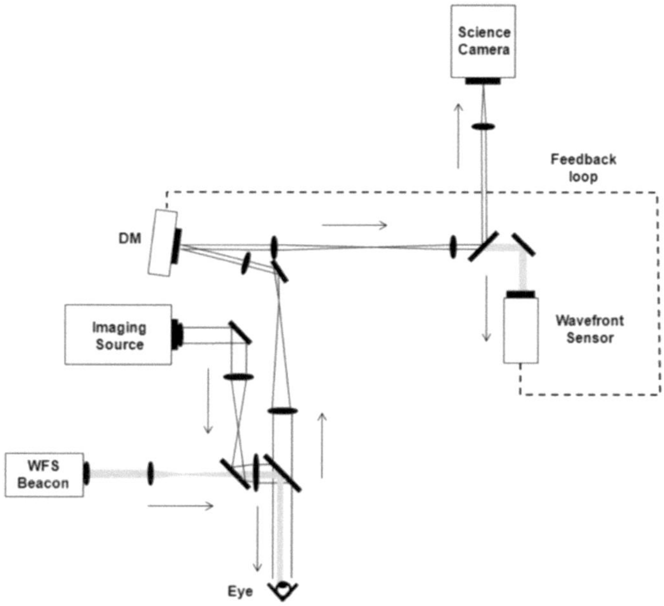

Fig. 6. Schematic of the flood illuminated (flash) AO fundus camera (Headington et al, 2011) in use at the New England College of Optometry.

Figure 6 shows the optical layout of the New England College of Optometry (NECO) AO flood illuminated (flash) fundus camera (Headington et al, 2011). The WFS beacon is used to measure the ocular aberration, a small incident beam (1mm diameter at the cornea) from a superluminescent diode (SLD) at 820 nm is focused to a ~10 µm diameter spot on the retina. The scattered light exits through the dilated pupil (6mm in diameter) and is redirected by the DM, through the dichroic beamsplitter into the WFS. The aberrations are sampled at 20 Hz and the required correction profile is sent to the DM. The system is fast enough to track and correct dynamic ocular aberration changes at a frequency of ~1 Hz. As with all AO systems used in vision, the DM and the WFS are placed in optical planes approximately conjugate to the pupil of the eye. Once the aberrations have been corrected, typically below 0.1 µm rms over the 6mm diameter pupil, the retinal image is acquired. The imaging source is usually an arclamp and delivers a 4-6 msec retinal exposure. The particular imaging wavelength (between 500-800 nm) is chosen to highlight a particular retinal feature. The imaging light follows the corrected path through the AO system and is redirected to the science camera via a dichroic beamsplitter. Typical retinal image sizes are 1-3° (0.3-0.9 mm diameter).

The system described in Figure 6 can be easily modified for functional vision testing. The science camera can be replaced by a visual test pattern, such as a visual acuity or a contrast sensitivity chart. The projected test image is then pre-distorted by the conjugate ocular aberration before being incident on the retina - see later section titled AO for Vision Testing.

2.2 Light level considerations

It is essential in the operation of any ophthalmic device that the light levels used are safe. The wide range of imaging modalities, frame rates, wavelengths, field sizes and exposure durations mean that the maximum permissible exposures (MPE) must be calculated on a case by case basis. Several reference standards are used in such calculations (ANSI, 2007; Delori et al, 2007).

2.3 Wavefront corrector requirements

Independent of the particular imaging modality, the benefit of AO in ophthalmic systems fundamentally relies on its ability to measure, track and correct the ocular aberrations. It is therefore imperative that the WFS, and in particular the wavefront corrector, have optimal operating characteristics. As DMs are the most commonly used form of wavefront corrector their performance is described in detail here. LC-SLMs (Li et al, 1998) can be modeled as piston-only DMs.

DMs can be divided into two broad categories, continuous surface and segmented (Fig 7). In both cases, there is a set of actuators that physically deform the mirrored surface. Examples of actuation mechanisms can be electrostatic, piezoelectric, magnetic, thermal or voicecoil. Refer to Tyson (2010) for more details on the various types of DMs and their actuation mechanisms. Figure 7A shows a cross section through a continuous surface DM. A two dimensional array of actuators deforms the surface. The greater the number of actuators the higher the spatial frequency correction capability. Light would be incident from the top of the figure. Figure 7B shows a segmented piston/tip/tilt (PTT) DM. In this case, each segment has three degrees of freedom. A common variant is a piston only DM in which an individual segment can only move in the vertical direction. For a continuous surface DM adjusting one actuator causes a deformation of the top mirrored surface and the degree of localization is termed the influence function. Certain DM types, such as membrane and bimorph, have very broad influence functions meaning that activation of one actuator causes a deformation over a large area of the DM. Segmented DMs however have much narrower influence functions; moving a piston only segment only changes that segment's mirror position and not that of its neighbours. The shape of this influence function, along with the number of actuators and the dynamic range, define the corrective ability of a DM.

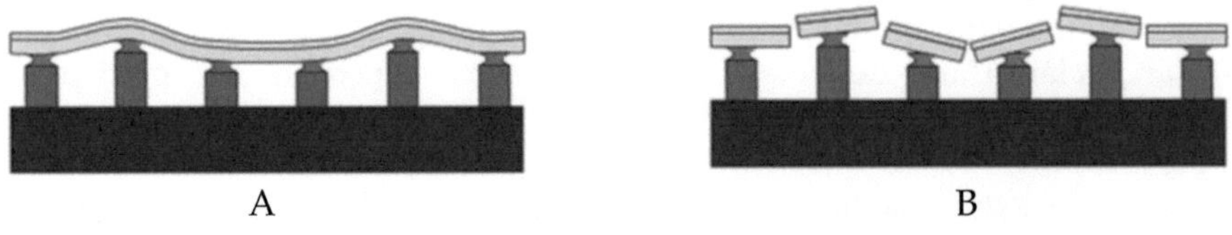

Fig. 7. Deformable mirror (DM) types. (A) Continuous surface mirror, deformed by an underlying array of actuators. (B) Segmented surface - the mirror is composed of a discrete array of segments each of which has three degrees of freedom: piston, tip and tilt (PTT). Piston only segmented DMs are also possible.

Figure 8 shows the DM correction performance as a function of the number of actuators or segments across a 7.5 mm pupil for a 0.6 μm wavelength. The Rochester population dataset described earlier (Fig 5), was analyzed after zeroing the defocus coefficient. For continuous surface DMs approximately 15 actuators are required to give diffraction limited performance (SR > 0.8), with 12 segments giving the same performance for PTT devices (with the caveat that three times as many control voltages are required to move one segment as compared to a single actuator). Piston only DMs require many more segments to achieve good correction with over 100 being necessary; however, these numbers are easily achievable with newer LC-SLMs.

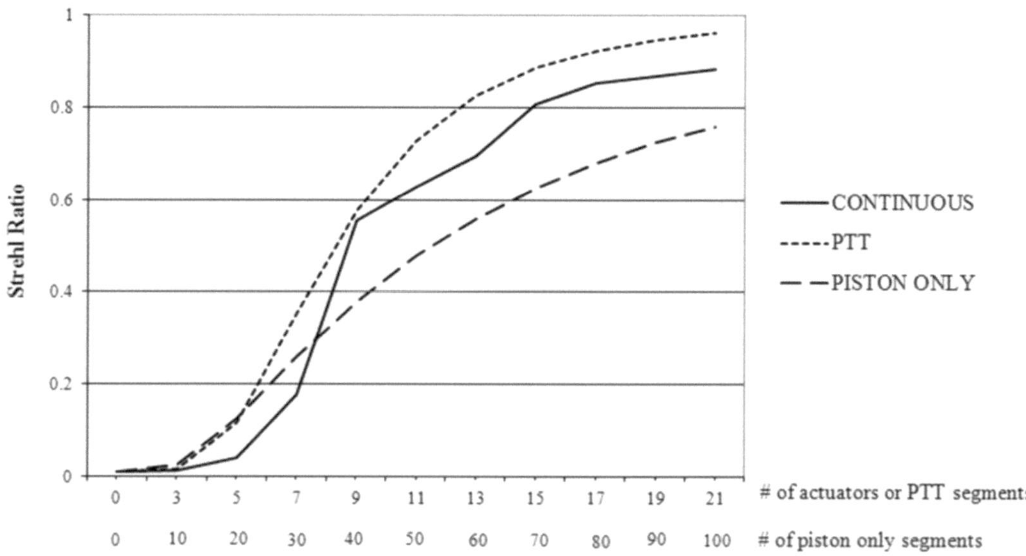

Fig. 8. DM correction performance as a function of the number of actuators or segments after zeroing the defocus coefficient. Continuous (solid line) or PTT segmented (short dashed) DMs have comparable performance with 12-15 actuators or segments being required to achieve a SR of 0.8.

3. AO modalities and applications in ophthalmic imaging

AO improves the capabilities of any ophthalmic instrument where the optics of the eye are involved, from fundus cameras to phoropters. With the high lateral resolution achievable through the use of AO, as described above, it is now possible to detect the earliest changes caused by retinal pathologies. Small structures, such as the smallest microaneurism (early signs of diabetic retinopathy), blood cells, photoreceptor cells, ganglion cells, RPE cells, the smallest capillaries and cells' organelles can now be observed with the high resolution achievable by AO imaging (e.g. Roorda & Williams, 1999; Roorda et al, 2007; Chen et al, 2011; Wang et al, 2011; Zhong et al, 2011). A number of laboratories and clinical centers have begun to evaluate eye diseases using AO imaging. Choi et al (2006) and Wolfing et al (2006) first reported *in vivo* images of photoreceptors in a patient with rod-cone dystrophy, which revealed a reduction in cone density. It has also been shown that retinitis pigmentosa and rod-cone dystrophy show a different pattern of cone degeneration (Duncan et al, 2007). Congenital color deficiencies have been studied using AO imaging (e.g. Carroll et al, 2004; Rha et al, 2010)). Choi and colleagues found that AO imaging is a reliable technique for assessing and quantifying the changes in photoreceptors in a number of optic neuropathies (Choi et al, 2008) and glaucoma (Choi et al, 2011). AO imaging has also proven to be useful in patients with inherited Stargardt's disease (Chen et al, 2011). Marmor et al (2008) used AO as well as conventional OCT to evaluate the visual significance of the foveal pit and found that it is not required for the specialization of foveal cones. More recently, McAllister et al (2010) have found variation in the degree of foveal hypoplasia and the corresponding variation in foveal cone photoreceptor specialization.

3.1 Flood illuminated (Flash) AO fundus cameras

The first AO retinal imaging systems were flood illuminated designs as depicted in Figure 6. They are extremely versatile and may be configured for a variety of imaging and vision testing experiments. Their disadvantage for imaging is that they are susceptible to ocular scatter as all of the reflected light is imaged onto the science camera, thus reducing imaging contrast, and they have essentially zero axial resolution. In addition, they tend to be slow with sub-Hertz image acquisition rates, although video rate systems have also been built (Rha et al, 2006).

In 1996, Miller et al (1996) obtained the first *in vivo* images of the cone receptors using a high resolution flood illuminated fundus camera (coupled with a precise second-order refraction). The introduction of a full AO system by Liang and Williams (1997) further enhanced the contrast and quality of the cone images. Since then many other flood AO systems have been built (e.g. Hofer, 2001b; Larichev et al, 2002; Glanc et al, 2004; Choi et al, 2006; Rha et al, 2006; Headington et al, 2011). They have utilized improved AO components and imaged a variety of retinal structure and function in both normal (e.g. Roorda & Williams, 1999, 2002; Pallikaris et al, 2003; Putnam et al, 2005; Jonnal et al, 2007; Doble et al, 2011) and diseased eyes (e.g. Carroll et al, 2004; Choi et al, 2006, 2008, 2011; Wolfing et al, 2006; Carroll, 2008).

3.2 Adaptive Optics Confocal Scanning Laser Ophthalmoscopes (AO-cSLO)

In a cSLO, a point of light is scanned rapidly across the retina in a two-dimensional transverse pattern. The reflected light passes through a pinhole that is confocal to a particular retinal layer. This light is then incident on a point detector such as a photomultiplier tube (PMT) or avalanche photodiode (APD). The two dimensional image can then be reconstructed from the detector output. This approach has two major advantages: (i) reduced scattering as only light from a particular point and retinal layer passes through the confocal pinhole (all other light is blocked) and all other light is blocked, and (ii) it allows for video rate imaging of retinal structure and processes.

The two dimensional scan is achieved in modern systems through the use of a fast-mirrored resonant scanner and a slower galvanometric frame scanner; typical frame rates are 20-30 Hz. The field of view is similar to flood-based AO systems (1-3°). Standard cSLOs have transverse and axial resolutions of approximately 5 μm and 200 μm respectively, but through the use of AO the resolution is improved to 2.5 μm transversely and <80 μm axially; with these numbers being dependent on the pupil size and imaging wavelength used.

In 1980, Webb and his colleagues (1980) demonstrated the first SLO, which was followed by the work of Dreher et al (1989) who used a DM in conjunction with an SLO. Wade and Fitzke (1998) used an SLO and post-processing to visualize the cone photoreceptors. The first closed-loop AO-cSLO was developed by Roorda et al (2002). Newer systems have pushed the performance even further allowing for the use of dual DMs (Chen et al, 2007a), increased fields of view (Ferguson et al, 2010) and the ability to visualize the rod photoreceptors (Dubra et al, 2011; Merino et al, 2011). For further details the reader is directed to the chapter on AO-cSLOs.

3.3 Adaptive Optics Optical Coherence Tomography (AO-OCT)

Optical coherence tomography (OCT) (Huang et al, 1991; Fercher et al, 1993; Swanson et al, 1993) is a non-invasive imaging modality that exploits the coherence properties of light to form an image. A light source is split into a reference channel and a sample channel via a modified Michelson interferometer. The retinal surface located in the sample channel is rapidly scanned and the reflected light interferes with that of the reference arm. Axial or 'A-scans' are generally acquired first allowing for the subsequent construction of the commonly displayed 'B-scans'. Three-dimensional volume scans can then be created. A major advantage of OCT is the axial and lateral resolutions are decoupled. The axial resolution depends on the bandwidth of the light source, the broader the bandwidth the higher the resolution, although compensation of the chromatic aberration of the eye is then required (Fernandez et al, 2005; Zawadzki et al, 2008). Axial resolutions of a few micrometers are possible. The lateral resolution is a function of the pupil size which in turn influences the level of aberration, hence the need for AO.

The first systems combined AO with OCT (Miller et al, 2003; Hermann et al, 2004; Zhang et al, 2005), more recently we are seeing the emergence of systems with multimodal imaging capabilities, AO-OCT-SLO systems (Iftimia et al, 2006; Merino et al, 2006; Miller et al, 2011; Zawadzki et al, 2011).

4. AO for vision testing

Human visual performance is limited by neural (retinal or at higher visual level) and optical factors. The evaluation of the contribution of each of these factors has proven difficult in the past, since the study of one was confounded by the other and vice versa. Prior attempts to separate the effect of these limiting factors included interferometry and the detection of contrast in images embedded in noise, however both methods have significant procedural limitations.

The use of AO for correcting aberrations in the human eye opened the possibility of evaluating the effect of neural factors since the optical effects are compensated for. Soon after the first applications of AO to obtain high-resolution images of the retina (Liang et al, 1997), a number of laboratories began exploring the use of AO to produce aberration-free retinal images to improve vision and evaluate visual function. It was shown that the correction of aberrations improved the contrast sensitivity function (CSF), showing sensitivity at frequencies up to 55 cycles per degree, not possible without AO correction (Liang et al, 1997), and improved the visual system's resolution or VA (Yoon & Williams, 2002). AO has also been used to modify the aberrations of the eye to study visual performance (Artal et al, 2004; Chen et al, 2007b). The limiting factor once aberrations are corrected with AO is the sampling of the photoreceptors. Positive effects of the correction of aberrations have been shown in other functions such as face recognition (Sawides et al, 2010) and even some improvement in the periphery of the visual field, where optics do not play such an important role (Roorda, 2011). These results seemed to imply that correcting aberrations, e.g. with refractive surgery, would significantly improve visual performance. However, since the eye is a living organ and the visual system ever changing, there are potentially significant limitations to the benefit of high order aberration correction to visual function as we describe below.

i. Accommodation

Accommodation is the process of changing the power of the eye by modifying the shape and position of the crystalline lens. The amount of accommodation required to form a clear retinal image is controlled by a number of cues, most of which are related to the retinal stimulus quality, such as blur and chromatic aberration, although some, e.g. retinal disparity and proximity, are not.

AO has allowed further evaluation of the stimuli that drive accommodation and disaccommodation, which seem to be non-parallel processes. Among the possible cues that may indicate the sign of defocus to drive accommodation are higher order monochromatic and chromatic aberrations. A number of authors have studied accommodation and disaccommodation with the manipulation of high order aberrations (Hampson et al, 2006; Chin et al, 2009a; Hampson et al, 2010) and suggested that aberrations play a role in the accommodation control of dynamic (stepwise and sinusoidal) stimuli. However, there is controversy as to the role of aberrations in accommodation since other authors have found improvement rather than a reduction of accommodation when correcting aberrations (Gambra et al, 2009). Further work is needed to determine the role of higher order aberrations in accommodation. Aberrations may play a role in the time response of accommodation rather than its accuracy (Fernandez & Artal, 2005).

Presbyopia, the decreased ability of the eye to accommodate as it ages, may also benefit from AO aberration correction (see section below on the correction of refractive error using AO).

ii. Refraction and Refractive Technologies Using Wavefront Sensing and AO Correction

Autorefractors are computer-controlled instruments that provide objective measurement of the eye's refractive error by measuring the vergence of the light reflected from the retina. They are often used by eye care professionals as a starting point for a subjective refraction. The use of clinically available aberrometers (e.g. Ophthonix Z-View aberrometer; Huvitz HRK-7000AW autorefractor/aberrometer) that measure higher order as well as lower order aberrations instead of the traditional autorefractor are becoming more popular as they have shown greater accuracy (Cooper et al, 2011). A spectacle correction based solely on the measures obtained from these instruments is not appropriate in most patients as they tend to overcorrect astigmatism - and some myopia - and give high errors when determining the axis for low astigmatic magnitudes. A phoropter is an instrument that contains lenses typically used by eye care professionals for subjective refraction of the eye typically used during an eye examination, i.e. correction of the lower aberrations - defocus and astigmatism. Phoropters incorporating AO would correct for higher order aberrations in addition to defocus and astigmatism using the wavefront pattern obtained with the aberrometer as a base and a subjective refraction as an endpoint.

AO vision simulators may also be useful tools to help finding the best refractive prescription for patients. Rocha et al (2010) used a crx1 AO Visual Simulator (Imagine Eyes SA) to correct and modify the wavefront aberrations in keratoconic eyes and symptomatic postoperative refractive surgery (LASIK) eyes. The AO visual simulator correction improved visual acuity by an average of two lines compared to their best spherocylinder correction. The AO technology may be of clinical benefit when counseling patients with highly aberrated eyes regarding their maximum subjective potential for vision correction.

Lower order aberrations, i.e. defocus and astigmatism, have been measured for hundreds of years and are typically corrected with spectacles, contact lenses, intraocular lenses or refractive surgery. The emerging AO technologies and the improvement in visual performance found with the correction of higher order aberrations has brought excitement to the field of refractive error correction. During the last decade there has been considerable debate concerning the visual impact of correcting the higher order aberrations of the eye.

First attempts to correct higher order aberrations used spectacle lenses and contact lenses. Both of these designs do not benefit significantly from correction of high order aberrations as aberrations are not constant but change with off-axis viewing, movement of the device, pupil size changes and accommodation, among other factors (Lopez-Gil et al, 2007). The use of scleral contact lenses helps with the stability problem of conventional contact lenses. A practical use of these lenses (Sabesan et al, 2007; Katsoulos et al, 2009; Sabesan & Yoon, 2010) for correction of the particularly elevated aberrations found in keratoconus seems plausible. Most of the difficulties found with spectacle lenses and contact lenses may be compensated if other methods of aberration correction, such as refractive surgery or intraocular lenses, are used.

AO has been applied in multifocal intraocular lenses for the correction of presbyopia as it extends the depth of focus by varying the amount of spherical aberration for axial (small pupil) rays for near vision and peripheral rays (larger pupils) for distance. Spherical aberration can be significantly reduced with aspheric intraocular lenses; however, there is a limited reduction in the total high order aberrations, even in perfectly positioned custom aspheric intraocular lenses, which may be influencing the unclear results in the studies assessing the potential benefits on visual performance of these lenses (Einighammer et al, 2009). AO has also been used to show that accommodative intraocular lenses for the correction of presbyopia actually work via pseudoaccommodative rather than accommodative mechanisms (Klaproth et al, 2011).

Refractive surgery has traditionally corrected lower order aberrations. Conventional refractive surgery may disrupt the compensation mechanism of corneal and internal ocular aberrations, creating a larger total amount of high order aberrations (Benito et al, 2009). With the advances in wavefront sensing technology, customized refractive surgery is common nowadays. Compared with conventional treatments, wavefront-guided ablations can achieve a reduction in preexisting higher-order aberrations and less induction of new higher-order aberrations, resulting in improved outcomes for contrast sensitivity and visual symptoms under mesopic and scotopic conditions (Kim & Chuck, 2008). However, concerns regarding the clinical applicability of customized wavefront correction have emerged, and the possibility of achieving supernormal vision in all patients has been challenged (e.g. Yeh & Azar, 2004).

The results of the reviewed studies suggest that many, but not all, observers with normal vision would perceive improvements in their spatial vision with customized (AO) vision correction, at least over a range of viewing distances, particularly when their pupils are larger than 3 mm. Keratoconic patients and patients suffering from high spherical aberration, e.g. as a result of conventional refractive surgery, would particularly benefit (Rocha et al, 2010). A recently developed technique combining customized (topography-guided) refractive surgery with riboflavin/UVA cross-linking seems a promising development for the treatment of progressive keratoconus (Krueger & Kanellopoulos, 2010).

iii. Using an AO-cSLO as a High-Frequency Eye Tracker

While the first attempts were made to correct the distortions found in the SLO frames due to eye movements, it was realized that these data are a record of the movements that had occurred during acquisitions. Therefore, the eyes can be tracked with an accuracy and frequency that would not be achieved with the best eye trackers available (Roorda, 2011).

iv. Simultaneous Stimuli Presentation and Image Delivery

With a cSLO the stimulus can be directly encoded in the rastered image, giving a real-time exact position of the stimulus in relation to the surrounding cones. Furthermore, with the incorporation of AO, the stimuli may be delivered to precise regions of the retina, to the level of individual photoreceptors (Sincich et al, 2009). Using a different channel for imaging and stimulus delivery, an infrared light may be used to image the retina while a visible light is used to present the stimuli and be used to record processes occurring at the retinal level as explained in the visual function section below. This technology can also be applied to fluorescence imaging to allow evaluation of sensitivity of non-absorbing structures such as axonal and dendritic structures of primate ganglion cells *in vivo* (Gray et al, 2008). The axial and lateral resolution achieved with AO was high enough to visualize individual dendrites and axons and was able to distinguish between ganglion cell types and function.

Still images and animations can also be delivered into a specific locus at the retina. Surprisingly, moving and stationary targets seem to generate different fixation loci and neither is correlated with the point of maximum cone density (Putnam et al, 2005; Stevenson et al, 2007).

AO-cSLO technology has also the potential for presenting stimuli at the level of a single photoreceptor and performing microperimetry - i.e. measure sensitivity - at an unprecedented degree of retinotopic precision (Makous et al, 2006; Tuten et al, 2011). Such technology would be very useful in studies determining the preferred locus of fixation (Putnam et al, 2005; 2011) and its relation with photoreceptor density. The preferred locus of fixation is an important parameter to obtain in patients with eccentric fixation and in patients with low vision who have central vision loss, e.g. caused by macular degeneration.

v. Visual Acuity and Contrast Sensitivity

By modifying an existing AO imaging device (fundus camera, SLO), stimuli may be imaged onto the retina without aberrations or with a controlled amount/type of aberrations. In addition to work by Liang et al (1997), Yoon and Williams (2002) also used AO to measure VA and contrast sensitivity (CS) through an aberration-corrected eye, with the limiting factor being the sampling of the photoreceptors, and found improved VA and CS at spatial frequencies of 16 and 24 cycles/deg. Most but not all observers showed improvement in VA and CS (e.g. Elliott & Chapman, 2009). Rossi et al (2007) found that myopes do not perform as well after AO correction as emmetropes; it was suggested that this difference is not due to larger cone spacing, as axial myopes (longer eyes) do not show smaller sampling than emmetropes (Li et al, 2010).

vi. AO Imaging Correlations with Visual Function

A number of reports have shown correlation between AO retinal imaging and visual function tests. Choi et al (2006) first reported that disruption of the cone photoreceptors mosaic in patients with various forms of retinal dystrophies is correlated to functional vision losses (visual fields, contrast sensitivity and and multifocal electroretinography - mfERG) (Fig 9). In toxic maculopathy caused by hydroxychloroquine (antimalarial drug used extensively in the treatment of autoimmune diseases), AO ophthalmoscopy also shows disruption of the cone photoreceptor mosaic in areas corresponding to visual field defects, and shows additional areas of irregularities in cone photoreceptor density in areas with otherwise normal visual field findings, suggesting that AO imaging is detecting changes earlier than visual field tests (Stepien et al, 2009). AO-cSLO imaging of the cone mosaic explained visual performance, a unilateral ring-like paracentral distortion that could otherwise not be explained using common clinical imaging instruments (Joeres et al, 2008).

More recently, Talcott et al (2011) has found that the use of AO imaging was more useful than the standard of care tests for evaluation of the effect of treatment in three patients with retinal degeneration. Furthermore, AO-cSLO has been used to evaluate cone spacing in familial mitochondrial DNA mutation. Visual function was affected with various levels of severity depending on the cone spacing pattern; it was improved in patients with a contiguous and regular cone mosaic. Patients expressing high levels of the mitochondrial DNA mutation T8993C showed abnormal cone structure, suggesting normal mitochondrial DNA is necessary for normal waveguiding by cones (Yoon et al, 2009). High resolution AO imaging and AO for visual function evaluation of photoreceptors have also been proven useful techniques in selection of patients for therapeutic trials of congenital achromatopsia and for monitoring the therapeutic response in these trials (Genead et al, 2011). AO fundus imaging has also been used to investigate photoreceptor structural changes in eyes with occult macular dystrophy (Kitaguchi et al, 2011). Furthermore, high resolution imaging with AO-cSLO has contributed significantly to our understanding of Stargardt's disease, a disease that severely affects central vision of young, otherwise healthy, individuals. AO-cSLO imaging showed abnormal cone spacing in regions of abnormal fundus autofluorescence and reduced visual function, although the earliest cone spacing abnormalities were observed in regions of homogeneous autofluorescence and normal visual function (Chen et al, 2011). In addition, visual resolution decreases rapidly outside the foveal center towards the peripheral retina, which seems to be more related to sampling of midget retinal ganglion cells than photoreceptor sampling (Rossi & Roorda, 2010b).

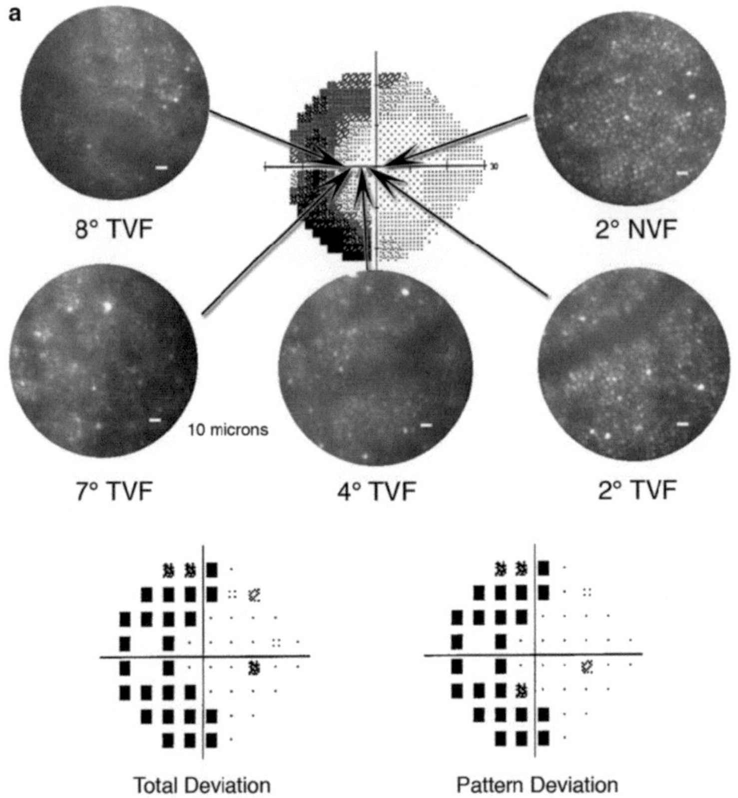

Fig. 9. AO-corrected retinal images for a subject with a retinal dystrophy at different locations of the retina shown with the corresponding visual field maps. Reprinted from Choi et al (2006), with permission from the journal Investigative Ophthalmology and Visual Science.

These studies indicate that AO fundus imaging is a reliable technique for assessing and quantifying the changes in the photoreceptor layer as the disease progresses. Furthermore, AO imaging correlates with visual function tests and may be useful in cases where visual function tests provide borderline or ambiguous results, as it allows visualization of individual photoreceptors. Some caution is warranted as there may be greater higher-order wavefront aberrations in eyes with macular disease than in control eyes without disease (Bessho et al, 2009). It has been suggested that portion of the aberration measurements may result from irregular or multiple reflecting retinal surfaces.

vii. Visual Perception at the Photoreceptor Level

AO systems are unique in that they provide structural and functional information of the visual system and therefore allow a new class of experiments with great scientific potential. For example, testing of image perception at the level of an individual photoreceptor is now possible with the use of AO. Roorda and Williams (1999) were the first to show *in vivo* images of the arrangement of the human trichromatic arrangement. By bleaching the retina with three different wavelengths they were able to classify the individual cones according to their photopigment type. The relative numbers of L and M cones varied significantly among subjects, even though all had the same color perception.

Using AO-cSLO, Sincich et al (2009) were able to deliver micron-scale spots of light to the centers of the receptive fields of neurons in the macaque LGN and resolve the contribution of single cone photoreceptors to the response of central visual neurons. They imaged and directly stimulated individual cones in the macaque in vivo, while neuron receptive fields were recorded in the LGN. It is therefore possible to now study the properties of different photoreceptors and their influence in visual perception.

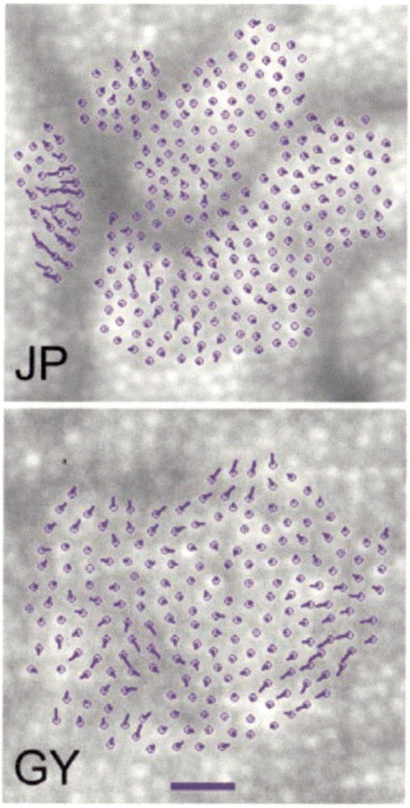

Fig. 10. Cone directionality plots for two subjects, from Roorda and Williams (2002). The circles represent cone locations and the lines the direction and magnitude of the departure of each cone's pointing direction in the pupil plane from the average of the ensemble. Reprinted with permission from the Journal of Vision.

As mentioned in the first section of this chapter, Roorda and Williams (2002) imaged the angular tuning properties (SCE) of individual photoreceptors in living human eyes. They found that the disarray in macular cones is very small, implying that the optical waveguide properties of ensembles of cones are similar to those of the single cones that compose them (Fig 10).

5. Ongoing challenges in applying AO for vision

Controlling and correcting aberrations allows the possibility of exploring their effects on human vision and their correction may improve visual performance. However, questions regarding the real benefits of totally eliminating aberrations remain (e.g. Chen et al, 2007b).

The subjective image quality depends on intrinsic optical (e.g. aberrations) and neural factors (Campbell & Green, 1965), as well as the prior experience of the observer. One of the neural limits is adaptation to blur (e.g. Georgeson & Sullivan, 1975; Webster et al, 2002; Vera-Diaz et al, 2010). Adaptation to blur is well known by eye care professionals as their myopic patients often report clearer vision after a period of not wearing their spectacles. Adaptation seems to occur not only for defocus and astigmatism (de Gracia et al, 2011), and other types of induced blur, but the visual system seems to also be adapt to the eye's own high order aberrations profile (Artal et al, 2004; Sawides et al, 2011), somehow removing the effects of blur induced by the optics of the eye. Best subjective image quality is obtained when some amount (~12 %) of high order aberrations are left uncorrected (Chen et al, 2007b). Further, observers seem to prefer a small amount of positive spherical defocus for best visual performance (Werner et al, 2009). Adaptation may have important implications in the correction of aberration with customized refractive surgery or contact lenses, as the benefits of that correction may be overcome with neural adaptation. However, the effect may disappear after adaptation to the new level of aberrations. Rossi and Roorda (2010a) have suggested that adaptation to aberrations does not hinder the correction with AO from giving immediate visual benefit since AO provided a significant improvement in visual resolution, and visual resolution is a low level task that is not expected to improve with training. It is yet to be determined whether the improvement found in static laboratory measures of VA and CS corresponds to an improvement in real life visual tasks. Furthermore, the improvement in these measures seems to be only found with larger pupil sizes (Elliott & Chapman, 2009) which do not occur in natural viewing. It also seems that the presence of aberrations is utilized as a cue by the visual system for accommodation (Kruger et al, 1993; Chen et al, 2006; Chin et al, 2009b) and perhaps for control of other oculomotor functions of the eye. Therefore, correcting these aberrations may negatively interfere with these visual functions.

In addition to neural limitations, the image quality presented to the retina has other optical dependencies besides aberrations; for example, ocular media density and light scatter, as well as diffraction (the eye is a diffraction-limited optical system at small pupil diameters). These are neither measured nor corrected by AO. Visual function, for example VA and CS at mid- and high-spatial frequencies, is compromised with normal aging. Elliott and Chapman (2009) found that increased high order aberrations that occur with aging (due to coma induced by asymmetric corneas, spherical aberration caused by changes in the lens, etc.) cannot completely account for the decline in spatial vision with aging. A larger role of the aforementioned optical factors and, perhaps more importantly, neural factors, exists in changes in spatial vision found with aging. Elliott et al (2007) and Vera-Diaz et al (2010) found no difference in the strength of adaptation to transient changes in image blur for younger and older observers, suggesting that cortical mechanisms of adaptation remain largely intact with age and could provide a mechanism for long-term adaptation to the increasing degree of high order aberrations with age.

Furthermore, ocular aberrations are dynamic, as explained above, and vary with eye movements, accommodation, etc., which bring subsequent limitations to the benefit of aberration correction.

Other potential limitations to the use of AO for vision science and clinical care are that these devices are not yet user-friendly and are cumbersome, although compact devices are being developed (Mujat et al, 2010). It has also been recently suggested that there are potential differences between AO fundus camera images and AO-cSLO images (Carroll et al, 2010), which requires further investigation. In addition, there are a number of technical limitations of AO systems that require further work. For example, during the time it takes to capture an image with an AO-cSLO device (about 30 msecs), the eye moves significantly and in unpredictable ways causing distortions in each frame. These distortions can be corrected for, but this is a work in progress.

6. The future of AO in vision science

In spite of the limitations described in the previous section, ophthalmic systems that employ AO have the potential to play a crucial role in the clinic. Although the use of AO in clinical settings is still to come, primarily because of the cost and the time-consuming nature of the testing and data processing, the recent substantial and rapid advances of these technologies suggest that their commercialization for ophthalmic clinical use is imminent. A number of prototypes are currently being tested. The AO fundus camera developed by Imagine Eyes © (rtx1™ Adaptive Optics Retinal Camera) is currently commercially available, for research use only, not for sale as a diagnostic device, and clinical trials began in France in 2009 and it may be available soon for clinical use. Physical Sciences Inc. has developed AO imaging devices and an AO-cSLO system, again for research only.

Noteworthy applications of AO imaging are the evaluation of the longitudinal progression of a disease and the evaluation of treatment efficacy. Further insight on the progression of retinal diseases is fundamental for understanding the molecular basics of these diseases. Likewise, the evaluation of the disease response to novel treatments with accurate objective methods such as AO imaging is far more informative than current subjective evaluation methods (e.g. VA and CS), and is critical for determining treatment efficacy. Retinal diseases are in general of slow progression, with some taking several years before changes in functional vision may be appreciated; however, disease progression has been shown with AO imaging techniques. With the use of AO it is now possible to image the smallest structures of the eye *in vivo*; not only retinal cells, but organelles and other microscopic structures are being imaged. As a result of the high transverse resolution of OCT and lateral resolution of SLO with AO, these evaluation measures are quickly being adopted as clinical trials' endpoints.

Talcott et al (2011) has recently completed the first longitudinal study of cone photoreceptors during retinal degeneration (in patients with inherited retinal degeneration, such as retinitis pigmentosa and Usher syndrome) and evaluated the response to a treatment with a ciliary neurotrophic factor. Changes in functional tests such as VA, CS and mfERG were monitored over 2 years. AO-cSLO images showed reduced cone loss in patients treated compared to the contralateral sham eyes (Sincich et al, 2009). Longitudinal studies on healthy eyes are necessary to create a database of normative data against which to compare disease data.

AO vision simulators may become useful tools to help choose the best refraction for patients. Clinicians will be able to show their patients what their vision would be like if they undergo a particular method of refractive surgery compared to another or compare between various kinds of intraocular lenses. Patient education could benefit from these simulators, particularly with the increased interest in presbyopic surgery, or the use of multifocal contact lenses, as patients may better understand and experiment with the visual benefits of various treatment methods.

7. Conclusion

AO is an extremely valuable tool in the study of the human visual system. Through bypassing the limitations of the optics of the eye, AO has enabled scientists to visualize single retinal cells *in vivo* and to probe the limits of human visual performance. Today, the field has moved on from technology development (although challenges still remain) to that of answering fundamental questions on retinal disease development and progression, human visual perception and aiding in the development of a range of new refractive technologies.

8. Acknowledgements

This work was supported in part by grant EY020901 from the National Institute of Health, Bethesda, MD, USA.

9. References

American National Standard Institute for the Safe Use of Lasers (ANSI). In Standard, A. N. (Ed.), (Vol. ANSI Z136.1 -- 2007): Laser Institute of America.

American National Standard Institute for the Safe Use of Lasers (ANSI). Ophthalmics Methods For Reporting Optical Aberrations Of Eyes. In (ANSI), A. N. S. I. (Ed.), (Vol. Z80.28-2010).

Artal, P., Chen, L., Fernandez, E. J., Singer, B., Manzanera, S. & Williams, D. R. (2004). Neural compensation for the eye's optical aberrations. *J Vis, 4*, 281-287.

Babcock, H. W. (1953). The possibility of compensating astronomical seeing. *Publ Astron Soc Pac, 65*, 229.

Bartsch, D. U., Zhu, L., Sun, P. C., Fainman, S. & Freeman, W. R. (2002). Retinal imaging with a low-cost micromachined membrane deformable mirror. *J Biomed Opt, 7*, 451-456.

Benito, A., Redondo, M. & Artal, P. (2009). Laser in situ keratomileusis disrupts the aberration compensation mechanism of the human eye. *Am J Ophthalmol, 147*, 424-431.

Bessho, K., Bartsch, D. U., Gomez, L., Cheng, L., Koh, H. J. & Freeman, W. R. (2009). Ocular wavefront aberrations in patients with macular diseases. *Retina, 29*, 1356-1363.

Burns, S. A., Wu, S., He, J. C. & Elsner, A. E. (1997). Variations in photoreceptor directionally across the central retina. *J Opt Soc Am A Opt Image Sci Vis, 14*, 2033-2040.

Campbell, F. W. & Green, D. G. (1965). Optical and retinal factors affecting visual resolution. *J Physiol, 181*, 576-593.

Carroll, J. (2008). Adaptive optics retinal imaging: applications for studying retinal degeneration. *Arch Ophthalmol, 126*, 857-858.

Carroll, J., Neitz, M., Hofer, H., Neitz, J. & Williams, D. R. (2004). Functional photoreceptor loss revealed with adaptive optics: an alternate cause of color blindness. *Proc Natl Acad Sci U S A, 101*, 8461-8466.

Carroll, J., Rossi, E. A., Porter, J., Neitz, J., Roorda, A., Williams, D. & Neitz, J. (2010). Deletion of the X-linked opsin gene array locus control region (LCR) results in disruption of the cone mosaic. *Vision Res, 50*, 1989–1899.

Chen, D. C., Jones, S. M., Silva, D. A. & Olivier, S. S. (2007a). High-resolution adaptive optics scanning laser ophthalmoscope with dual deformable mirrors. *J Opt Soc Am A Opt Image Sci Vis, 24*, 1305-1312.

Chen, L., Artal, P., Gutierrez, D. & Williams, D. R. (2007b). Neural compensation for the best aberration correction. *J Vis, 7*, 9 1-9.

Chen, L., Kruger, P. B., Hofer, H., Singer, B. & Williams, D. R. (2006). Accommodation with higher-order monochromatic aberrations corrected with adaptive optics. *J Opt Soc Am A Opt Image Sci Vis, 23*, 1-8.

Chen, Y., Ratnam, K., Sundquist, S. M., Lujan, B., Ayyagari, R., Gudiseva, V. H., Roorda, A. & Duncan, J. L. (2011). Cone photoreceptor abnormalities correlate with vision loss in patients with Stargardt disease. *Invest Ophthalmol Vis Sci, 52*, 3281-3292.

Chin, S. S., Hampson, K. M. & Mallen, E. A. (2009a). Effect of correction of ocular aberration dynamics on the accommodation response to a sinusoidally moving stimulus. *Opt Lett, 34*, 3274-3276.

Chin, S. S., Hampson, K. M. & Mallen, E. A. (2009b). Role of ocular aberrations in dynamic accommodation control. *Clin Exp Optom, 92*, 227-237.

Choi, S. S., Doble, N., Hardy, J. L., Jones, S. M., Keltner, J. L., Olivier, S. S. & Werner, J. S. (2006). In vivo imaging of the photoreceptor mosaic in retinal dystrophies and correlations with visual function. *Invest Ophthalmol Vis Sci, 47*, 2080-2092.

Choi, S. S., Enoch, J. M. & Kono, M. (2004). Evidence for transient forces/strains at the optic nerve head in myopia: repeated measurements of the Stiles-Crawford effect of the first kind (SCE-I) over time. *Ophthalmic Physiol Opt, 24*, 194-206.

Choi, S. S., Zawadzki, R. J., Keltner, J. L. & Werner, J. S. (2008). Changes in cellular structures revealed by ultra-high resolution retinal imaging in optic neuropathies. *Invest Ophthalmol Vis Sci, 49*, 2103-2119.

Choi, S. S., Zawadzki, R. J., Lim, M. C., Brandt, J. D., Keltner, J. L., Doble, N. & Werner, J. S. (2011). Evidence of outer retinal changes in glaucoma patients as revealed by ultrahigh-resolution in vivo retinal imaging. *Br J Ophthalmol, 95*, 131-141.

Chui, T. Y., Song, H. & Burns, S. A. (2008a). Adaptive-optics imaging of human cone photoreceptor distribution. *J Opt Soc Am A Opt Image Sci Vis, 25*, 3021-3029.

Chui, T. Y., Song, H. & Burns, S. A. (2008b). Individual variations in human cone photoreceptor packing density: variations with refractive error. *Invest Ophthalmol Vis Sci, 49*, 4679-4687.

Cooper, J., Citek, K. & Feldman, J. M. (2011). Comparison of refractive error measurements in adults with Z-View aberrometer, Humphrey autorefractor, and subjective refraction. *Optometry, 82*, 231-240.

Curcio, C. A. & Allen, K. A. (1990). Topography of ganglion cells in human retina. *J Comp Neurol, 300*, 5-25.

Curcio, C. A., Sloan, K. R., Kalina, R. E. & Hendrickson, A. E. (1990). Human photoreceptor topography. *J Comp Neurol, 292*, 497-523.

de Gracia, P., Dorronsoro, C., Marin, G., Hernandez, M. & Marcos, S. (2011). Visual acuity under combined astigmatism and coma: optical and neural adaptation effects. *J Vis, 11*.

Delori, F. C., Webb, R. H. & Sliney, D. H. (2007). Maximum permissible exposures for ocular safety (ANSI 2000), with emphasis on ophthalmic devices. *J Opt Soc Am A Opt Image Sci Vis, 24*, 1250-1265.

Diaz-Santana, L., Torti, C., Munro, I., Gasson, P. & Dainty, C. (2003). Benefit of higher closed-loop bandwidths in ocular adaptive optics. *Opt Express, 11*, 2597-2605.

Do, M. T. & Yau, K. W. (2010). Intrinsically photosensitive retinal ganglion cells. *Physiol Rev, 90*, 1547-1581.

Doble, N., Choi, S. S., Codona, J. L., Christou, J., Enoch, J. M. & Williams, D. R. (2011). In vivo imaging of the human rod photoreceptor mosaic. *Opt Lett, 36*, 31-33.

Doble, N., Miller, D. T., Yoon, G. & Williams, D. R. (2007). Requirements for discrete actuator and segmented wavefront correctors for aberration compensation in two large populations of human eyes. *Appl Opt, 46*, 4501-4514.

Doble, N., Yoon, G., Chen, L., Bierden, P., Singer, B., Olivier, S. & Williams, D. R. (2002). Use of a microelectromechanical mirror for adaptive optics in the human eye. *Opt Lett, 27*, 1537-1539.

Dreher, A. W., Bille, J. F. & Weinreb, R. N. (1989). Active optical depth resolution improvement of the laser tomographic scanner. *Appl Opt, 28*, 804-808.

Dubra A, Sulai Y, Norris JL, Cooper RF, Dubis AM, Williams DR, Carroll J. (2011). Noninvasive imaging of the human rod photoreceptor mosaic using a confocal adaptive optics scanning ophthalmoscope. *Biomed Opt Express*,2(7):1864-1876.

Duncan, J. L., Zhang, Y., Gandhi, J., Nakanishi, C., Othman, M., Branham, K. E., Swaroop, A. & Roorda, A. (2007). High-resolution imaging with adaptive optics in patients with inherited retinal degeneration. *Invest Ophthalmol Vis Sci, 48*, 3283-3291.

Einighammer, J., Oltrup, T., Feudner, E., Bende, T. & Jean, B. (2009). Customized aspheric intraocular lenses calculated with real ray tracing. *J Cataract Refract Surg, 35*, 1984-1994.

Elliott, D. B. & Chapman, G. J. (2009). Adaptive gait changes due to spectacle magnification and dioptric blur in older people. *Invest Ophthalmol Vis Sci, 51*, 718-722.

Elliott, S. L., Hardy, J. L., Webster, M. A. & Werner, J. S. (2007). Aging and blur adaptation. *J Vis, 7*, 8.

Enoch, J. M. (1963). Optical properties of the retinal receptors. *Journal of the Optical Society of America A, 53*, 71-85.

Fercher, A. F., Hitzenberger, C. K., Drexler, W., Kamp, G. & Sattmann, H. (1993). In vivo optical coherence tomography. *Am J Ophthalmol, 116*, 113-114.

Ferguson, R. D., Zhong, Z., Hammer, D. X., Mujat, M., Patel, A. H., Deng, C., Zou, W. & Burns, S. A. (2010). Adaptive optics scanning laser ophthalmoscope with integrated wide-field retinal imaging and tracking. *J Opt Soc Am A Opt Image Sci Vis, 27*, A265-277.

Fernandez, E. J. & Artal, P. (2005). Study on the effects of monochromatic aberrations in the accommodation response by using adaptive optics. *J Opt Soc Am A Opt Image Sci Vis, 22*, 1732-1738.

Fernandez, E. J., Iglesias, I. & Artal, P. (2001). Closed-loop adaptive optics in the human eye. *Opt Lett, 26*, 746-748.

Fernandez, E. J., Vabre, L., Hermann, B., Unterhuber, A., Povazay, B. & Drexler, W. (2006). Adaptive optics with a magnetic deformable mirror: applications in the human eye. *Opt Express, 14*, 8900-8917.

Fernández, E. & Drexler, W. (2005). Influence of ocular chromatic aberration and pupil size on transverse resolution in ophthalmic adaptive optics optical coherence tomography. *Opt Express, 13*, 8184-8197.

Gambra, E., Sawides, L., Dorronsoro, C. & Marcos, S. (2009). Accommodative lag and fluctuations when optical aberrations are manipulated. *J Vis, 9*, 4 1-15.

Genead, M. A., Fishman, G. A., Rha, J., Dubis, A. M., Bonci, D. M., Dubra, A., Stone, E. M., Neitz, M. & Carroll, J. (2011). Photoreceptor Structure and Function in Patients with Congenital Achromatopsia. *Invest Ophthalmol Vis Sci.*

Georgeson, M. A. & Sullivan, G. D. (1975). Contrast constancy: deblurring in human vision by spatial frequency channels. *J Physiol, 252*, 627-656.

Glanc, M., Gendron, E., Lacombe, F., Lafaille, D., Le Gargasson, J. F. & Léna , P. (2004). Towards wide-field retinal imaging with adaptive optics. *Optics Comm, 230*, 225-238.

Gray, D. C., Wolfe, R., Gee, B. P., Scoles, D., Geng, Y., Masella, B. D., Dubra, A., Luque, S., Williams, D. R. & Merigan, W. H. (2008). In vivo imaging of the fine structure of rhodamine-labeled macaque retinal ganglion cells. *Invest Ophthalmol Vis Sci, 49*, 467-473.

Gruppetta, S., Koechlin, L., Lacombe, F. & Puget, P. (2005). Curvature sensor for the measurement of the static corneal topography and the dynamic tear film topography in the human eye. *Opt Lett, 30*, 2757-2759.

Hampson, K. M., Chin, S. S. & Mallen, E. A. (2010). Effect of temporal location of correction of monochromatic aberrations on the dynamic accommodation response. *Biomed Opt Express, 1*, 879-894.

Hampson, K. M., Paterson, C., Dainty, C. & Mallen, E. A. (2006). Adaptive optics system for investigation of the effect of the aberration dynamics of the human eye on steady-state accommodation control. *J Opt Soc Am A Opt Image Sci Vis, 23*, 1082-1088.

Hardy, J. W. (1998). *Adaptive Optics for Astronomical Telescopes.* Oxford, UK: Oxford University Press.

He, J. C., Burns, S. A. & Marcos, S. (2000). Monochromatic Aberrations in the Accommodated Human Eye. *Vision Res, 40*, 41-48.

Headington, K., Choi, S. S., Nickla, D. & Doble, N. (2011). Single Cell, In vivo Imaging of the Chick Retina with Adaptive Optics. *Current Eye Research*, 36:947-957.

Hermann, B., Fernandez, E. J., Unterhuber, A., Sattmann, H., Fercher, A. F., Drexler, W., Prieto, P. M. & Artal, P. (2004). Adaptive-optics ultrahigh-resolution optical coherence tomography. *Opt Lett, 29*, 2142-2144.

Hofer, H., Artal, P., Singer, B., Aragon, J. L. & Williams, D. R. (2001a). Dynamics of the eye's wave aberration. *J Opt Soc Am A Opt Image Sci Vis, 18*, 497-506.

Hofer, H., Chen, L., Yoon, G. Y., Singer, B., Yamauchi, Y. & Williams, D. R. (2001b). Improvement in retinal image quality with dynamic correction of the eye's aberrations. *Opt Express, 8*, 631-643.

Huang, D., Swanson, E. A., Lin, C. P., Schuman, J. S., Stinson, W. G., Chang, W., Hee, M. R., Flotte, T., Gregory, K., Puliafito, C. A. & et al. (1991). Optical coherence tomography. *Science, 254*, 1178-1181.

Iftimia, N. V., Hammer, D. X., Bigelow, C. E., Ustun, T., de Boer, J. F. & Ferguson, R. D. (2006). Hybrid retinal imager using line-scanning laser ophthalmoscopy and spectral domain optical coherence tomography. *Opt Express, 14*, 12909-12914.

Iglesias, I., Ragazzoni, R., Julien, Y. & Artal, P. (2002). Extended source pyramid wave-front sensor for the human eye. *Opt Express, 10*, 419-428.

Joeres, S., Jones, S. M., Chen, D. C., Silva, D., Olivier, S., Fawzi, A., Castellarin, A. & Sadda, S. R. (2008). Retinal imaging with adaptive optics scanning laser ophthalmoscopy in unexplained central ring scotoma. *Arch Ophthalmol, 126*, 543-547.

Jonas, J. B., Schneider, U. & Naumann, G. O. H. (1992). Count and density of human retinal photoreceptors. *Graefe's Archive of Clinical and Experimental Ophthalmology, 230*, 505.

Jonnal, R. S., Besecker, J. R., Derby, J. C., Kocaoglu, O. P., Cense, B., Gao, W., Wang, Q. & Miller, D. T. (2010). Imaging outer segment renewal in living human cone photoreceptors. *Opt Express, 18*, 5257-5270.

Jonnal, R. S., Rha, J., Zhang, Y., Cense, B., Gao, W. & Miller, D. T. (2007). In vivo functional imaging of human cone photoreceptors. *Opt Express, 15*, 16141-16160.

Kanis, M. J. & van Norren, D. (2008). Delayed recovery of the optical Stiles-Crawford effect in a case of central serous chorioretinopathy. *British Journal of Ophthalmology, 92*, 292-292.

Katsoulos, C., Karageorgiadis, L., Vasileiou, N., Mousafeiropoulos , T. & Asimellis, G. (2009). Customized hydrogel contact lenses for keratoconus incorporating correction for vertical coma aberration. *Ophthalmic Physiol Opt, 29*, 321-329.

Kaufman, P. L. & Alm, A. (2002). *Adler's Physiology of the Eye* (10 ed.): Elsevier Health Sciences.

Kim, A. & Chuck, R. S. (2008). Wavefront-guided customized corneal ablation. *Curr Opin Ophthalmol, 19*, 314-320.

Kitaguchi, Y., Kusaka, S., Yamaguchi, T., Mihashi, T. & Fujikado, T. (2011). Detection of photoreceptor disruption by adaptive optics fundus imaging and Fourier-domain optical coherence tomography in eyes with occult macular dystrophy. *Clin Ophthalmol, 5*, 345-351.

Klaproth, O. K., Titke, C., Baumeister, M. & Kohnen, T. (2011). [Accommodative intraocular lenses - principles of clinical evaluation and current results]. *Klin Monbl Augenheilkd, 228*, 666-675.

Krueger, R. R. & Kanellopoulos, A. J. (2010). Stability of simultaneous topography-guided photorefractive keratectomy and riboflavin/UVA cross-linking for progressive keratoconus: case reports. *J Refract Surg, 26*, S827-832.

Kruger, P. B., Mathews, S., Aggarwala, K. R. & Sanchez, N. (1993). Chromatic aberration and ocular focus: Fincham revisited. *Vision Res, 33*, 1397-1411.

Larichev, A. V., Ivanov, P. V., Iroshnikov, N. G., Shmalhauzen , V. I. & Otten, L. J. (2002). Adaptive system for eye-fundus imaging. *Quantum Electronics, 32*, 902-908.

Li, F. H., Mukohzaba, N., Yoshida, N., Igasaki, Y., Toyoda, H., Inoue, T., Kobayashi, Y. & Hara, T. (1998). Phase modulation characteristics analysis of optically-addressed parallel-aligned nematic liquid crystal phase-only spatial light modulator combined with a liquid crystal display. *Opt Review, 5*, 174-178.

Li, K. Y., Tiruveedhula, P. & Roorda, A. (2010). Intersubject variability of foveal cone photoreceptor density in relation to eye length. *Invest Ophthalmol Vis Sci, 51*, 6858-6867.

Liang, C. & Williams, D. R. (1997). Aberrations and retinal image quality of the normal human eye. *J Opt Soc Am A Opt Image Sci Vis, 14*, 2873-2883.

Liang, J., Grimm, B., Goelz, S. & Bille, J. F. (1994). Objective measurement of wave aberrations of the human eye with the use of a Hartmann-Shack wave-front sensor. *J Opt Soc Am A Opt Image Sci Vis, 11*, 1949-1957.

Liang, J., Williams, D. R. & Miller, D. T. (1997). Supernormal vision and high-resolution retinal imaging through adaptive optics. *J Opt Soc Am A Opt Image Sci Vis, 14*, 2884-2892.

Lopez-Gil, N., Rucker, F. J., Stark, L. R., Badar, M., Borgovan, T., Burke, S. & Kruger, P. B. (2007). Effect of third-order aberrations on dynamic accommodation. *Vision Res, 47*, 755-765.

Makous, W., Carroll, J., Wolfing, J. I., Lin, J., Christie, N. & Williams, D. R. (2006). Retinal microscotomas revealed with adaptive-optics microflashes. *Invest Ophthalmol Vis Sci, 47*, 4160-4167.

Marmor, M. F., Choi, S. S., Zawadzki, R. J. & Werner, J. S. (2008). Visual insignificance of the foveal pit: reassessment of foveal hypoplasia as fovea plana. *Arch Ophthalmol, 126*, 907-913.

Martin, J. A. & Roorda, A. (2005). Direct and noninvasive assessment of parafoveal capillary leukocyte velocity. *Ophthalmology, 112*, 2219-2224.

McAllister, J. T., Dubis, A. M., Tait, D. M., Ostler, S., Rha, J., Stepien, K. E., Summers, C. G. & Carroll, J. (2010). Arrested development: high-resolution imaging of foveal morphology in albinism. *Vision Res, 50*, 810-817.

Merino, D., Dainty, C., Bradu, A. & Podoleanu, A. G. (2006). Adaptive optics enhanced simultaneous en-face optical coherence tomography and scanning laser ophthalmoscopy. *Opt Express, 14*, 3345-3353.

Merino, D., Duncan, J. L., Tiruveedhula, P. & Roorda, A. (2011). Observation of cone and rod photoreceptors in normal subjects and patients using a new generation adaptive optics scanning laser ophthalmoscope. *Biomed Opt Express, 2*, 2189-2201.

Miller, D. T., Kocaoglu, O. P., Wang, Q. & Lee, S. (2011). Adaptive optics and the eye (super resolution OCT). *Eye (Lond), 25*, 321-330.

Miller, D. T., Qu, J., Jonnal, R. S. & Thorn, K. (2003). Coherence Gating and Adaptive Optics in the Eye. *Proc. SPIE* 65-72.

Miller, D. T., Williams, D. R. & Morris, G. M. (1996). Images of cone photoreceptors in the living human eye. *Vision Res, 36*, 1067-1079.

Mujat, M., Ferguson, R. D., Patel, A. H., Iftimia, N., Lue, N. & Hammer, D. X. (2010). High resolution multimodal clinical ophthalmic imaging system. *Opt Express, 18*, 11607-11621.

Netter, F. H. (2006). *Netter's Atlas of Human Anatomy.* (4 ed.): Saunders-Elsevier.

Neuroscience Online. In Byrne, J. H. (Ed.): The University of Texas Health Science Center at Houston (UTHealth). 1997.

Noll, R. J. (1976). Zernike polynomials and atmospheric turbulence. *J Opt Soc Am A Opt Image Sci Vis, 66*, 207–211.

Pallikaris, A., Williams, D. R. & Hofer, H. (2003). The reflectance of single cones in the living human eye. *Invest Ophthalmol Vis Sci, 44*, 4580-4592.

Porter, J., Guirao, A., Cox, I. G. & Williams, D. R. (2001). Monochromatic aberrations of the human eye in a large population. *J Opt Soc Am A Opt Image Sci Vis, 18*, 1793-1803.

Porter, J., Queener, H., Lin, J., Thorn, K. E. & Awwal, A. (2006). *Adaptive Optics for Vision Science.* Hoboken, NJ Wiley-Interscience.

Prieto, P., Fernandez, E., Manzanera, S. & Artal, P. (2004). Adaptive optics with a programmable phase modulator: applications in the human eye. *Opt Express, 12*, 4059-4071.

Putnam, N. M., Hammer, D. X., Zhang, Y., Merino, D. & Roorda, A. (2010). Modeling the foveal cone mosaic imaged with adaptive optics scanning laser ophthalmoscopy. *Opt Express, 18,* 24902-24916.

Putnam, N. M., Hofer, H. J., Doble, N., Chen, L., Carroll, J. & Williams, D. R. (2005). The locus of fixation and the foveal cone mosaic. *J Vis, 5,* 632-639.

Putnam, N. M., Tiruveedhula, P. & Roorda, A. (2011). Characterization Of The Preferred Retinal Locus Of Fixation And The Locus Of Perceived Fixation In Relation To The Photoreceptor Mosaic *Association for Research and Vision in Ophthalmology.* Fort Lauderdale, Florida.

Ragazzoni, R. (1996). Pupil plane wavefront sensing with an oscillating prism. *J of Mod. Opt, 43,* 289-293.

Rha, J., Dubis, A. M., Wagner-Schuman, M., Tait, D. M., Godara, P., Schroeder, B., Stepien, K. & Carroll, J. (2010). Spectral domain optical coherence tomography and adaptive optics: imaging photoreceptor layer morphology to interpret preclinical phenotypes. *Adv Exp Med Biol, 664,* 309-316.

Rha, J., Jonnal, R. S., Thorn, K. E., Qu, J., Zhang, Y. & Miller, D. T. (2006). Adaptive optics flood-illumination camera for high speed retinal imaging. *Opt Express, 14,* 4552-4569.

Rocha, K. M., Vabre, L., Chateau, N. & Krueger, R. R. (2010). Enhanced visual acuity and image perception following correction of highly aberrated eyes using an adaptive optics visual simulator. *J Refract Surg, 26,* 52-56.

Roddier, F. (1988). Curvature sensing and compensation: a new concept in adaptive optics. *Appl Opt, 27,* 1223.

Roorda, A. (2011). Adaptive optics for studying visual function: a comprehensive review. *J Vis, 11.*

Roorda, A., Romero-Borja, F., Donnelly Iii, W., Queener, H., Hebert, T. & Campbell, M. (2002). Adaptive optics scanning laser ophthalmoscopy. *Opt Express, 10,* 405-412.

Roorda, A. & Williams, D. R. (1999). The arrangement of the three cone classes in the living human eye. *Nature, 397,* 520-522.

Roorda, A. & Williams, D. R. (2002). Optical fiber properties of individual human cones. *J Vis, 2,* 404-412.

Roorda, A., Zhang, Y. & Duncan, J. L. (2007). High-resolution in vivo imaging of the RPE mosaic in eyes with retinal disease. *Invest Ophthalmol Vis Sci, 48,* 2297-2303.

Rossi, E. A. & Roorda, A. (2010a). Is visual resolution after adaptive optics correction susceptible to perceptual learning? *J Vis, 10,* 11.

Rossi, E. A. & Roorda, A. (2010b). The relationship between visual resolution and cone spacing in the human fovea. *Nat Neurosci, 13,* 156-157.

Rossi, E. A., Weiser, P., Tarrant, J. & Roorda, A. (2007). Visual performance in emmetropia and low myopia after correction of high-order aberrations. *J Vis, 7,* 14.

Sabesan, R., Ahmad, K. & Yoon, G. (2007). Correcting highly aberrated eyes using large-stroke adaptive optics. *J Refract Surg, 23,* 947-952.

Sabesan R, Yoon G. (2010). Neural compensation for long-term asymmetric optical blur to improve visual performance in keratoconic eyes. *Invest Ophthalmol Vis Sci,* 51(7):3835-3839.

Sawides, L., de Gracia, P., Dorronsoro, C., Webster, M. & Marcos, S. (2011). Adapting to blur produced by ocular high-order aberrations. *J Vis, 11.*

Sawides, L., Gambra, E., Pascual, D., Dorronsoro, C. & Marcos, S. (2010). Visual performance with real-life tasks under adaptive-optics ocular aberration correction. *J Vis, 10*, 19.

Shack, R. B. & Platt, B. C. (1971). Production and use of a lenticular Hartmann screen. *J Opt Soc Am A Opt Image Sci Vis, 61*, 656.

Sincich, L. C., Zhang, Y., Tiruveedhula, P., Horton, J. C. & Roorda, A. (2009). Resolving single cone inputs to visual receptive fields. *Nat Neurosci, 12*, 967-969.

Snell, R. S. & Lemp, M. A. (1998). *Clinical Anatomy of the Eye* (2 ed.). Oxford: Backwell Science.

Stepien, K. E., Han, D. P., Schell, J., Godara, P. & Carroll, J. (2009). Spectral-domain optical coherence tomography and adaptive optics may detect hydroxychloroquine retinal toxicity before symptomatic vision loss. *Trans Am Ophthalmol Soc, 107*, 28-33.

Stevenson, S., Kumar, G. & Roorda, A. (2007). Psychophysical and oculomotor reference points for visual direction measured with the adaptive optics scanning laser ophthalmoscope. *J Vis, 7*, 137.

Stiles, W. S. & Crawford, B. H. (1933). The luminous efficiency of rays entering the eye pupil at different points. *Proceedings of the Royal Society of London. Series B: Biological Sciences, 112*, 428-450.

Swanson, E. A., Izatt, J. A., Hee, M. R., Huang, D., Lin, C. P., Schuman, J. S., Puliafito, C. A. & Fujimoto, J. G. (1993). In vivo retinal imaging by optical coherence tomography. *Opt Lett, 18*, 1864-1866.

Talcott, K. E., Ratnam, K., Sundquist, S. M., Lucero, A. S., Lujan, B. J., Tao, W., Porco, T. C., Roorda, A. & Duncan, J. L. (2011). Longitudinal study of cone photoreceptors during retinal degeneration and in response to ciliary neurotrophic factor treatment. *Invest Ophthalmol Vis Sci, 52*, 2219-2226.

Tam, J., Martin, J. A. & Roorda, A. (2010). Noninvasive visualization and analysis of parafoveal capillaries in humans. *Invest Ophthalmol Vis Sci, 51*, 1691-1698.

Thibos, L. N. & Bradley, A. (1997). Use of liquid-crystal adaptive-optics to alter the refractive state of the eye. *Optom Vis Sci, 74*, 581-587.

Thibos, L. N., Hong, X., Bradley, A. & Cheng, X. (2002). Statistical variation of aberration structure and image quality in a normal population of healthy eyes. *J Opt Soc Am A Opt Image Sci Vis, 19*, 2329-2348.

Tuten, W. S., Tiruveedhula, P. & Roorda, A. (2011). Adaptive Optics Scanning Laser Ophthalmoscope-based Microperimetry, *Association for Research and Vision in Ophthalmology*. Fort Lauderdale, Florida.

Tyson, R. K. (2010). *Principles of Adaptive Optics.* (3rd Edition ed.). USA: CRC Press.

Vargas-Martin, F., Prieto, P. M. & Artal, P. (1998). Correction of the aberrations in the human eye with a liquid-crystal spatial light modulator: limits to performance. *J Opt Soc Am A Opt Image Sci Vis, 15*, 2552-2562.

Vera-Diaz, F. A., Woods, R. L. & Peli, E. (2010). Shape and individual variability of the blur adaptation curve. *Vision Res, 50*, 1452-1461.

Vohnsen, B. (2007). Photoreceptor waveguides and effective retinal image quality. *J Opt Soc Am A Opt Image Sci Vis, 24*, 597-607.

Wade, A. & Fitzke, F. (1998). A fast, robust pattern recognition asystem for low light level image registration and its application to retinal imaging. *Opt Express, 3*, 190-197.

Wang, Q., Kocaoglu, O. P., Cense, B., Bruestle, J., Jonnal, R. S., Gao, W. & Miller, D. T. (2011). Imaging retinal capillaries using ultrahigh-resolution optical coherence tomography and adaptive optics. *Invest Ophthalmol Vis Sci, 52*, 6292-6299.

Webb, R. H., Hughes, G. W. & Pomerantzeff, O. (1980). Flying spot TV ophthalmoscope. *Appl Opt, 19*, 2991-2997.

Webster, M. A., Georgeson, M. A. & Webster, S. M. (2002). Neural adjustments to image blur. *Nat Neurosci, 5*, 839-840.

Webvision: The Organization of the Retina and Visual System. In Kolb, H., Nelson, R., Fernandez, E. & Jones, B. (Eds.): WorldPress. 2011.

Werner, J. S., Elliott, S. L., Choi, S. S. & Doble, N. (2009). Spherical aberration yielding optimum visual performance: evaluation of intraocular lenses using adaptive optics simulation. *J Cataract Refract Surg, 35*, 1229-1233.

Westheimer, G. (1967). Dependence of the magnitude of the Stiles-Crawford effect on retinal location. *J Physiol, 192*, 309-315.

Westheimer, G. (2008). Directional sensitivity of the retina: 75 years of Stiles-Crawford effect. *Proc Biol Sci, 275*, 2777-2786.

Wolfing, J. I., Chung, M., Carroll, J., Roorda, A. & Williams, D. R. (2006). High-resolution retinal imaging of cone-rod dystrophy. *Ophthalmology, 113*, 1019 e1011.

Yeh, S. I. & Azar, D. T. (2004). The future of wavefront sensing and customization. *Ophthalmol Clin North Am, 17*, 247-260.

Yoon, G. Y. & Williams, D. R. (2002). Visual performance after correcting the monochromatic and chromatic aberrations of the eye. *J Opt Soc Am A Opt Image Sci Vis, 19*, 266-275.

Yoon, M. K., Roorda, A., Zhang, Y., Nakanishi, C., Wong, L. J., Zhang, Q., Gillum, L., Green, A. & Duncan, J. L. (2009). Adaptive optics scanning laser ophthalmoscopy images in a family with the mitochondrial DNA T8993C mutation. *Invest Ophthalmol Vis Sci, 50*, 1838-1847.

Zawadzki, R. J., Cense, B., Zhang, Y., Choi, S. S., Miller, D. T. & Werner, J. S. (2008). Ultrahigh-resolution optical coherence tomography with monochromatic and chromatic aberration correction. *Opt Express, 16*, 8126-8143.

Zawadzki, R. J., Jones, S. M., Pilli, S., Balderas-Mata, S., Kim, D. Y., Olivier, S. S. & Werner, J. S. (2011). Integrated adaptive optics optical coherence tomography and adaptive optics scanning laser ophthalmoscope system for simultaneous cellular resolution in vivo retinal imaging. *Biomed Opt Express, 2*, 1674-1686.

Zhang, Y., Rha, J., Jonnal, R. & Miller, D. (2005). Adaptive optics parallel spectral domain optical coherence tomography for imaging the living retina. *Opt Express, 13*, 4792-4811.

Zhong, Z., Song, H., Chui, T. Y., Petrig, B. L. & Burns, S. A. (2011). Noninvasive measurements and analysis of blood velocity profiles in human retinal vessels. *Invest Ophthalmol Vis Sci, 52*, 4151-4157.

7

Adaptive Optics Confocal Scanning Laser Ophthalmoscope

Jing Lu, Hao Li, Guohua Shi and Yudong Zhang
Institute of Optics and Electronics, Chinese Academy of Sciences
China

1. Introduction

Scanning laser ophthalmoscope (SLO) was first presented by R. H. Webb (R. H. Webb, et al, 1987) in 1987, which is the same as a scanning laser microscope except that human eye is used as the objective lens and retina is usually the sample being imaged. With its high contrast real-time imaging and axial sectioning capability, SLO has many advantages and applications in retina imaging (Austin. Roorda, et al, 2002), eye tracking (D. X. Hammer, et al, 2005), hemodynamic (R. Daniel Ferguson, Daniel X. Hammer, 2004), tomography (Fernando Romero-Borja, et al, 2005), etc. However, due to ocular aberrations, resolution of image is dramatically degraded (R. H. Webb, et al, 1987), which results in that conventional SLOs always have large field (usually 10°~20°) and low resolution (lateral resolution>30μm, axial resolution>300μm for 6mm pupil).

Rochester University first used Adaptive Optics (AO) to compensate for monochromatic aberrations of human eye in a fundus camera (J. Liang, et al, 1997). Then Institute of Optics and Eleconics of China developed a table-top adaptive optical system for human retinal imaging (Yudong Zhang, et al, 2002, Ning Ling, Yudong Zhang, et al, 2002, Ning Ling, Xuejun Rao, et al, 2002, Ning Ling, Yudong Zhang, et al, 2004). They obtained a high resolution retina image with 6°×6° field of view (Ling Ning, et al, 2005). The first AOCSLO was reported by Austin Roorda (Austin. Roorda, et al, 2002), it yielded the first real-time images of photoreceptors and blood flow in living human retina at video rates, with lateral resolution<2.5μm, axial resolution<100μm. In 2005, Indiana University developed Tracking AOCSLO (D. X. Hammer, et al, 2005), and David Merino combined AOCSLO and AOOCT (D. Merino, Chris Dainty, 2006) to improve both lateral resolution and axial resolution.

2. Basic concept

The schematic of AOCSLO system is shown in Fig.1. A beam of light emitted from SLD is collimated, passes through various optical components, and then enters the eye which focuses it to a small spot on retina. While the eye's pupil plane is conjugate to the two scanning mirrors (horizontal scanner: 16 KHz resonant scanner, vertical scanner: 30Hz galvometric scanner), the angle of incidence of the incoming beam is continuously changed with line-of-sight as a pivot, which provide raster scanning of the focused spot of the retina. Light scattered back from the eye passes through the same optical path as the incoming beam then is collected and focused on a pinhole which is conjugate to the focused spot on retina. A photomultiplier tube is (PMT) placed after the pinhole to record the intensity of the scattered light for each position of the spot. By synchronizing the signals from PMT and two scanning mirrors, we can consecutively build the image of retina.

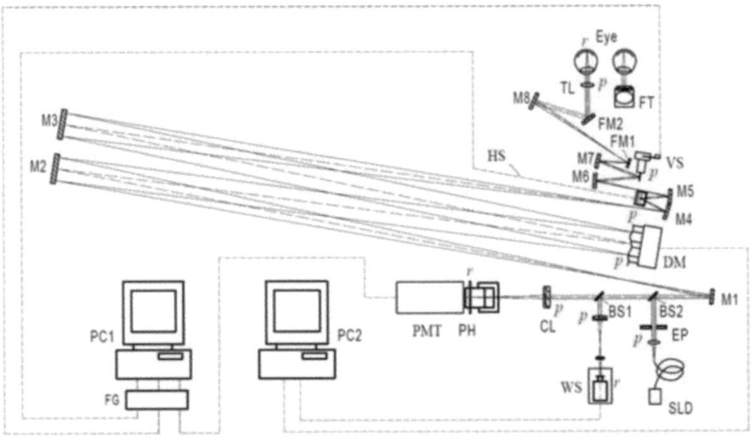

Fig. 1. The schematic of AOCSLO system. HS, horizontal scanner; VS, vertical scanner; WS, wavefront sensor; DM, deformable mirror; PMT, photomultiplier tubes; BS1, BS2, beam splitter; CL, collecting lens; FM1, FM2, fold mirrors; SLD,680nm; M1~M8, spherical mirrors; PH, pinhole; TL, trial lens; FT, fixing target.

Light scattered back from retina is split into two parts: one enters PMT for imaging and the other is captured by a Shack-Hartmann wavefront sensor. This arrangement avoids chromatic aberration caused by using different wavelength for imaging and wavefront sensing. In order to compensate for ocular aberrations, a deformable mirror is placed at the pupil conjugate plane.

2.1 Optical design

The main difference in optical path between SLO and AOCSLO is conjugate planes. There are seven pupil conjugate planes in AOCSLO optical system: actual pupil, HS, VS, DM, lenslet array of WS, and another 2 positions. It's essential to preserving conjugate planes, as WS should measure the exact aberrations of the eye. If the pupil is not conjugate to the scanning mirrors, beam would wander across the pupil as it is scanned, and WS would not see a stable aberration pattern (nor would the DM).

The light delivery path is a double-pass system including light source, two scanners, WS, DM and PMT. It can be thought of as a series of telescopes that relay the pupil to the various conjugate elements. Spherical mirrors are used for pupil relay to prevent back reflection (which is particularly harmful in a double-pass system) and chromatic aberrations. The drawback of using spherical mirrors is that they have to be used off-axis, which generates off-axis aberrations like astigmatism and coma. Coma could be minimized by changing angles and distanced between the spherical mirrors. Astigmatism is compensated by trial lens.

The beam is scanned on the retina with a horizontal scanner (HS) and a vertical scanner (VS), which continuously changing system aberration. As AO system could not catch up with the speed of scanning mirror (16KHz for HS). The optical system should be designed for diffraction limited performance over the entire field while DM holds flat. For high resolution retinal imaging, 1x1 degree field is used. Zooming out to a larger field, the system is capable of 3x3 degrees. At each position of HS and VS, a completely new optical system configuration is encountered yielding a different wavefront aberration. A continuous x-y raster scan actually contains an infinite number of system configurations. To realistically represent this, tilt positions of the scanning mirrors are divided up into a seventeen configuration points spanning the area of a grid on retina, which is shown in Fig.2.

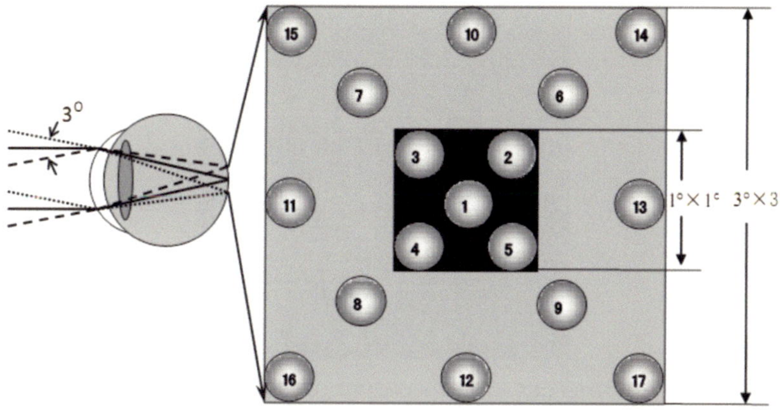

Fig. 2. Multi-configuration of rater scanning

Tilting angles and distances between mirrors are set as variables, and RMS wavefront errors of seventeen configurations are the target of optimization. Configurations of 1x1 degree are more weighted because this field of view is used in high resolution imaging, which is more critical in image quality. After optimization, spot diagram of the seventeen configurations is shown in Fig.3, in which each circle is the Airy disk of corresponding configuration. Most of the spot radius are smaller than Airy radius. In the peripheral configurations, coma and astigmatism are much larger than central configuration.

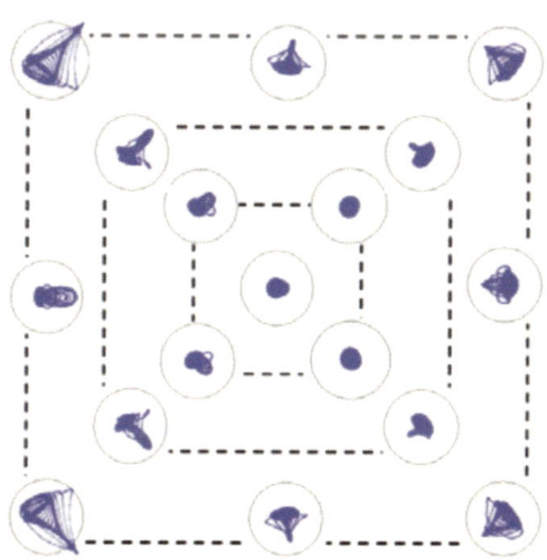

Fig. 3. Spot diagram of the seventeen configurations.

2.2 Adaptive optics in AOCSLO

AO system is used to compensate for the aberrations of human eye's optics. Commonly we use a Shack-Hartmann wavefront sensor. The lenslet array is made up of 5-mm focal length lenslets, each with an actual diameter of 200μm. In our AOCSLO system, the wavefront sensor has square sub-apertures, 11 across the diameter and a total or 97 lenslets inside a 6-mm pupil (see Fig. 4). The centroid of the focused spot from each lenslet is estimated by calculating the first moment of intensity at every spot location.

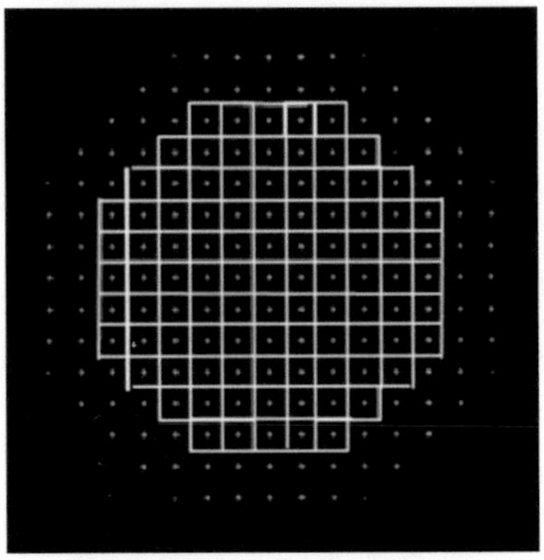

Fig. 4. Shack-Hartmann wavefront sensor's output image.

A 37-channel deformable mirror (DM) is placed in the optical path to compensate for the aberrations. DM is placed before the raster scanners to minimize the size of mirrors required for relaying the light. The clear diameter of DM is 40mm, and therefore the pupil has to be magnified to fill the mirror aperture. The size of the DM is the primary reason for the large size of the instrument. Aberrations are corrected both in the in-going and the out-going light paths. Correcting aberrations on the way into the eye helps to present a smaller focal spot on the retina and result in higher resolution of features in the retina. Correcting the aberrations on the way out helps to focus the light to a more compact spot in the pinhole plane, resulting in increased axial resolution. Figure 5 shows the wavefront sensor geometry superimposed on the DM actuator position geometry.

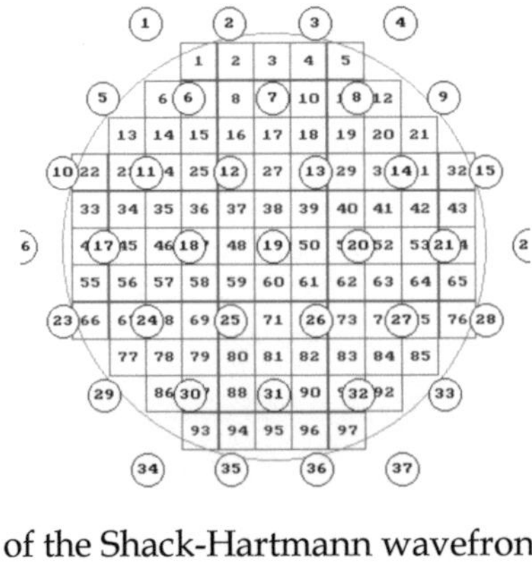

Fig. 5. Centers of the lenslets of the Shack-Hartmann wavefront sensor projected onto the DM actuator array.

2.3 Sectioning analysis

AOCSLO has a relatively high axial resolution, which makes it possible for retina sectioning. A method easy to implement is to use the deformable mirror (DM), which is shown in Fig.4. In Fig.6 (a), beam is focused on the plane of photoreceptor as DM is only used to correct for human eye's aberrations. Then extra defocus is added on DM, which lets the beam focus on retina at a different depth as shown in Fig.6 (b). Light scattered back from this plane follows the same path as the incoming beam, and passes through the DM with extra defocus. As a result, focal position is stable in detecting arm, which means there is no need to move the pinhole. This method has many advantages, including decrease in moving elements, enhancement in speed and accuracy. According to the eye model Liou and Brenan set up in 1997, Fig.6 (c) shows the relationship between the defocus of DM and the depth of focal movement. Human retina is about 300μm thick, so an extra defocus of 4μm is enough to section the whole retina.

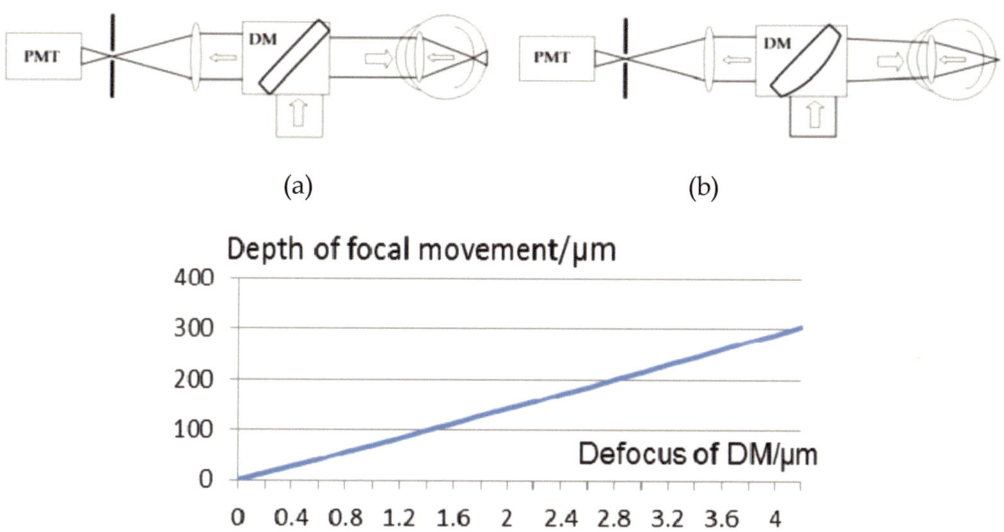

(c)

Fig. 6. Use deformable mirror (DM) for retina sectioning.

3. Result

3.1 Retina imaging

The AOSLO system was tested in adult volunteers, whose pupils were dilated by one drop of Mydrin-p to 6mm. Laser power on pupil plane was 100μW, which is more than 10 times lower than the maximum permissible exposure specified by the American National Standards Institute for the safe use of lasers (Laser Institute of America, 2000). Pinhole size is 40μm. Fig.7 shows retina images taken from the same area before and after close-loop. With AO correction, contrast and resolution were dramatically improved to reveal cone mosaic of human retina.

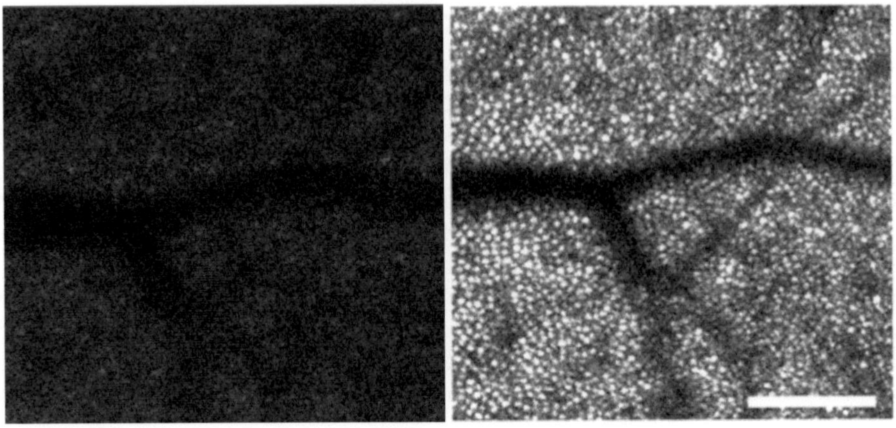

Fig. 7. Retina image before and after close loop. Scale bar is 100µm

3.2 Retina sectioning

The axial resolution of AOCSLO was measured experimentally and compared with the theoretically ideal axial resolution (See Fig.8). We used a model eye (6mm clear aperture, 100mm focal length doublet lens and a ground glass with 1% reflectivity) to measure axial resolution. The ground glass was moved through different positions near focus. At each of these defocus location, the intensity behind pinhole was registered. Then the point spread function (PSF) through focus could be measured.

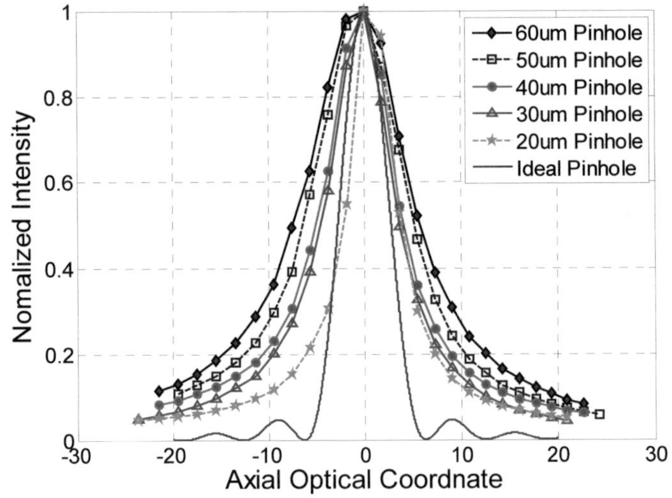

Fig. 8. Axial resolution measured experimentally in different pinhole size. The maximum intensity of each PSF is normalized.

As discussed in Section 1.3, DM could be used for retina sectioning. In practice, the defocus was adjusted over a range of values ranging from -1.5 to +1.5µm, which corresponded to an axial range of 200µm in retina (each step size 0.375µm corresponded to about 25µm). Figure 9 is a montage showing an image obtained at each axial location.

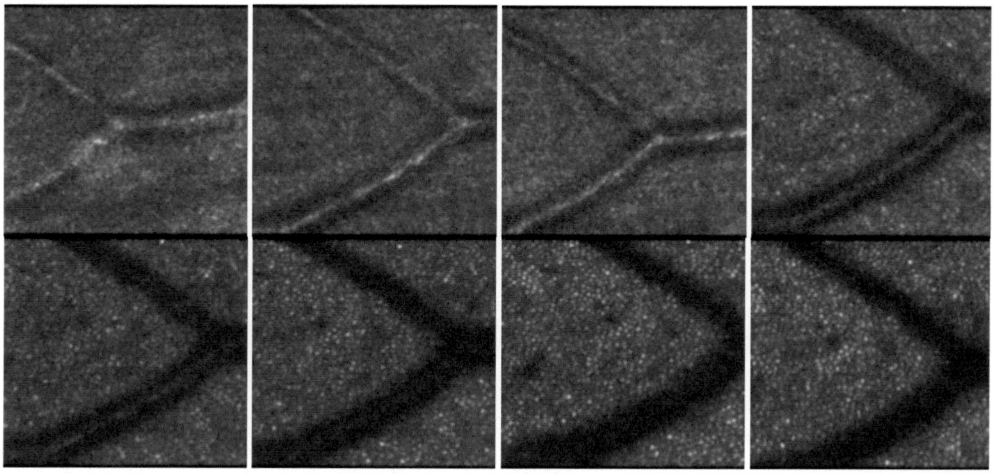

Fig. 9. Different layers of human retina. The sectioning passes through Nerve Fiber Layer, Blood Vessels and Photoreceptor Layer.

4. Real-time deconvolution of AOCSLO image

The ocular aberrations compensation with AO isn't perfect, and it still leaves some residual aberration. On one hand, the point-spread function (PSF) measured by wavefront sensor is suffered from various kinds of noises and it is just the PSF of the optical system. The PSF of the total system is affected by several factors, such as the image detector. On the other hand, wavefront measurements with Shack-Hartmann sensor lack information on the ocular scatter which may give rise to wavefront measure errors (Fernando Díaz-Doutón, et al, 2006, Justin M. Wanek, et al, 2007). Thus, image deconvolution is necessary to improve the quality of the images. We used a blind deconvolution algorithm named Incremental Wiener filter (Mou-yan Zou, Unbehauen Rolf, 1995) to recover the images

In the Increment Wiener Filter, we use an error array as

$$S(\mu,\nu) = Y(\mu,\nu) - \hat{X}(\mu,\nu)\hat{H}(\mu,\nu), \tag{1}$$

where $Y(\mu,\nu)$ is the frequency of the origin image; $\hat{X}(\mu,\nu)$ and $\hat{H}(\mu,\nu)$ are the estimations of frequency of the undistorted image and PSF. The estimations given by the Incremental Wiener filters are

$$\hat{X}_{new}(\mu,\nu) = \hat{X}_{old}(\mu,\nu) + \frac{\hat{H}^*(\mu,\nu)S(\mu,\nu)}{\left|\hat{H}(\mu,\nu)\right|^2 + \gamma_x}, \tag{2}$$

$$\hat{H}_{new}(\mu,\nu) = \hat{H}_{old}(\mu,\nu) + \frac{\hat{X}^*(\mu,\nu)S(\mu,\nu)}{\left|\hat{X}(\mu,\nu)\right|^2 + \gamma_h}. \tag{3}$$

Where * represents conjugation; γ_x and γ_h are small positive constants, they can be denoted by

$$\gamma_x = 0.2\left|\hat{H}(0,0)\right|^2, \tag{4}$$

$$\gamma_h = 0.2\left|\hat{X}(0,0)\right|^2. \tag{5}$$

Eq.(2) and Eq.(3) are iterated to estimate undistorted image and PSF. The Increment Wiener Filter is often used in conjunction with the object domain constraints. The constraint that both image and PSF should be positive is usually used.

The initial value of $\hat{H}(\mu,\nu)$ can be calculated from the wavefront measurement, and the initial value of $\hat{X}(\mu,\nu)$ can be calculated by Wiener Filter. The Wiener Filter is defined by

$$\hat{X}(\mu,\nu) = \frac{\hat{H}^*(\mu,\nu)Y(\mu,\nu)}{\left|\hat{H}(\mu,\nu)\right|^2 + \lambda}, \tag{6}$$

where λ is the regularization parameter. The optimum value of λ for each frequency can be calculated if the undistorted image and noise power spectra are known. However, in many practical situations the information is not available and λ is an empirical value.

Recently, advances in Graphics Processing Units (GPU) enable dramatic increases in computing performance. GPUs with programmable processing units can be used to accelerate many compute intensive applications. The standard C like language interface is also provided by NVIDIA's GPU with compute unified device architecture(CUDA) computing environment, and make it much more accessible to software developers (NVIDIA, 2009). GPUs provide very high memory bandwidth and tremendous computational horsepower. For example, the NVIDIA GTX280 can achieve a sustained memory bandwidth of 141.7 GB/s and can achieve 933 billions of floating-point operations per second (GFLOPS). With help of GPU, the deconvolution algorithms Incremental Wiener filter was realized in real-time.

Experiments on retinal images of original photoreceptor cells and capillary blood vessels are shown in Fig.10. Fig.10 (a) is an original photoreceptor cells image, and Fig.10(b) is an original capillary blood vessels image. Both Fig.10 (a) and (b) were taken with aberration corrected by AO. Fig.10 (c) and (d) are the images restored by Incremental Wiener filter. The initial value of the PSF in our deconvolution algorithm was estimated from the wavefront sensor, and λ in Eq.(6) was set to 2. It is enough to recover the images by twelve iterations of our deconvolution algorithm, and more iterations don't improve the image quality evidently. The images in Fig. 10 are 256x256 pixels, and the field of view is 0.5 degree (about 140 μm). We compared the calculation time of the Incremental Wiener filter programmed through CUDA and Matlab, and the results are shown in Table 1. Our host computer's CPU is E4600, and its GPU is NVIDIA GTX280. As is shown in Table 1, our algorithm can be realized in real-time on GPU, and the calculation speed of the CUDA program is about 282 times faster than that of Matlab program.

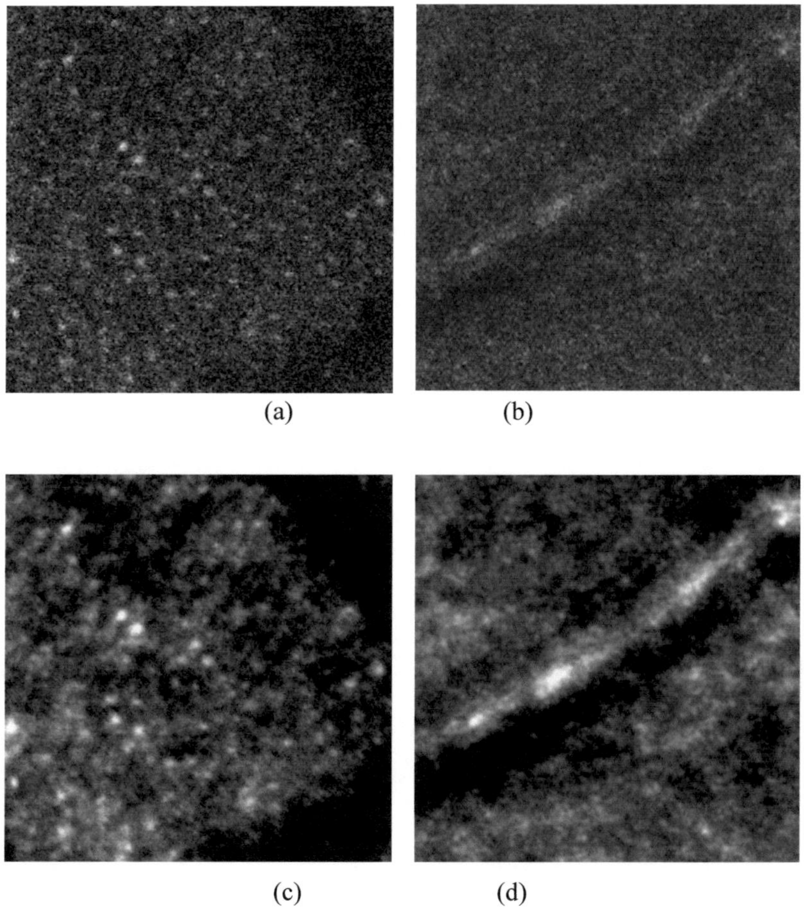

Fig. 10. Experiments on original photoreceptor cells and capillary blood vessels images of retina. (a) an original photoreceptor cells image. (b) an original capillary blood vessels image. (c) and (d) are the images restored by Incremental Wiener filter.

Image	CUDA program	Matlab program
Fig.3(c)	5.13ms	1444.10ms
Fig.3(d)	5.11ms	1440.05ms

Table 1. Comparison between the calculation time of the Incremental Wiener filter programmed through CUDA and matlab

To measure the quality of the retinal images, we also define the contrast of the image I as

$$C = \sqrt{\frac{\sum_{i}^{M}\sum_{j}^{N}\left[I(i,j)-\overline{I}\right]^{2}}{MN}}, \tag{7}$$

where M and N denote the width and the height of image I, and $\overline{I}$ represents the mean of image I. Large value of C indicates that the image quality is improved. The contrasts of the four images in Fig.10 are (a) 32.7 (b) 21.1 (c) 39.4 (d) 39.6, respectively. It shows that our method can improve the image quality efficiently.

The original PSFs of Fig.10 (a) and (b) are shown in Fig.11 (a) and (b), respectively. The PSFs are measured by wave front sensor, and they are initial values of Incremental Wiener filter. The restored PSFs of Fig.10 (a) and (b) by Incremental Wiener filter are shown in Fig. 11(c) and (d), respectively. As shown in Fig.11, there are significant differences between the PSFs measured by wave front sensor and restored by our algorithm. It indicates that it isn't accurate to use the PSF calculated from the wave front measurement as the PSF of the total system.

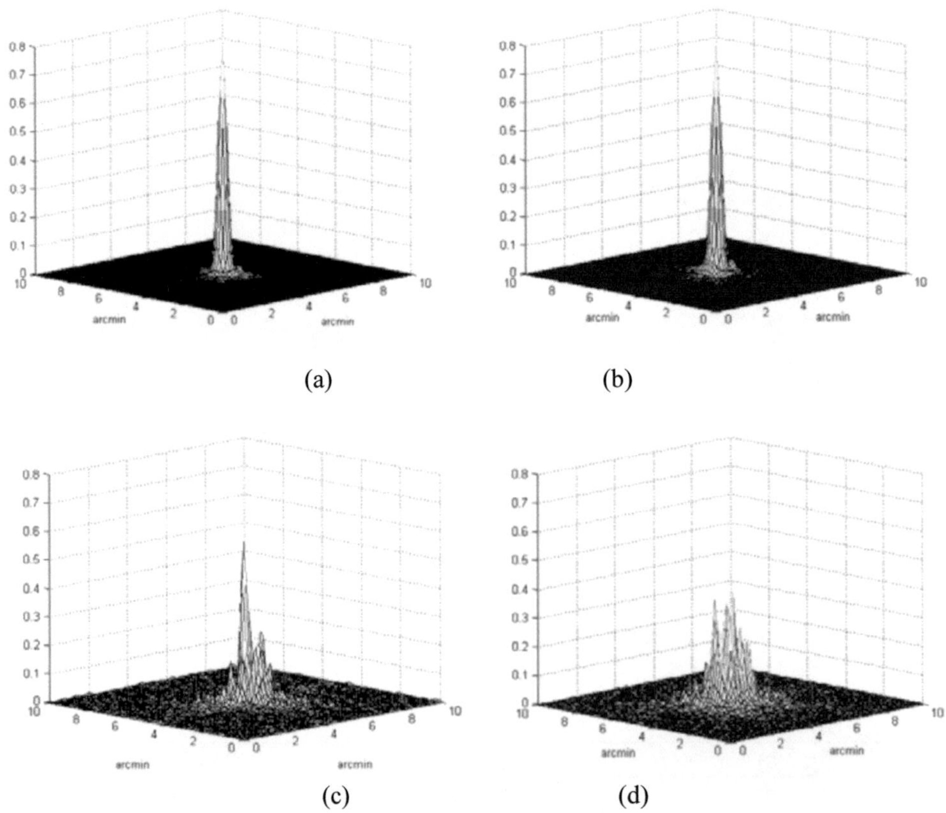

Fig. 11. (a) and (b) are the original PSFs of Fig. 5(a) and (b), respectively. (c) and (d) are the restored PSFs of Fig5(c) and (d) by Incremental Wiener filter, respectively.

5. Tracking features in retinal images of AOCSLO

The high resolution of the AO retinal imaging systems has been compromised by the eye motion. This eye motion includes components of various frequencies from low to a relatively high frequency (50-100Hz) (David W. Arathorn, et al, 2007). When the observed object appears to be fixed in the field of view, the motion sweep any projected point of the object across many cone diameters. With a small field size (about 1 deg field, 280μm), the motion affects quality severely.

In order to remove the eye motion, a computational technique known as Kanade-Lucas-Tomasi (KLT) is applied to track the features in the retinal image (Jianbo Shi and Carlo Tomasi, 1994). KLT is a tracking algorithm with low computational complexity and high accuracy. The tracked features can be provided by Scale-Invariant Feature Transform (SIFT). SIFT is an image registration algorithm (David G. Lowe, et al, 2004), and it can automatically abstract point features which are blob-like structures with subpixel resolution from two images. With the tracked points, the second-order polynomial transformation can be used to remove distortions of the retina image. And if more complex transformation such as three-order polynomial transformation is used, more general distortions can also be removed. The second-order polynomial transformation of a distorted image point $[x\ y]^T$ to the corrected image point $[x'\ y']^T$ can be written in the following form:

$$\begin{bmatrix} x' \\ y' \end{bmatrix} = \begin{bmatrix} a_{00}\ a_{10}\ a_{01}\ a_{11}\ a_{20}\ a_{02} \\ b_{00}\ b_{10}\ b_{01}\ b_{11}\ b_{20}\ b_{02} \end{bmatrix} \begin{bmatrix} 1 \\ x \\ y \\ xy \\ x^2 \\ y^2 \end{bmatrix}, \tag{8}$$

where the distortion is represented by a and b parameters. The second-order polynomial transformation allows mappings of all lines to curves. Given a set of matched points, the parameters can be determined as follows. Assume k points match in the reference and distorted images. A matrix R is defined which contains the coordinates of the k matched points in the reference image:

$$R = \begin{bmatrix} x'_1\ x'_2 \cdots x'_k \\ y'_1\ y'_2 \cdots y'_k \end{bmatrix}. \tag{9}$$

The D matrix contains the corresponding points in the distorted image:

$$D = \begin{bmatrix} 1 & 1 & \cdots & 1 \\ x_1 & x_2 & \cdots & x_k \\ y_1 & y_2 & \cdots & y_k \\ x_1y_1 & x_2y_2 & \cdots & x_ky_k \\ x^2_1 & x^2_2 & \cdots & x^2_k \\ y^2_1 & y^2_2 & \cdots & y^2_k \end{bmatrix}. \tag{10}$$

An A matrix is defined contains the transformation parameters:

$$A = \begin{bmatrix} a_{00} \, a_{10} \, a_{01} \, a_{11} \, a_{20} \, a_{02} \\ b_{00} \, b_{10} \, b_{01} \, b_{11} \, b_{20} \, b_{02} \end{bmatrix}. \tag{11}$$

The linear regression estimate for A is given by:

$$A = RD^T (DD^T)^{-1}. \tag{12}$$

To implement the second-order polynomial transformation at least six corresponding point pairs are needed.

The point features abstracted by SIFT in the reference image are shown in Fig.12(a), and the points matched by KLT in an image are shown in Fig.12(b). The image with distortions removed by the second-order polynomial transformation is shown in Fig.12(c).

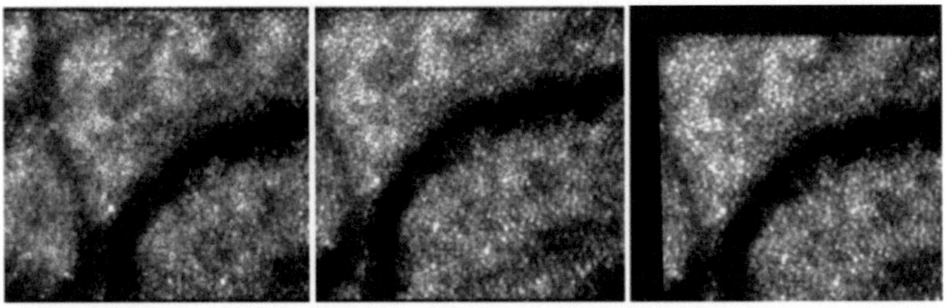

(a) (b) (c)

Fig. 12. Removing distortions by the second-order polynomial transformation. (a) Shows the point features abstracted by SIFT in the reference image. (b) Shows the points matched by KLT in a distorted image. (c) Shows the corrected image with distortions removed by the second-order polynomial transformation.

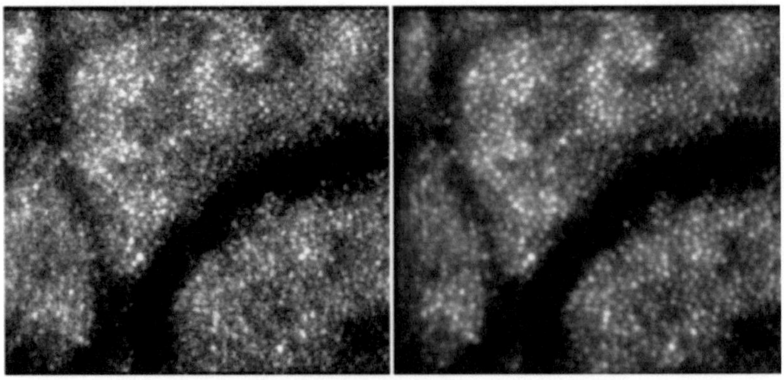

(a) (b)

Fig. 13. Co-added 10 successive frames. (a) the reference frame. (b) the average image without distortions

Ten successive frames with distortions removed are co-added, and the result is shown in Fig.13 (b). The reference frame is shown in Fig.13 (a). If more frames are selected, the superposition is small and the image quality isn't improved evidently. The power spectra of Figs.13 (a) and (b) are compared in Fig.14. It can be seen that at the high frequencies which probably correspond to noise, the power spectrum of Fig.13 (b) is decreased. It indicates that the noises are effectively suppressed and the image quality is improved.

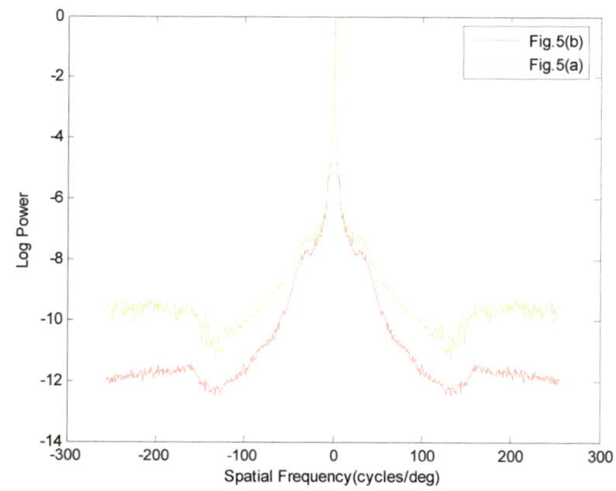

Fig. 14. Average power spectra of Figs.8 (a) and (b).

6. Conclusion

In this chapter, we discuss the design of an Adaptive Optics Confocal Scanning Laser Ophthalmoscope (AOCSLO) and provide high resolution retina image *in vivo*. Then in order to improve the quality of AOCSLO image, we introduce real-time deconvolution and tracking features in retinal images.

7. Acknowledgement

This work was supported by The Main Direction Program of Knowledge Innovation of Chinese Academy of Sciences (No. KGCX2-Y11-920)

8. References

Austin. Roorda, F. Romero-Borja, W. J. Donnelly, H. Queener, T. J. Hebert and M. C. W. Campbell. (2002). Adaptive optics scanning laser ophthalmoscopy. *Optics Express.*, Vol. 10, pp 405-412

D. Merino, Chris Dainty. (2006). Adaptive optics enhanced simultaneous en-face optical coherence tomography and scanning laser ophthalmoscopy. *Optics Express.*, Vol. 14, pp 3345-3353

D. X. Hammer, R. D. Ferguson, N. V. Iftimia, T. Ustun and S. A. Burns. (2005). Tracking adaptive optics scanning laser ophthalmoscope (TAOCSLO). *Investigative Ophthalmology & Visual Science.*, Vol. 46

David G. Lowe. (2004). Distinctive Image Features from Scale-Invariant Keypoints. *International Journal of Computer Vision 60.*, pp 91-110

David W. Arathorn, Qiang Yang, Curtis R. Vogel, Yuhua Zhang, Pavan Tiruveedhula, and Austin Roorda. (2007). Retinally stabilized cone-targeted stimulus delivery. *Opt. Express.*, Vol. 15 pp 13731-13744

Fernando Díaz-Doutón, Antonio Benito, Jaume Pujol, Montserrat Arjona, José Luis Güell, and Pablo Artal. (2006). Comparison of the retinal image quality with a hartmann-shack wavefront sensor and a double-pass instrument. *Investigative Ophthalmology & Visual Science.,*Vol. 47, pp 1710-1716

Fernando Romero-Borja, Krishnakumar Venkateswaran, Austin Roorda, Thomas Hebert. (2005). Optical slicing of human retinal tissue in vivo with the adaptive optics scanning laser ophthalmoscope. *Applied Optics.,* Vol. 44, pp 4032-4040

Hwey-Lan Liou and Nobel A. Brenan (1997). Anatomically accurate, finite model eye for optical modeling. *J. Opt. Soc. Am. A,* Vol. 14, pp 1684-1695

Jianbo Shi and Carlo Tomasi. (1994). Good Features to Track. *IEEE Conference on Computer Vision and Pattern Recognition.,* pp 593-600

Jing Lu, Yudong Zhang, Xuejun Rao. (2007). Optical design of a confocal scanning laser ophthalmoscope based on adaptive optics. *Proc. of SPIE.,* Vol. 6624-127

Justin M. Wanek, Marek Mori, and Mahnaz Shahidi. (2007). Effect of aberrations and scatter on image resolution assessed by adaptive optics retinal section imaging. *Opt. Soc. Am. A.,* Vol. 24, pp 1296-1304

Liang, D. R. Williams, and D. Miller. (1997). Supernormal vision and high-resolution retinal imaging through adaptive optics. *Opt. Soc. Am. A., Vol. 14,* pp 2884-2892

Ling Ning, Zhang Yudong，Rao Xuejun, Cheng Wang, Yiyun HU, Wenhan Jiang. (2005). High resolution mosaic image of capllaries of human retina by adaptive optics. *Chinese Optics Letters.,* Vol. 3, No. 4, pp 225～226

Mou-yan Zou, Unbehauen Rolf. (1995). New algorithms of two-dimensional blind deconvolution. *Optical Engineering.,* Vol. 34, pp 2945-2956

Ning Ling, Xuejun Rao, Zheping Yang, Chen Wang. (2002). Wave front sensor for the measurement of vivid human eye. *Proceeding of the 3rd International Workshop on Adaptive Optics for Industry and Medicine.,* Ed. S. R. Restaino & S. W. Teare, pp 85-90.

Ning Ling, Yudong Zhang, Xuejun Rao, Cheng Wang, Yiyun Hu, Wenhan Jiang, Chunhui Jiang. (2002). Experiments of high-resolution retinal imaging with adaptive optics. *SPIE Proc.,* Vol. 5639, pp.37-45

Ning Ling, Yudong Zhang, Xuejun Rao, Xinyang Li, Chen Wang, Yiyun Hu, Wenhan Jiang. (2002). Small table-top adaptive optical systems for human retinal imaging. *SPIE Proc.,* Vol. 4825, pp. 99-105

NVIDIA. (2009). NVIDIA CUDA Compute Unified Device Architecture Programming Guide Version 2.3.1

R. Daniel Ferguson, Daniel X. Hammer. (2004). Wide-field retinal hemodynamic imaging with the tracking scanning laser ophthalmoscope. *Optics Express.,* Vol. 12, pp 5198-5208

R. H. Webb, G. W. Hughes, and F. C. Delori. (1987). Confocal scanning laser ophthalmoscope. *Apply Optics.,* Vol. 26, pp 1492-1499J.

Yu Tian. (2009). Deconvolution from Wavefront Sensing and Adaptive Optics Image Post-processing. *PHD Dissertation.,* Institute of Optics and Electronics, Chinese Academy of Science

Yudong Zhang, Ling Ning, Xuejun Rao, Xingyang Li, Cheng Wang, Xuean Ma, Wenhan Jiang. (2002). A small adaptive optical system for human retinal imaging. *Proceeding of the 3rd International Workshop on Adaptive Optics for Industry and Medicine.,* Ed. S. R. Restaino & S. W. Teare, pp.97-104

Yuhua Zhang, Siddharth Poonja, Austin Roorda. (2006). MEMS-based adaptive optics scanning laser ophthalmoscopy. *Optics Letters.,* Vol. 31, pp 1268-1270

Part 4

Wavefront Sensors and Deformable Mirrors

8

Advanced Methods for Improving the Efficiency of a Shack Hartmann Wavefront Sensor

Akondi Vyas, M. B. Roopashree and B. Raghavendra Prasad
Indian Institute of Astrophysics, II Block, Koramangala, Bangalore
India

1. Introduction

Wavefront sensor is a device that measures the optical wavefront aberration. The Shack Hartmann wavefront sensor (SHWS), named after Johannes Franz Hartmann and Roland Shack, is one of the most often used optical wavefront sensor. It is made up of an array of microlenses (all having the same focal length and aperture size) and a detector placed at the focal plane of these microlenses. Johannes Franz Hartmann developed a device that consisted of an opaque screen with multiple holes to test the quality of imaging in large telescope systems and fine-tune the telescope focus (Hartmann, 1900). In the process of developing an adaptive optics system to improve the resolution of satellite images taken from the earth, Roland Shack came up with a feasible model of the sensor by using an array of tiny lenses instead of holes (Shack & Platt, 1971). The technological advancements in the field of optical fabrication allowed the industry to make lenses as small as 100 μm using materials like fused silica, ZnS, ZnSe, Si, Ge, etc. Individual microlenses are also called subapertures and the SHWS spots formed at the focal plane where the detector is placed are referred to as "spots" in this chapter.

SHWS is widely used in diverse wavefront sensing applications. It is very commonly used in astronomical adaptive optics systems (Gilles & Ellerbroek, 2006), lens testing (Birch et al., 2010), ophthalmology (Wei & Thibos, 2010) and microscopy (Cha et al., 2010). It is used in the correction of errors due to non-flatness of spatial light modulators in holographic optical tweezer applications (López-Quesada et al., 2009). A few modifications over the simple SHWS were also suggested in the literature. A hexagonal arrangement of the lenslet array can increase the sensitivity and dynamic range of the detector (Wu et al., 2010). A differential SHWS was proposed which measures the wavefront slope differentials (Zou et al., 2008). In order to make a dynamic microlens array with greater dynamic range, digital SHWS were developed with the help of Liquid Crystal-Spatial Light Modulators (LC-SLMs) (Zhao et al., 2006). It is also possible to characterize atmospheric turbulence using the SHWS via the structure function measurements, C_N^2 profile measurements, wind velocity profile estimation and measurement of deviations from the theoretical model of Kolmogorov turbulence (Sergeyev & Roggemann, 2011; Silbaugh et al., 1996). Also, using the SHWS is advantageous over the curvature sensor when large numbers of sensing elements (subapertures) are to be used (Kellerer & Kellerer, 2011) and hence is the best option in large telescope adaptive optics.

The simplest model of a SHWS is shown in Fig. 1. When a plane wavefront is incident on the SHWS, it produces equidistant focal spots at the detector plane. Any wavefront distortion

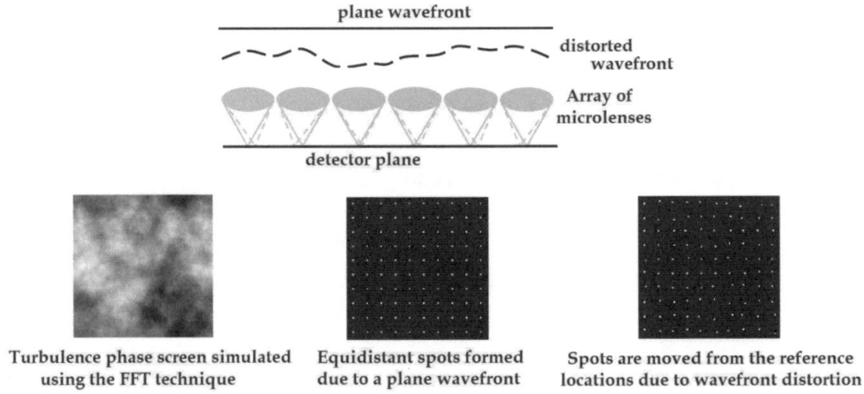

Fig. 1. Description of SHWS: Phase screen (left) following the Kolmogorov turbulence model of the atmosphere, simulated using the Fast Fourier Transform (FFT) technique. Simulated spot pattern with no turbulence assuming a SHWS with 10×10 subapertures is shown in the center and the distorted spot pattern image simulated by taking the turbulence into account is shown on the right.

introduced in the incident wavefront will displace the spots from their original locations. The distance moved by individual spots contains the information of the local wavefront slopes of the incoming distorted wavefront. The retrieval of the wavefront shape in a SHWS is a two step process. In the first step, the positions of the spots corresponding to individual subapertures is determined through spot centroiding techniques and the local slopes, (β_x, β_y) of the wavefront are subsequently determined by calculating the shift in the spots from a reference location (calculated from image of the spot pattern captured when no distortion is present). As a second step, the shape of the wavefront is reconstructed from the measured local slope values.

The simplest method of locating the spot is to identify the pixel with maximum intensity. In this technique of peak identification, the finite number of detector pixels per subaperture limits the accuracy of locating the position of the spot. In order to accurately determine the location of the spots, we need to calculate the location of the spot in subpixel units or increase the resolution of imaging. Dealing with high resolution images means more readout time and greater computational effort and hence should be avoided in applications requiring high-speed operation. The slopes are related to the centroid locations in the following manner,

$$\beta_x = \frac{x_p - x_r}{f} = \frac{\Delta x}{f}, \qquad \beta_y = \frac{y_p - y_r}{f} = \frac{\Delta y}{f} \tag{1}$$

where $R = (x_r, y_r)$ is the reference coordinate and $P = (x_p, y_p)$ is the position within a single subaperture to which the spot is moved after distortion of the wavefront, $\Delta x = x_p - x_r$ and $\Delta y = y_p - y_r$ represent the magnitude of shifts along 'x' and 'y' directions and 'f' is the focal length of the microlenses.

Since most of the wavefront reconstruction error arises from inaccurate centroiding, it is a very important step. Centroiding in the case of astronomical adaptive optics is even more challenging because the science objects under observation are weak light sources and hence need longer exposure times (several seconds to a few minutes). This even makes them inappropriate for wavefront sensing, which needs to be done at a much faster rate. For this reason, another star (called Natural Guide Star or NGS) in the close vicinity of the science target is used as a reference source. In the absence of a NGS, an artificial star (called Laser Guide Star or LGS) is generated by shining a high power laser into the atmosphere, and is used as a reference source.

LGS has become an integral part of adaptive optics systems for better sky coverage (Fugate et al., 1991; Primmerman et al., 1991). LGS can be of two types - Rayleigh or sodium beacon. Rayleigh beacon is formed by the light scattered from molecules at lower altitudes, ranging from 10 km to 15 km depending on the power of the laser and the site conditions (Thompson & Castle, 1992). This low altitude reference source would lead to under sampling and hence the problem of focal anisoplanatism arises (Fried, 1982). A high power 589 nm laser can be used to excite the meteoric origin sodium atom layers in the mesosphere which are present at a mean altitude of 92 km and with a mean thickness of ~10 km (Hart et al., 1995). The de-excitation of the atoms from the upper states ($3p_{\frac{3}{2}}$ and $3p_{\frac{1}{2}}$) to the lower state ($3s_{\frac{1}{2}}$) via spontaneous decay results in resonant backscattering and hence an artificial star. Increasing the power of the laser enhances the chances of stimulated emission against spontaneous decay and thereby reducing the intensity of the desired backscattered photons, eventually leading to population inversion and medium saturation. To avoid this problem, an optimum laser power is used. The limit on the power of LGS also limits the number of available photons for wavefront sensing. The number of available photons, p_N decides the centroiding accuracy and the minimum exposure time which in turn controls the adaptive optics servo bandwidth (Hardy, 1998).

The other noise concerns include background noise due to Rayleigh scattering of laser light and under sampling of the spot due to servo bandwidth constraints. In the case of laser guide star (LGS) based sensing, the elongation of the spots in large telescopes and the variability in sodium density profile cause further errors. Readout noise of the detectors in addition to the photon noise may also seriously degrade centroiding accuracy in a few cases. The following section discusses a few basic centroiding algorithms discussed in the literature.

2. Centroiding algorithms

2.1 Center of gravity

The method of center of gravity (CoG) calculates the centroid location as the weighted mean of the position coordinates, (x, y) and the weight being the spot intensity as a function of position coordinates, $I(x, y) = I(X, Y)$. Here, X, Y are used to denote discreteness. The centroid, (x_s, y_s) of a single subaperture spot pattern, (I) is evaluated using,

$$(x_s, y_s) = \left(\frac{\sum I_{ij} X_{ij}}{\sum I_{ij}}, \frac{\sum I_{ij} Y_{ij}}{\sum I_{ij}} \right) \tag{2}$$

where i, j are row and column indices running from 1 to M, where $M \times M$ is the size of a single subaperture matrix. This is the simplest of all the centroiding techniques and is best suited to situations where the light intensity levels are sufficiently high and the signal to noise ratio (SNR) is good enough.

2.2 Weighted center of gravity

If the shape of the spot pattern is known beforehand, then this method takes advantage of this additional information for accurate determination of the location of the centroid (Fusco et al., 2004). The mathematical form that is assumed for the shape of the spot is called the weighting function and is multiplied with the intensity function before applying the CoG algorithm as previously discussed. The estimated centroid location becomes:

$$(x_s, y_s) = \left(\frac{\sum W_{ij} I_{ij} X_{ij}}{\sum W_{ij} I_{ij}}, \frac{\sum W_{ij} I_{ij} Y_{ij}}{\sum W_{ij} I_{ij}} \right) \tag{3}$$

The weighting function $W(x, y)$ is generally assumed to be a Gaussian function when a natural guide star (NGS) is used as the reference in wavefront sensing:

$$W(x, y) = \text{Amplitude} \times \exp\left[-\frac{(x - x_0)^2}{2\sigma_x^2} - \frac{(y - y_0)^2}{2\sigma_y^2} \right] \tag{4}$$

The spots **cannot** be assumed to be Gaussian in a digital SHWS due to the domination of diffractive noise and instability of the light source. Hence, an adaptive thresholding and dynamic windowing method is used in this case for accurate centroid detection (Yin et al., 2009). Also, in the case of LGS adaptive optics system, the spot **cannot** be assumed to be Gaussian for large telescope systems due to the problem of spot elongation. In this case, the weighting function, $W(x, y)$ can be simulated by assuming ideal conditions (no turbulence) (Vyas et al., 2010c).

This algorithm is best suited in the closed loop adaptive optics systems where the shift in the spots over consecutive temporal measurements is small. It is not suitable for large shift in the spots and hence inappropriate to open loop systems and large phase errors (Vyas et al., 2009b).

2.3 Intensity Weighted Centroiding

Intensity Weighted Centroiding (IWC) is similar to WCoG with a difference that the weighting function, W_{ij} is the intensity distribution of the spot pattern, I_{ij}. Hence, in IWC, the estimated centroid position becomes,

$$(x_c, y_c) = \left[\frac{\sum_{ij} I_{ij}^2 x_{ij}}{\sum_{ij} I_{ij}^2}, \frac{\sum_{ij} I_{ij}^2 y_{ij}}{\sum_{ij} I_{ij}^2} \right] \tag{5}$$

In comparison to the CoG method, this algorithm performs a better job under low light level conditions and low background and readout noise.

2.4 Iteratively weighted center of gravity

The problem of inaccurate centroiding in the case of large shift in the spots can be overcame by using the iteratively weighted center of gravity (IWCoG) algorithm where the centroid location is computed iteratively (Baker & Moallem, 2007). The weighting function is modified after each iteration and is centered around the new centroid location, (x_s^n, y_s^n) identified by the nth iteration. The centroid location in the nth $(n \geq 2)$ iteration is defined as,

$$(x_s^n, y_s^n) = \left\{ \frac{\sum_{ij} W_{ij}^n I_{ij} x_{ij}}{\sum_{ij} W_{ij}^n I_{ij}}, \frac{\sum_{ij} W_{ij}^n I_{ij} y_{ij}}{\sum_{ij} W_{ij}^n I_{ij}} \right\} \tag{6}$$

where

$$W^n(x,y) = \text{Amplitude} \times \exp\left[-\left\{ \frac{(x - x_s^{n-1})^2}{2\sigma_x^2} + \frac{(y - y_s^{n-1})^2}{2\sigma_y^2} \right\} \right] \tag{7}$$

The first (initial) iteration uses the weighting function defined in Eq. 4. The width of the Gaussian can also be modified after each iteration, but in general has a very little effect on the accuracy of centroid estimation if the same optimal width is used in all the iterations. Any iterative process carries along with it problems like - saturation of performance, non uniform convergence and speed. These issues are answered by a hybrid algorithm (Vyas et al., 2010b), a combination of the IWCoG and the correlation technique, which is discussed in the coming sections.

2.5 Matched Filter algorithm

The Matched Filter Centroiding (MFC) algorithm measures the centroid location by maximizing the cross correlation of the spot with an assumed reference spot (Leroux & Dainty, 2010). Interpolation is performed on the resultant cross correlation matrix to locate the centroid with sub-pixel accuracy (Poyneer, 2003).

3. Wavefront reconstruction from SHWS slope measurements

3.1 Vector Matrix Multiply (VMM) method

The wavefront phase can be obtained by assuming the Southwell's sampling geometry (as shown in Fig. 2) which defines the relation between the wavefront phase and the local slope measurements. Other sampling geometries (Fried, 1977; Hudgin, 1977) can also be adopted instead of Southwell's. From the grid geometry shown in the Fig. 2, it can be shown that (Southwell, 1980) for a SHWS with a pitch of 'h' and $N \times N$ subapertures,

$$\begin{aligned} \frac{S_{i+1,j}^x + S_{ij}^x}{2} &= \frac{\phi_{i+1,j} - \phi_{ij}}{h}, \quad i = 1,2,...N-1 \quad \text{and} \quad j = 1,2,...N \\ \frac{S_{i+1,j}^y + S_{ij}^y}{2} &= \frac{\phi_{i,j+1} - \phi_{ij}}{h}, \quad i = 1,2,...N \quad \text{and} \quad j = 1,2,...N-1 \end{aligned} \tag{8}$$

The application of least square curve fitting model, on the SHWS slope data, reduces the problem into a matrix form, $DS = A\phi$, where D and A are sparse matrices of sizes $2N^2 \times 2N^2$ and $2N^2 \times N^2$ respectively. S is a vector containing the measured slope values, 'x' slopes followed by 'y' slopes. The wavefront phase can therefore be evaluated using the following expression,

$$\phi = (A^{\dagger}A)^{-1}A^{\dagger}DS \tag{9}$$

To solve for ϕ using Matlab, we can apply the in-built command, LSQR on the linear system of equations, $A\phi = B$, where $B(= DS)$ is a vector of size $2N^2 \times 1$.

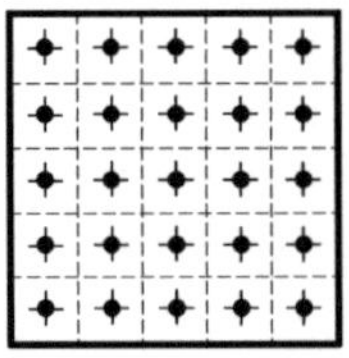

Fig. 2. Southwell slope geometry (for case of 5 × 5 SHWS) which defines the relation between the wavefront phase and the local slope measurements. The dots represent points where the phase is being estimated and the horizontal and vertical lines over the dots represent the 'x' and 'y' slopes measured by the SHWS. The dotted lines show the separation between the lenslets.

3.2 Fast Fourier Transform (FFT) based reconstructor

This is a modal reconstruction process where complex exponentials are used as basis functions and the wavefront can be reconstructed from its local discrete slopes by a simple multiplicative filtering operation in the spatial frequency domain (Freischlad & Koliopoulos, 1986). Taking the case of Hudgin geometry where the first differences, of the phase values, are the measured slope values,

$$\begin{aligned} S^x_{i,j} &= \phi_{i+1,j} - \phi_{i,j} \\ S^y_{i,j} &= \phi_{i,j+1} - \phi_{i,j} \end{aligned} \tag{10}$$

Applying the Fourier transform and using the shift property of the discrete Fourier transform on Eq. 10 gives,

$$\begin{aligned} \mathbf{S}^x_{k,l} &= \Phi_{k,l} \left[exp\left(\frac{\hat{i}2\pi k}{N}\right) - 1 \right] \\ \mathbf{S}^y_{k,l} &= \Phi_{k,l} \left[exp\left(\frac{\hat{i}2\pi l}{N}\right) - 1 \right] \end{aligned} \tag{11}$$

where $\mathbf{S}$ and Φ represent the Fourier transforms of vectors containing slopes and phase values respectively. Φ can now be solved and an inverse Fourier transform performed to arrive at the wavefront phase profile, ϕ. The suitability of different geometries for FFT technique is studied in comparison with the VMM method (Correia et al., 2008).

3.3 Monte Carlo simulations

Monte Carlo simulations are used for testing adaptive optics systems. It involves the generation of atmosphere like turbulence phase screens, simulation of the SHWS spot pattern and the retrieval of the wavefront shape using centroiding and wavefront reconstruction algorithms. The performance of the SHWS and the wavefront sensing algorithm is quantified by calculating the correlation coefficient between the simulated (initial) phase screen and the reconstructed wavefront.

3.3.1 Temporally evolving phase screens

Atmospheric statistics are determined by two important parameters, turbulence strength determined by the profile of the refractive index structure constant, $C_N^2(h)$ and vertical wind velocity profile, $v_w(h)$. These metrics are functions of altitude from the surface of the Earth, $'h'$. Hence, it can be viewed as if the atmosphere is made up of infinite number of turbulence layers. In most situations, it is enough to consider a finite number of discrete layers[1] to closely depict the atmosphere. Hence, we limit ourself to a few layers while simulating atmospheric wavefronts. FFT techniques are generally used for the generation of large phase screens in quick time. The limitations like inefficient representation of low frequency turbulence characteristics in the FFT method of phase screen generation can be minimized by the addition of low frequency subharmonics (Johansson & Gavel, 1994; Sedmak, 2004).

Random wavefronts with turbulence parameters corresponding to different layers can be generated and superposed to simulate a wavefront which closely represents the effect of the full column of the atmosphere. This method of simple superposition of wavefronts is well suited to obtain static and temporally uncorrelated wavefronts. To simulate temporally evolving wavefronts, frozen in turbulence approximation can be used on individual layers for simplicity. A very large wavefront (X^l of size $M \times M$ square pixels corresponding to layer l) with turbulence strength defined for a single layer may be simulated as a first step. A small portion of this very large wavefront can then be chosen as the initial phase screen ($P_1^l(t=0)$, 't' represents time) in the temporal evolution of this particular layer. Subsequent phase screens ($P_i^l(t)$, i=2,3,...n) at later times are formed by translating the initially selected portion on the very large wavefront X^l in a definite direction and well defined velocity read from a pre-assumed wind velocity profile (as shown in Fig. 3). The number of such phase screens generated is limited by the number of pixels on the large wavefront, X^l. If the size of $P_i^l(t)\ \forall\, i$ is $N \times N$ square pixels, then the maximum value 'n' can take is equal to $(M - N + 1)$. The time interval between two temporally adjacent phase screens is defined as: $\Delta T = d_l\ /\ v_l$, where v_l is the layer velocity and d_l is the distance moved on the phase screen in a time ΔT. In the simulation of temporally evolving phase screens, ΔT is kept a constant for all the layers and d_l (representative of the number of pixels moved within ΔT in accordance with the wind velocity of the layer 'l') is varied for different layers. The phase screens representing the evolution of atmospheric turbulence are finally obtained by superposing different evolving layers. The correlation of the phase screen $P(t)$ at time $t = 0$ with the phase screen at any other time, $t = t'$ reduces with increasing t' as shown in Fig. 4 for the simulated temporally evolving phase screens.

The simulation of temporally evolving phase screens described above has a serious problem when a finite number of pixels are used for their representation. The structure of the wind profile would not always allow to move the phase screen by integral number of pixels on the large wavefront, X^l in time intervals of ΔT.[2] Hence the question of moving the phase screen by sub-pixel value arises. The most commonly used mathematical tool for moving $P_i^l(t)$ on X^l is through bilinear interpolation. This method is although simple and well established, it does not completely retain the phase statistics of the simulated wavefronts. Therefore, methods like random mid point displacement (interpolates at the mid point of the pixel) and statistical

[1] A single atmospheric layer is defined as the depth over which the variations of $C_N^2(h)$ and $v_w(h)$ profiles are insignificant

[2] d_l is always not an integral multiple of pixel values

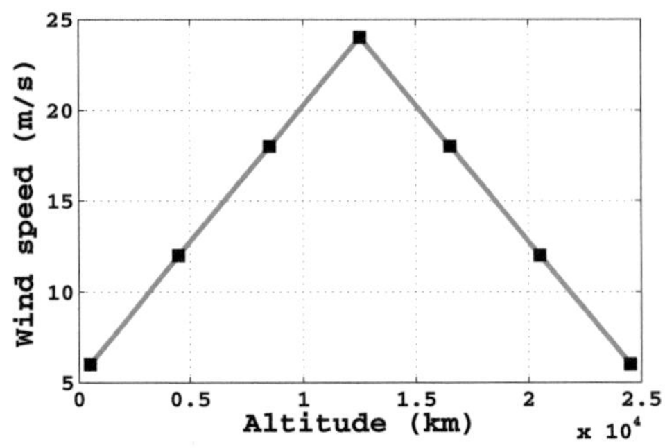

Fig. 3. Vertical wind profile used for simulation of multilayered evolving turbulence

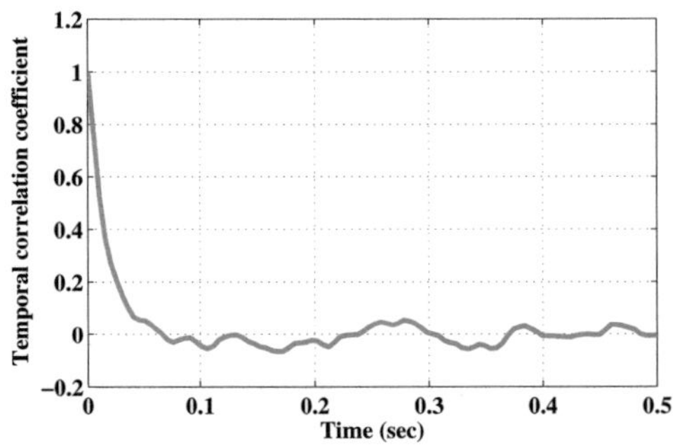

Fig. 4. Simulation of temporal evolution of turbulence based on a seven layer wind model - Drop in the correlation coefficient with time

interpolation (Wu et al., 2009) have been devised. Statistical interpolation is preferred over random mid point displacement method since it can interpolate to any fraction of pixel pitch. A comparison of the performance of statistical interpolation with other interpolation techniques in simulating evolving phase screens is reported by Roopashree et al. (2010).

3.3.2 Metric for comparison

Wavefront reconstruction accuracy is quantified through the correlation coefficient (**CC**) which is used as a metric for comparing the simulated phase screen and the reconstructed one,

$$\mathbf{CC} = \frac{E(XY) - E(X).E(Y)}{\sigma_X \sigma_Y} \tag{12}$$

where E represents the expectation value, X and Y are vectors containing the pixel values of the phase profiles under comparison and σ_X, σ_Y represents the standard deviations.

4. Optimizing the SHWS design parameters

The accuracy of wavefront sensing depends largely on the design parameters of the SHWS, which are highly interdependent and they include, the number of subapertures (N_S), subaperture size (d) and the focal length of the microlenses (f). The ability of the SHWS in precisely estimating the wavefront shape depends dominantly on the number of detector pixels corresponding to a single subaperture (N_P), the detector readout noise and its quantum efficiency (Hardy, 1998). As 'N_S' increases, the wavefront sensing accuracy increases at the cost of increased detector readout time and additional computational effort. Increasing the subaperture size will enhance the number of collected photons and in turn reduce the wavefront sensor error. The subaperture size cannot be increased beyond the Fried parameter, 'r_0' so that the angular resolution is not limited by the spatial scale of turbulence, but by the subaperture size. The focal length of the microlens array determines the maximum detectable tilt range of each subaperture. To increase the dynamic range of the SH sensor, detectors with large 'N_P' could be selected, but at the cost of reduced wavefront sampling frequency. There can be a significant degree of error due to ill-positioning of the imaging device at the focal plane of the microlenses. The experimental limits to the performance are light intensity levels, exposure time scales and operating speeds.

5. Working model

This section describes a working model of the low cost, simple laboratory SHWS at the AO Laboratory, Indian Institute of Astrophysics. A CCD camera (model Pulnix TM-1325CL, readout noise = 50 e^-, pitch = 6.45μm) placed at the focal plane of an array of microlenses (from Flexible Optical B.V., pitch = 200μm, f = 4 cm) makes our wavefront sensor. A continuous membrane deformable mirror (Multi-DM from Boston Micromachines, 140 actuators, pitch = 450μm, maximum stroke = 5.5μm) was used to generate distorted wavefronts. He-Ne laser (15mW, λ=632.8nm) was used as the light source. Also, 4-f geometry was used to remove high frequency noise. The number of pixels per subaperture is 31 in our case and for sensing the whole DM aperture, we used 25 $\times$ 25 microlenses. Low resolution higher order Zernike polynomials (matrix of size: 12 $\times$ 12) are simulated and corresponding voltage values to be addressed on the DM actuators are calculated. The maximum voltage value that is applied to a single DM actuator is 40V, in order to avoid crosstalk between the adjacent subapertures of the SHWS.

Fig. 5a shows the simulation of a low resolution Zernike polynomial (Z_m^n with radial index, $n = 10$ and azimuthal index, $m = 4$). The maximum gray scale in the image corresponds to 40V and minimum gray scale is 0V. The CCD image of the SHWS spot pattern that was captured by addressing the simulated Zernike polynomial, Z_4^{10} on the DM is shown in Fig. 5b. The wavefronts reconstructed using CoG and WCoG methods are shown in Figs. 5c and 5d respectively. Readout noise does not significantly affect wavefront reconstruction accuracy in our wavefront sensing experiment. This was tested by performing multiple trials of reconstructing the same wavefront distortion introduced on the DM. We find that the percentage error in the wavefront reconstruction accuracy due to readout noise is $\sim$0.24%. The existence of a good light level makes CoG a better option over WCoG (Roopashree et al., 2011).

6. Advanced centroiding algorithms

This section discusses the problems associated with the centroiding techniques described in the earlier sections and brings out advanced methods to deal with those issues in the case of astronomical sensing. The following subsections discuss the algorithms: improved iteratively

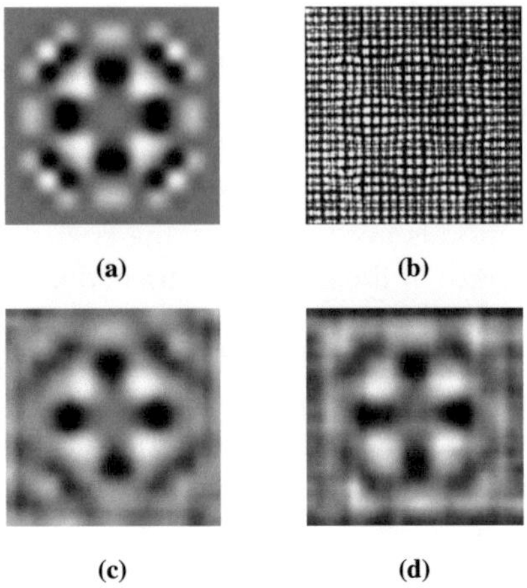

Fig. 5. (a) Simulated low resolution Zernike polynomial, Z_4^{10} addressed on the DM (b) Spot pattern on the SHWS: $I = 0.28\mu$W at the location of the photodetector, SLM in off state (c) Reconstructed phase using CoG, $c = 0.733$ (d) using WCoG, $c = 0.718$.

weighted center of gravity, thresholded Zernike reconstructor based centroiding and Gaussian pattern matching algorithm.

6.1 Improved iteratively weighted center of gravity

The advent of large telescopes made the sensing process more challenging by posing multiple problems due to the use of the LGS. To make it even more complicated, the field of view can only be widened by making use of more than one LGS. The finite thickness of the sodium layer makes it difficult to generate spots with an artificial source that is point-like. The situation becomes worse while dealing with large aperture telescopes ($\geq$10 m). The observed spots are elongated with the elongation length being a function of the distance of the spot from the laser launch point and the elongation axis in the direction of the line joining the subaperture center and the launch point. The spot elongation (ϵ) can be estimated by,

$$\epsilon \approx \frac{L\delta H}{H^2 + H.\delta H} \tag{13}$$

where L is the physical distance from the subaperture center to the launch point of LGS, δH is the sodium layer thickness and H is the altitude of the sodium layer. The result from the simulation of an elongated spot pattern (no atmosphere) with the launch point matching with the center of the telescope aperture is shown in Fig 6. Another problem along with the non uniform elongation is the temporal variability of the vertical sodium density profile (Davis et al., 2006). Hence, an advanced image processing tool is necessary for accurate detection of the centroid position in the case of LGS based Shack Hartmann sensor (Schreiber et al., 2009; Thomas et al., 2008).

Any iterative process like the IWCoG carries along with it problems like - saturation of performance, non uniform convergence and slow convergence, which can be confirmed from a detailed Monte Carlo analysis (Vyas et al., 2010b). The density of sodium peaks nearly at an

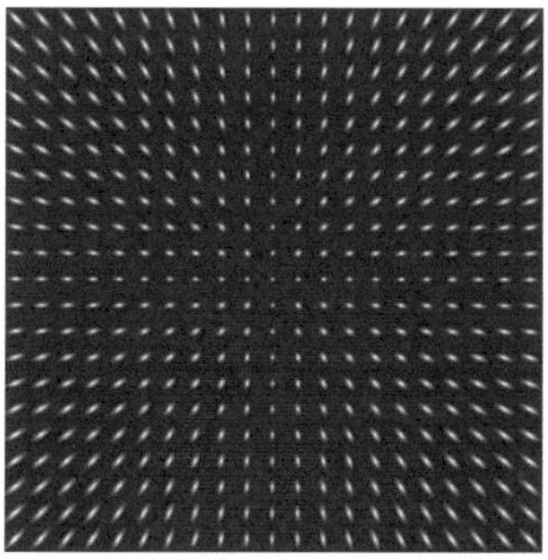

Fig. 6. Laser guide star spot pattern as seen at the focal plane of the lenslet array

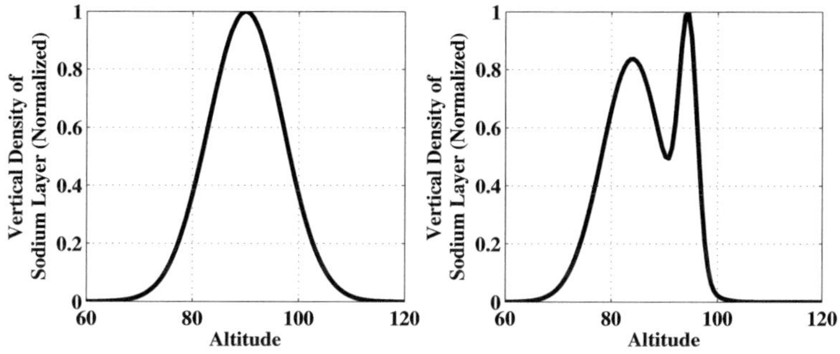

Fig. 7. Single mode and bimodal vertical line profiles of the column of Sodium layer in the mesosphere. Altitude in km units

altitude of 90 km. It was also experimentally observed that there exists two peaks instead of one and hence it is called as bimodal vertical sodium density profile (Drummond et al., 2004). The line profile can be approximated with a single mode (Gaussian distribution) or modeled as if it is bimodal (linear combination of two Gaussian functions) in nature as shown in Fig 7.

$$\varrho(h) = \begin{cases} \exp\left[-\frac{(\mathrm{h}-90)^2}{10^2}\right] & ;\ \text{single modal profile} \\ \frac{1}{1.2025}\left[\exp\left(-\frac{(\mathrm{h}-84)^2}{8.24^2}\right) + \exp\left(-\frac{(\mathrm{h}-94.5)^2}{2.35^2}\right)\right] & ;\ \text{bimodal profile} \end{cases} \tag{14}$$

Here, "h" represents the altitude from sea level and $\varrho(h)$ is the unit normalized density as a function of "h". A factor of $\frac{1}{1.2025}$ in bimodal profile arises due to the unit normalization of density profile when h $\in$ [65 115]. The single mode is centered on 90 km with FWHM of 10 km. The bimodal profile has two peaks centered on 84 km and 94.5 km with FWHM of 8.24 km and 2.35 km respectively. The effect of temporally varying sodium layer profile as shown in Fig. 8 can be included in the simulations of elongated laser guide star spots. To generate randomly varying single modal sodium layer profile, as a first iteration, pseudo random number picked from a normal distribution is added to the mean height of the profile (h=90 km) in Eq. 14. In the subsequent iterations, 'i' pseudo random numbers are added to the earlier iteration value,

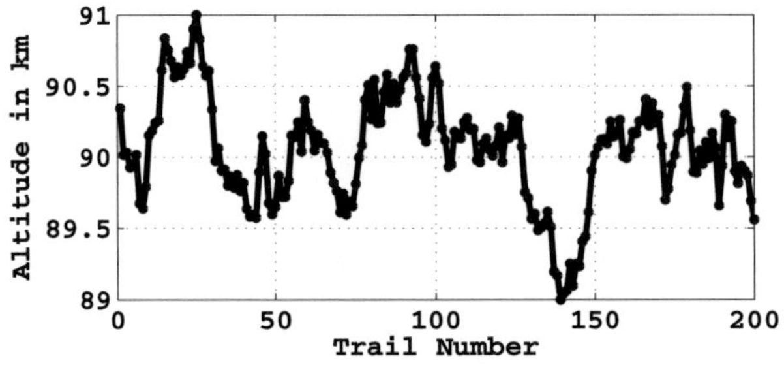

Fig. 8. The fluctuations in the altitude of the sodium layer

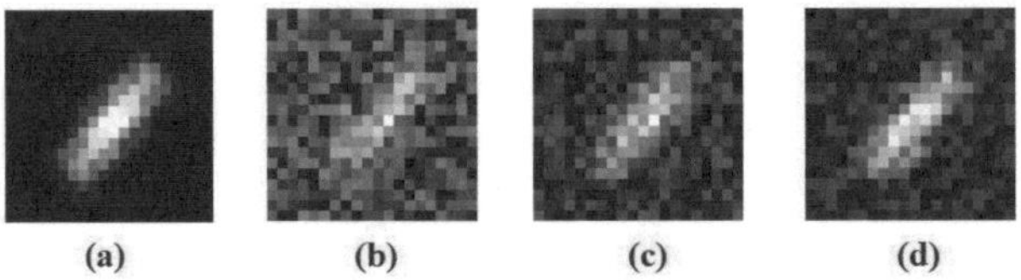

Fig. 9. Spot patterns (using single mode vertical sodium profile) with different SNRs (a)∞ (b) 1 (c) 2 (d) 3

'h_{i-1}'. In the case of bimodal profile, the peaks of the two Gaussian functions are modified by adding pseudo random numbers, similar to the case of single modal profile.

Instead of simulating the entire adaptive optics loop, we concentrate on simulating the case of centroiding the noisy spot pattern corresponding to a single subaperture. A single subaperture spot pattern is simulated by convolving the vertical sodium layer density profile (length proportional to the elongation length defined in Eqn. 13 with atmosphere like phase screen (Schreiber et al., 2009). The simulated phase screen includes the effect of the round trip that the laser has undertaken and the phase screens can be easily simulated based on a Fourier technique (Harding et al., 1999). The phase screens simulated are based on spatial and temporal superposition of multiple layers of the atmosphere (Roopashree et al., 2010).

Once the spot pattern is simulated, photon noise and readout noise are added. Poisson noise is added to the simulated spot pattern by replacing intensity values at individual pixels by pseudo Poisson random numbers picked up from a Poisson distribution generated with mean equal to the intensity value at those pixels. Readout noise is added by generating M^2 (as many as the number of pixels in the spot) pseudo Gaussian random numbers (zero mean and standard deviation defined by SNR) and adding to individual intensity values in the spot pattern. The resultant spot pattern images for different SNR are shown in Fig 9.

The Centroid Estimation Error (CEE) can be used as a performance metric in determining the accuracy of centroiding in the Monte Carlo simulations. It is defined as,

$$CEE = \sqrt{(x_{Act} - x_s)^2 + (y_{Act} - y_s)^2} \tag{15}$$

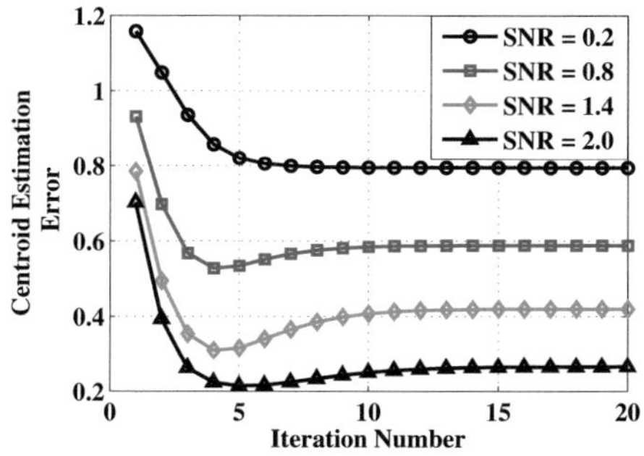

Fig. 10. Saturation and non uniform convergence of IWCoG algorithm at different SNR

where (x_{Act}, y_{Act}) is the actual centroid position of the spot that is although unknown in real situations, is predefined while performing Monte Carlo simulations.

6.1.1 Error saturation

We expect that increasing the number of iterations would reduce the CEE which does not happen in the case of iterative algorithms in general. The error tends to saturate after a finite number of iterations. In the IWCoG algorithm, in most cases, the CEE saturates within a few (3-10) iterations as shown in Fig 10. This error saturation effect can prove to become a serious problem since the CEE cannot be reduced further at the cost of increased computational time and effort, as a result leading to redundant calculations.

The saturation of CEE in IWCoG algorithm can be circumvented by implementing a simple and effective technique called the Iterative Addition of Random Numbers (IARN). After the nth iteration, the new weighting function for the $(n+1)^{th}$ iteration is calculated such that it is centered on a modified centroid location, (x_R^n, y_R^n) instead of (x_c^n, y_c^n) such that,

$$(x_R^n \, , \, y_R^n) = (x_c^n + N_x r_x^n \, , \, y_c^n + N_y r_y^n) \tag{16}$$

where, r_x^n and r_y^n are the generated pseudo random numbers with zero mean and unit variance, which assumes different values after each iteration. N_x and N_y are normalization constants (crudely they define the magnitude of randomness introduced; in our case $N_x = N_y = N$). The addition of these pseudo random numbers would not allow the saturation of CEE. This process although creates other problems like irregular convergence or no convergence as shown in Fig. 11, the minimum CEE that can be reached reduces by a large amount. This method is more significant at low SNR as depicted in Fig. 12.

6.1.2 Non uniform convergence

It can also be observed from Fig. 10 that at different SNR, the iteration number with minimum CEE is different. Due to the non uniform convergence as seen in Fig. 11, the CEE after the last iteration may not be the minimum CEE among all the iterations. For SNR of 0.2, the minimum occurs at n=10; and for SNR=2, minimum CEE occurs at n=5. This suggests that the optimum iteration number is a function of SNR. Hence, we need a technique to identify the iteration with minimum error.

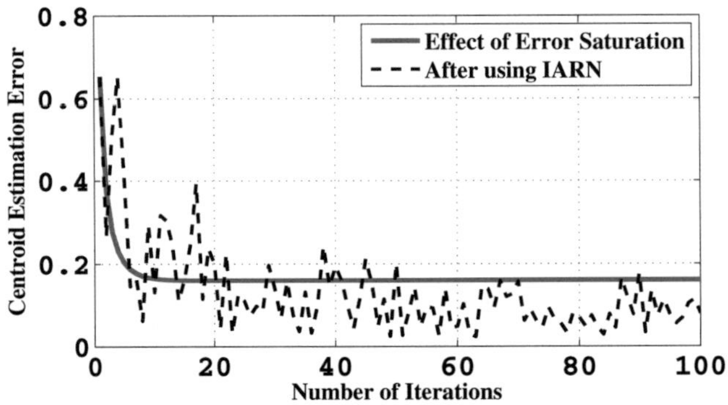

Fig. 11. The effect of error saturation can be overcome by using the IARN based IWCoG algorithm. This procedure largely reduces the CEE.

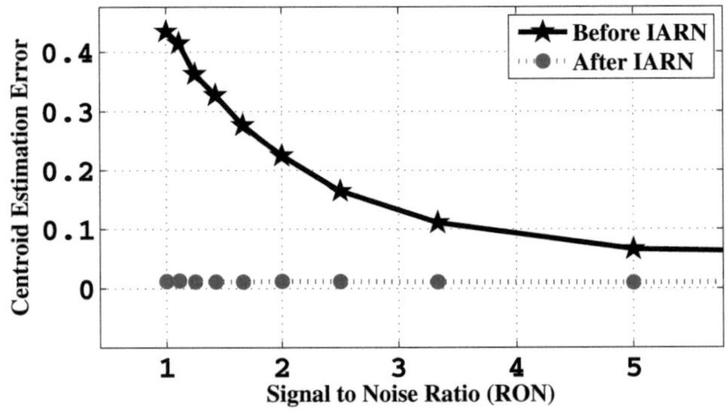

Fig. 12. IARN based IWCoG algorithm. Monte Carlo simulations were performed for 1000 different spot patterns at different SNR and the CEE is taken as a minimum among 1000 iterations.

A simple technique based on iterative computation of correlation can be used to identify the iteration with minimum error. After the nth iteration, an ideal spot pattern (simulated using PSF with atmospheric turbulence excluded and with zero noise) is calculated with its center at (x_c^n, y_c^n) and compared with the actual noisy spot pattern image using the correlation coefficient as the metric for comparison. There exists a strong negative correlation between this metric and the CEE as shown in Fig 13. At this point we make a hypothesis that the iteration number with minimum error corresponds to the one with maximum correlation coefficient. However, there exists a finite decorrelation between this metric and CEE which cannot be always neglected. To thoroughly validate the obtained result, the centroid location with minimum error is estimated multiple number of times based on this hypothesis.

This modified version of IARN based IWCoG method is a consistent performer. Simulations show that there is a significant improvement in the obtained CEE after the application of this modified IARN method on IWCoG as against simple IWCoG (Fig 14).

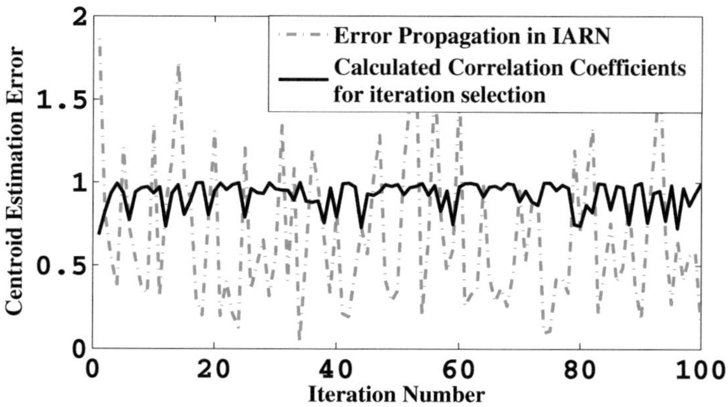

Fig. 13. Error propagation in the case of a single noisy LGS spot pattern. There is a negative correlation between the CEE and the correlation coefficient calculated by comparing the noisy spot pattern with a ideal spot pattern positioned at the new centroid estimated in a particular iteration of IWCoG algorithm, i.e. dips in CEE correspond to peaks in the correlation coefficient

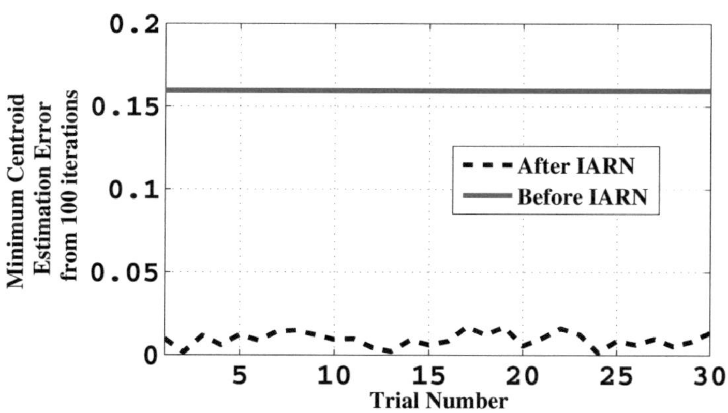

Fig. 14. Improved CEE after application of modified IARN based IWCoG algorithm

6.1.3 Speed of convergence

The maximum number of iterations (n_{max}) used although increases the accuracy in some cases, the number of iterations it needs to go below a predefined error threshold or minimum error sets a limit on the maximum speed of computation. IWCoG algorithm with $n_{max} > 1$ is the slowest among all the other centroiding techniques discussed here. The need to compute large number of iterations for cases of low SNR causes additional burden on the processor. An improved speed of convergence reduces the number of iterations to be performed and hence improves speed.

The time taken to compute the centroid position of a single subaperture in the lenslet array as a function of the number of iterations is shown in Fig 15. The Monte Carlo simulations were performed on a 1.4GHz Intel(R) Core(TM)2 Solo CPU with 2GB RAM.

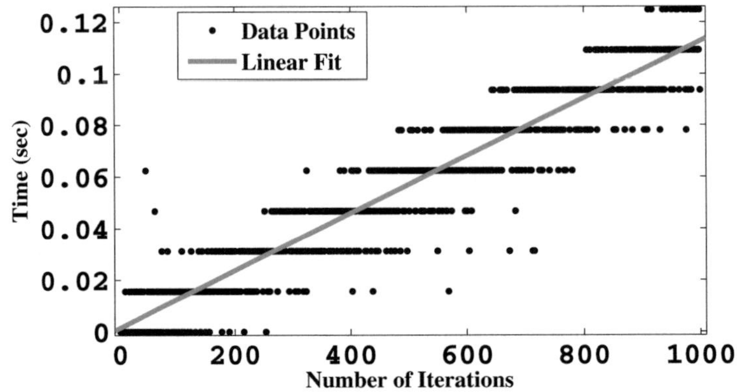

Fig. 15. Time taken for computations. The graph looks like a staircase. This is because of the discrete measurements made by the computer using MATLAB R2008a.

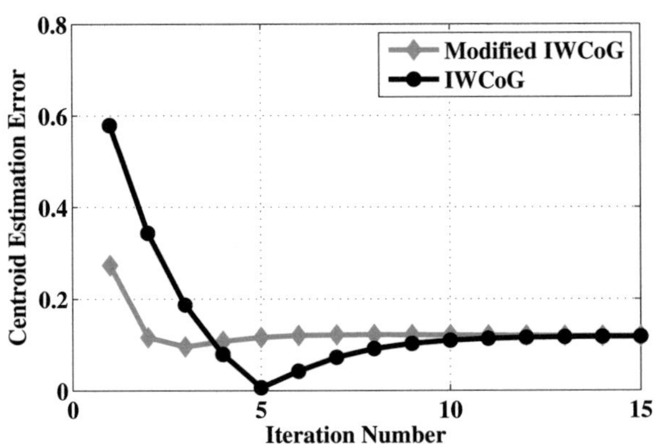

Fig. 16. Increased speed of convergence at low iteration number

A simple and efficient technique can improve the speed significantly. In the first (initial) iteration, the weighting function is centered on the point of maximum intensity (X_{I-Max} , Y_{I-Max}) in the spot pattern image. The improved speed of convergence at low iteration number in this faster IWCoG as against regular IWCoG method is shown in Fig 16 (Monte Carlo simulations performed using 1000 spot pattern images for SNR = 1). Although it seems as if only the first few (three) iterations converge faster using this modified IWCoG, for nearly 10% of the spot images the convergence improves by atleast five iterations.

A comparison of the improved IWCoG algorithm based on IARN is shown in Figs 17 and 18 for high and low SNR. It can be observed that the performance of improved IWCoG based on IARN is better at low SNR (≤ 5).

The novelty of this improved IARN based IWCoG algorithm is its consistency and constancy of CEE irrespective of the SNR. In addition, it is highly accurate and fast centroiding technique. The normalization constant N plays an important role and is a subject of further numerical understanding. Small value of N introduces small variations in the estimated

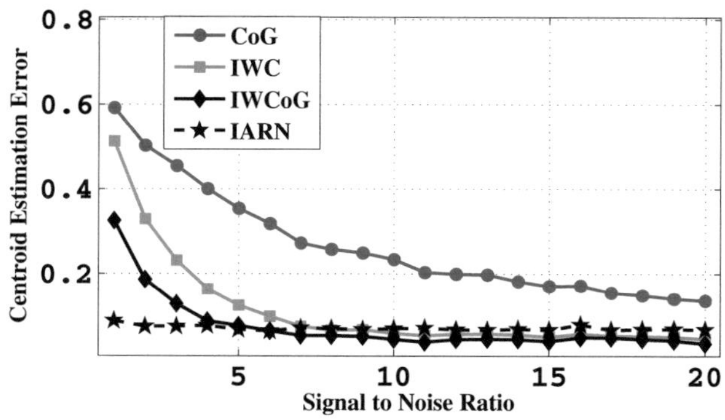

Fig. 17. A comparison of IARN based IWCoG with other algorithms. It can be observed that IARN is most appropriate at SNR ≤ 5. At high SNR, simple IWCoG performs better. A proper choice of "N" may allow IARN based IWCoG method to dominate at higher SNR also.

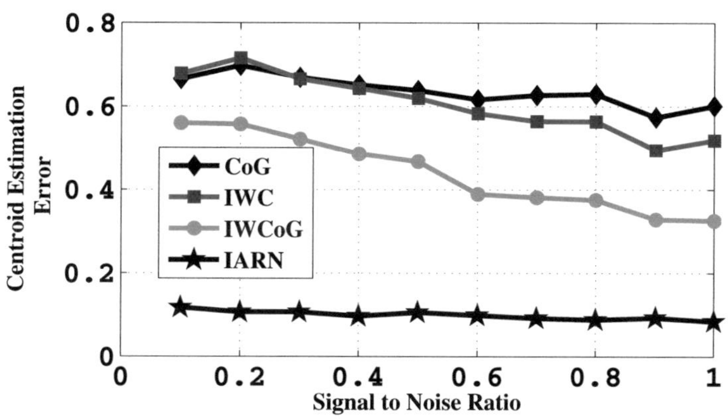

Fig. 18. The consistency of IARN method at very low SNR makes it a more reliable centroiding technique

centroid position and large N introduces large variations. Simulation results shown here use N=1, i.e. the centroid is moved within a pixel. The disadvantage of the correlation based identification of iteration with minimum error is that it is time-consuming. Also, the dependence of this minimum error iteration detection technique has to be investigated further for its performance at different SNR.

6.2 Thresholded Zernike reconstructor

In this section, a centroiding technique based on thresholded Zernike reconstructor is presented. This technique relies on removing the high spatial frequency noise present in the spots by reconstructing them via the calculation of lower order complex Zernike moments. The reconstructed spots are thresholded to remove unwanted noise features and the position of the centroid is then estimated using the weighted centroiding algorithm.

Fig. 19. Spot corresponding to a single subaperture with SNR = 1,2,3,4,5 from left to right.

The parameters of interest include the spot elongation, spot orientation (depends on the position of the laser launch point with respect to the subaperture), the amount of shift in the spot, spot size in non-elongated direction and the SNR. The sample spots at different SNR are shown in Fig. 19.

The proposed method of centroiding is based on image reconstruction using Zernike polynomials. A detailed description of the mathematical procedure for the calculation of Zernike moments for individual sub aperture images is given in Vyas et al. (2010c). The major parameters of interest include number of Zernike moments used for reconstruction and the thresholding percentage. Once the spot is reconstructed using Zernike polynomials, it is ready to be thresholded. A threshold percentage is chosen in the first place. Generally preferred between 50 and 90%; anything lower or higher would either retain noise features after thresholding or remove pixels that are very essential for centroiding purposes. Individual pixels on single subaperture spots are treated as 'object pixels' if the value of the pixel is greater than the threshold. Imposing an intensity threshold on the spot pattern allows us to remove most of the high spatial frequency events (or noise features). A sample spot image at different SNR that is reconstructed and threholded is shown in Fig 20. Also, an improper choice of the thresholding percentage can lead to erroneous features in the reconstructed spot images as shown in Fig. 21.

The Zernike reconstructed and thresholded spots are then subjected to weighted centroiding. In the case of NGS, a Gaussian weighting function is assumed and for the case of LGS, the weighting function used is a standard reference of the spot pattern without noise, by assuming the most probabilistic simple sodium layer profile. The thresholded Zernike reconstructor based centroiding has been numerically tested for the case of NGS and LGS based sensing (Vyas et al., 2010a;c).

Based on the Monte Carlo simulations, we could see a significant improvement in the centroid estimation accuracy when thresholded Zernike reconstructor is applied in conjugation with the weighted centroiding technique. A comparison of this technique with simple center of gravity (CoG), weighted center of gravity (WCoG), iterative addition of random numbers (IARN) based iteratively weighted center of gravity (Baker & Moallem, 2007; Vyas et al., 2010b) (IWCoG) is shown in Fig. 22. It can be observed that TZR algorithm performs equally in comparison with other algorithms at SNR between 1 and 2. At a SNR less than 1, IWCoG performs better than TZR. When TZR is combined with WCoG instead of CoG, it performs better than any other algorithm at low SNR.

The elongation (e) of the LGS spot pattern is assumed to be equal to the elongation of the equivalent closest ellipse.

$$e = \frac{a-b}{a} \tag{17}$$

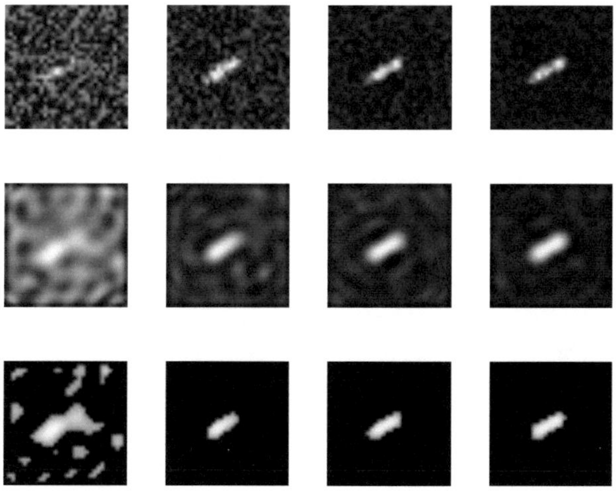

Fig. 20. The first row contains spots (32 × 32 pixels) with SNR = 1,2,3,4. The second row contains spots reconstructed using Zernike moments with $n \leq 24$. The third row contains the thresholded spots with intensity greater than 60% of maximum intensity.

Fig. 21. Left most spot pattern is the noisy spot of size 30 × 30 pixels with SNR = 1, the reconstructed spot is shown on its right using $n \leq 29$ and thresholded spots (65, 70, 75%) are shown in the same order.

where 'a' and 'b' are the major and minor radii of the ellipse. The elongation of the LGS spot has little effect on the centroiding estimation error after application of TZR as shown in Fig. 23. When compared against smaller spots or streaks, the performance of the algorithm is better in the case of larger ones.

The effect of spot orientation on centroid estimation accuracy is shown in Fig. 24. The estimate for the centroid location is best when the spots are oriented along the direction of the pixels (at 0 orientation angle). The performance of the algorithm can be improved by using a polar coordinate detector where the major axis of each rectangular pixel is aligned with the axis of elongation (Thomas et al., 2008).

The temporal fluctuations in the vertical sodium layer density profile is simulated and shown in Fig. 25. The centroiding estimation error in the case of fluctuating profile is shown in Fig. 25. It can be observed that the fluctuations in the sodium profile nearly doubles the centroid estimation error. It is hence important to continuously monitor the fluctuations and

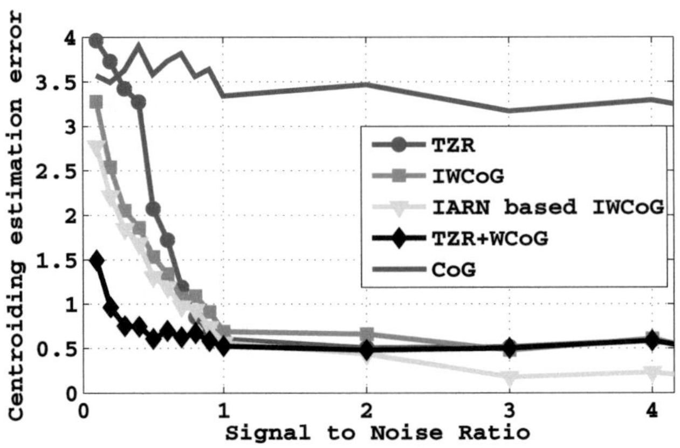

Fig. 22. A comparison of different centroiding algorithms with the TZR method in conjugation with WCoG. It can be seen that at low SNR, TZR performs much better than any other centroiding technique. The performance is also stable. Although when applied alone without WCoG, TZR is not better than IWCoG.

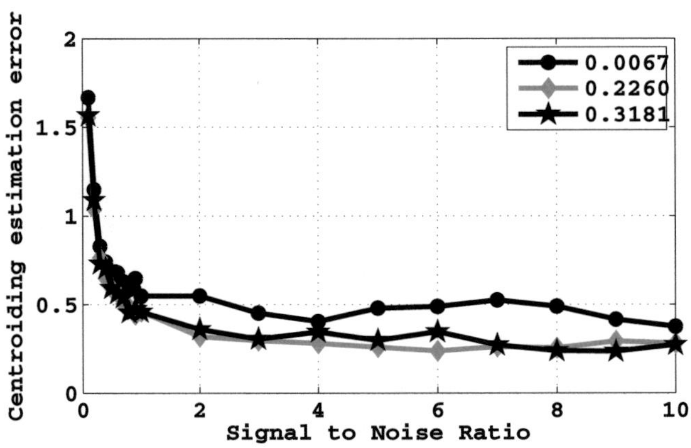

Fig. 23. TZR method performs well at different elongation lengths.

make corrections to the weighting function used in the centroiding of the spots. Numerical observations suggest that by doing so, the centroiding estimation error reduces by little above 10%, but does not reach the value that is obtained with zero fluctuations.

Using a very few Zernike moments for reconstruction weakens the spot of its details and having too many of them include the higher order effects which can be identified with noise. As the image size increases, it becomes important to include more and more Zernike terms to closely reconstruct the features of the spot. Percentage thresholding has significant effect on centroid estimation as can be seen in Fig. 26. Centroid estimation error dropped with reducing thresholding percentage up to 60%. Reducing the thresholding percentage further is risky and would retain noise features which may lead to large errors in the case of a spot of size 30×30 pixels with 29 Zernike modes.

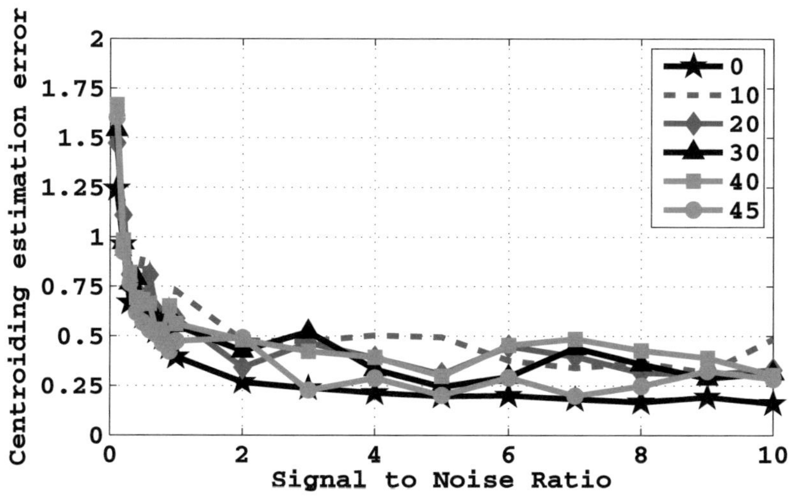

Fig. 24. Spot orientation affects the centroiding estimation accuracy. It is best when the streaks are oriented in the direction of the sides of the detector pixels.

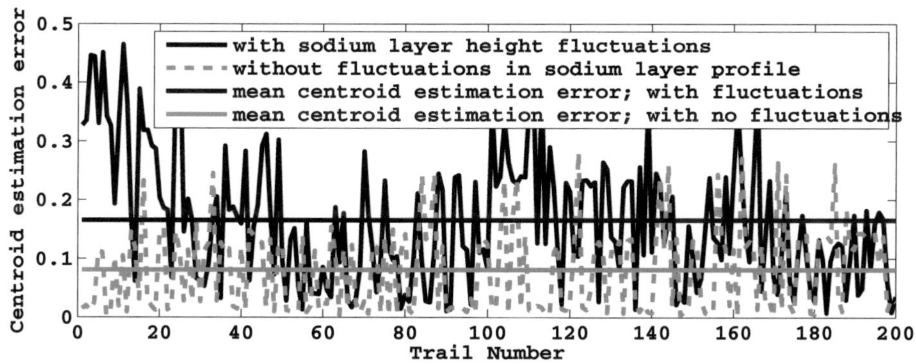

Fig. 25. A comparison of the performance of centroiding accuracy with and without sodium layer fluctuations in the case of TZR method. Since the weighting function was not modified with changing sodium layer profile, the mean centroiding accuracy reduces when fluctuations exist. Hence, it becomes important to track the profile fluctuations continuously.

The number of pixels used for a single subaperture can also influence centroiding estimation while using TZR. This is because image reconstruction in the case of smaller images becomes easier with fewer Zernike moments. The effect of changing image size is shown in Fig. 27. At a low SNR, reconstructing a 24×24 spot pattern using $n \leq 29$ leads to inefficient representation of all the required spot features, whereas representation of a 12×12 spot with the same number of Zernike modes is more efficient. Hence, in this case ($n \leq 29$ and 60% thresholding) at low SNR, smaller spots must be preferred. On the other hand, at a high SNR, fewer Zernike modes are sufficient to represent all the features of the spot. Centroiding accuracy is good for larger spot pattern at high SNR.

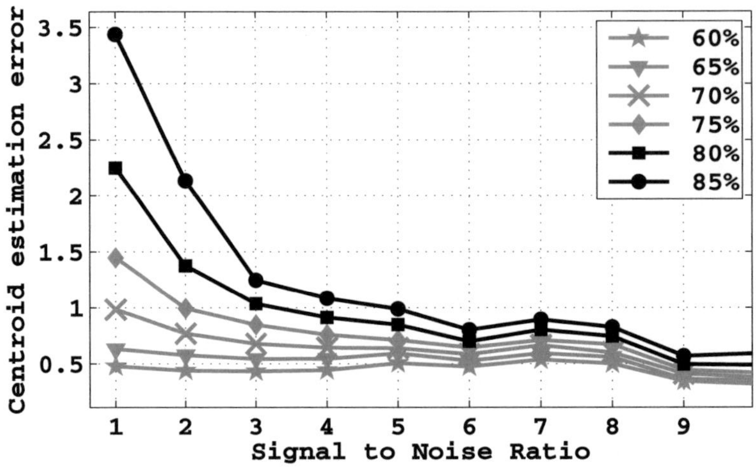

Fig. 26. The effect of centroiding at different thresholding percentage on spots reconstructed with n $\leq$ 29.

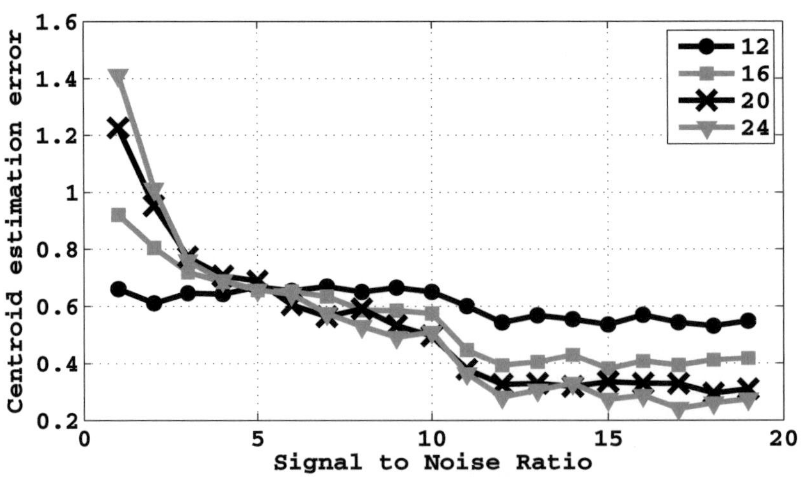

Fig. 27. Spots reconstructed using n $\leq$ 29 and 60% thresholding. Spot orientation was 30^0; with spot size along non-elongated direction being 3 pixels uniformly for all image sizes.

6.3 Gaussian pattern matching

At high noise conditions, the thresholded Zernike reconstructed images lead to multiple noisy features along with the actual spot pattern as depicted in Figs. 20 and 21. This is due to the fact that at high noise level conditions there can be large scale features (sometimes the features can be scaled to dimensions comparable to that of the spot) which might not be removed even after the denoising procedure is implemented.

To use this erroneous spot for accurate centroiding, we take the advantage of the fact that in the case of NGS, the spot finally formed at the focal plane of a lens must assume an airy pattern which can be approximated to a Gaussian like structure. The pattern matching algorithm is implementing in three steps: feature recognition, shape identification and profile identification.

6.3.1 Feature recognition

In this step, the features on the spot pattern image are counted and identified. Feature recognition can be performed by using many existing pattern recognition algorithms. In this algorithm, we used a simple Hough peak identification method to detect the features and number them in the order of peak height. To eliminate small scale features which may arise due to unavoidable scintillations, we impose threshold conditions on the size of the features.

6.3.2 Shape identification

Most features don't have a circular shape. The circularity or the extent of the feature being circular is measured for each of the features. This can be measured in many ways. In the proposed algorithm, the local centroid for the feature is calculated and the distance from the centroid at which intensity becomes zero is measured. This distance is called the distance parameter. The distance parameter is estimated at different angles (0-360) from the feature centroid position. We define circularity, C of a single feature as the inverse of the variance of the distance parameter computed over different angles from the centroid position. For an ideal circular feature, the circularity is infinity since the variance is zero. A lower cutoff for this parameter is chosen to eliminate features that are not close to a circular shape.

6.3.3 Profile identification

In the previous step, the shape of the spot was used for selective elimination. In this step, the intensity profile is used to choose the actual spot feature. The number of features are recounted and identified. The fall off in the intensity from the centroid of individual features is measured and compared with a standard Gaussian shape. In this process, the features that do not follow a Gaussian like structure are eliminated. This technique is more suitable to use along with the CoG and IWC methods (Vyas et al., 2010a).

7. Improved consistency: Dither based SHWS

It can be shown that the wavefront reconstruction accuracy is not a constant, but fluctuates about a mean value irrespective of the sensing geometry at place (see Figs. 28 and 29), by applying Monte Carlo simulations on VMM and FFT methods. These inconsistencies will have a serious effect on the stability of imaging and maintenance of a good Strehl

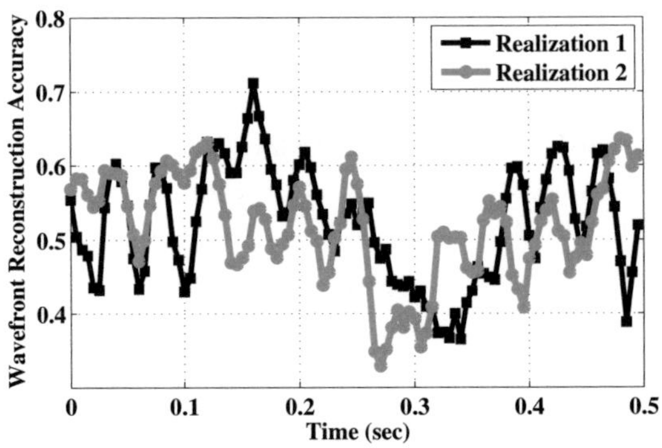

Fig. 28. Inconsistency of the wavefront reconstruction accuracy in the case of FFT technique when applied to evolving atmosphere like turbulence phase screens

ratio. In this section, a possible solution to this problem is addressed. The wavefront reconstruction accuracy in SHWS depends strongly on the way wavefront distortion points match with the points of phase estimation (position of the lenslets in the SHWS). A small dither signal which acts like a translating operator on the wavefront sensor with respect to the phase screen can be used to improve the wavefront reconstruction accuracy. For temporally evolving turbulence, the dither signal applied on the sensor can improve the consistency of the wavefront reconstruction accuracy.

For phase screens simulated using the Kolmogorov model, the wavefront reconstruction accuracy is calculated by applying dither such that the sensor shifts in all directions and by different magnitudes. Probabilistically the point of the best wavefront reconstruction is found to be near the center of the phase screen. We show through numerical simulations that the consistency and the accuracy of wavefront reconstruction can be significantly improved using this technique. In real-time systems, the dither signal to be applied can be obtained from the wavefront sensor data of the immediate past within the wavefront decorrelation time. The practicality of building such a sensor is also discussed.

A dither signal is applied on the wavefront sensor (ie. the lenslet array; the detector is placed at a fixed position) so that it is displaced to a new location such that the wavefront reconstruction accuracy is maximized (Fig. 30). This idea is evaluated through numerical simulations. Considering the case of a telescope with an effective diameter of 1m at a site with $r_0 \sim$ 10cm and a Shack Hartmann sensor with 100 subapertures Monte Carlo simulations were performed. The wavefront reconstruction accuracy depends on the way in which the wavefront distortion points match with the centers of subapertures. The center of the lenslet array was displaced to different discrete locations with respect to the center of the incoming wavefront within the distance of a single subaperture. The point of the best wavefront reconstruction is the point to which the center of the lenslet array has to be moved to. It can be seen that the application of dither signal not only increases the accuracy of reconstruction, but also significantly improves the consistency in the case of Fourier and VMM reconstruction procedures (Figs. 31 and 32). In a real situation, the position of best wavefront reconstruction must be obtained by looking at the strehl ratio. The position of the maximum intensity point varies with time as shown in Fig. 33.

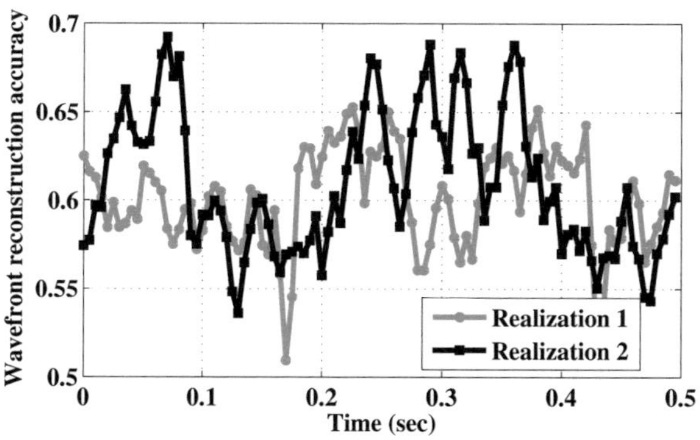

Fig. 29. Inconsistency of the wavefront reconstruction accuracy in the case of vector matrix multiply wavefront reconstruction technique when applied to evolving atmosphere like turbulence phase screens

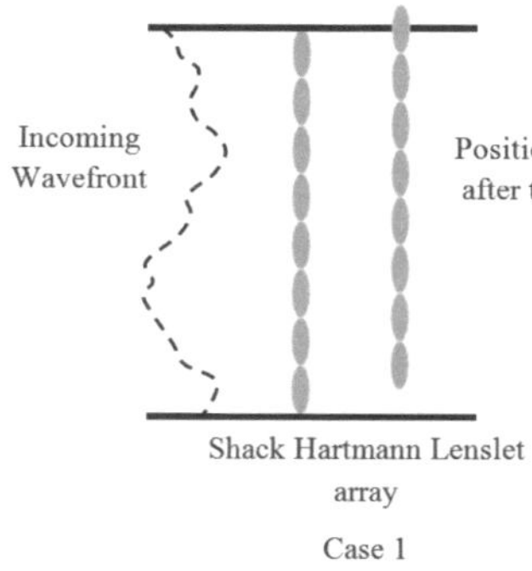

Fig. 30. Schematic demonstration of the dither based sensor

Probabilistically the point of the best wavefront reconstruction occurs close to the center of the wavefront as shown in Fig. 34 (computed for the case of a single phase screen). Hence it is enough to apply a dither close to the center of the wavefront. The choice of the spatial range over which the dither signal may be applied varies with the number of Shack Hartmann subapertures, the Fried parameter and the degree of freedom in the lenslet array.

The dither should be applied at a frequency three times the frequency of adaptive optics wavefront correction. This gives enough time to check the strehl ratio and make suitable corrections to the dither applied. The variance of wavefront reconstruction accuracy as a function of the time interval between the application of two dither signals is shown in Fig. 35.

It can be seen that by applying dither within shorter intervals of time gives smaller variance in the wavefront reconstruction accuracy and hence is more consistent.

This dither based sensor can be realized using a Liquid Crystal based Spatial Light Modulator (LC-SLM) by projecting a digital diffractive optical lenslet array on it (Vyas et al., 2009a;

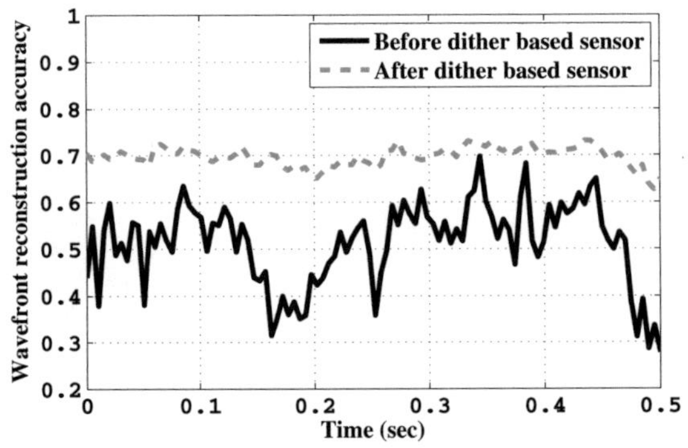

Fig. 31. Inconsistency of the wavefront reconstruction accuracy in the case of vector matrix multiply wavefront reconstruction technique when applied to evolving atmosphere like turbulence phase screens

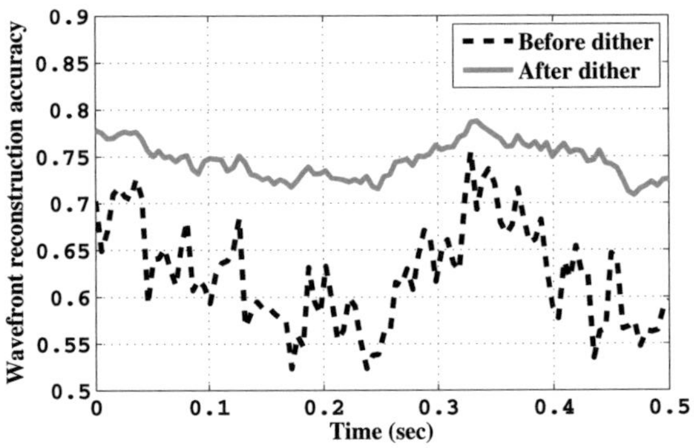

Fig. 32. Inconsistency of the wavefront reconstruction accuracy in the case of vector matrix multiply wavefront reconstruction technique when applied to evolving atmosphere like turbulence phase screens

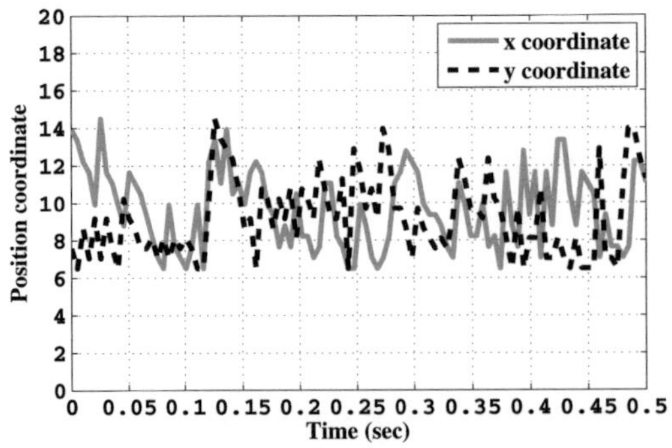

Fig. 33. The propagation of the position coordinate with the maximum wavefront reconstruction accuracy. The analysis performed at time intervals of 0.005 sec

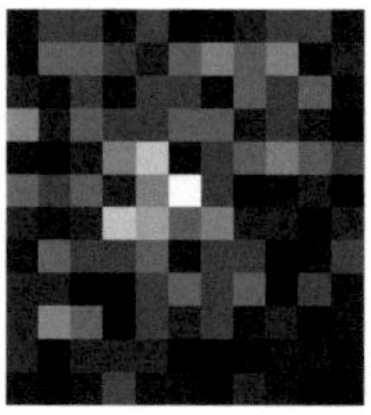

Fig. 34. Probability case of VMM

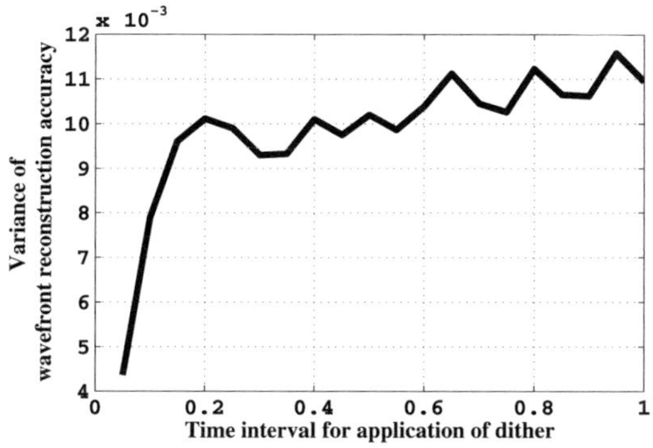

Fig. 35. Dependence of variance of wavefront reconstruction on the time interval between application of two dither signals on the wavefront sensor. This plot is a result of 20 sets of temporally evolving phase screens (100 in number each)

Zhao et al., 2006). The application of the dither signal is much simpler here since the physical movement of the sensor is not needed. The speed and physical dimensions of the SLM limits its application to this method in different situations. It has been recently shown that the application of multiple dither sensors improves the wavefront reconstruction accuracy and its consistency in the case of large telescope AO systems (Vyas et al., 2011).

8. References

Baker, K. L. & Moallem, M. M. (2007). Iteratively weighted centroiding for shack- hartmann wave-front sensors, *Opt. Express* 15(8): 5147–5159.
URL: *http://www.opticsexpress.org/abstract.cfm?URI=oe-15-8-5147*

Birch, G. C., Descour, M. R. & Tkaczyk, T. S. (2010). Hyperspectral shack–hartmann test, *Appl. Opt.* 49(28): 5399–5406.
URL: *http://ao.osa.org/abstract.cfm?URI=ao-49-28-5399*

Cha, J. W., Ballesta, J. & So, P. T. C. (2010). Shack-hartmann wavefront-sensor-based adaptive optics system for multiphoton microscopy, 15(4): 046022.
URL: *http://dx.doi.org/doi/10.1117/1.3475954*

Correia, C., Kulcsár, C., Conan, J.-M. & Raynaud, H.-F. (2008). Hartmann modelling in the discrete spatial frequency domain application to real-time reconstruction in adaptive

optics, 7015(1): 701551.
URL: *http://dx.doi.org/doi/10.1117/12.788455*

Davis, D. S., Hickson, P., Herriot, G. & She, C.-Y. (2006). Temporal variability of the telluric sodium layer, *Opt. Lett.* 31(22): 3369–3371.
URL: *http://ol.osa.org/abstract.cfm?URI=ol-31-22-3369*

Drummond, J., Telle, J., Denman, C., Hillman, P., Spinhirne, J. & Christou, J. (2004). Photometry of a sodium laser guide star from the starfire optical range. ii. compensating the pump beam, *Publications of the Astronomical Society of the Pacific* 116: 952–964.

Freischlad, K. R. & Koliopoulos, C. L. (1986). Modal estimation of a wave front from difference measurements using the discrete fourier transform, *J. Opt. Soc. Am. A* 3(11): 1852–1861.
URL: *http://josaa.osa.org/abstract.cfm?URI=josaa-3-11-1852*

Fried, D. L. (1977). Least-square fitting a wave-front distortion estimate to an array of phase-difference measurements, *J. Opt. Soc. Am.* 67(3): 370–375.
URL: *http://www.opticsinfobase.org/abstract.cfm?URI=josa-67-3-370*

Fried, D. L. (1982). Anisoplanatism in adaptive optics, *Monthly Notices of the Royal Astronomical Society* 72: 52–61.

Fugate, R. Q., Fried, D. L., Ameer, G. A., Boeke, B. R., Browne, S. L., Roberts, P. H., Ruane, R. E., Tyler, G. A. & Wopat, L. M. (1991). Measurement of atmospheric wavefront distortion using scattered light from a laser guide-star, *Nature* 353: 144–146.

Fusco, T., Nicolle, M., Rousset, G., Michau, V., Beuzit, J.-L. & Mouillet, D. (2004). Optimization of a shack-hartmann-based wavefront sensor for xao systems, 5490(1): 1155–1166.
URL: *http://dx.doi.org/doi/10.1117/12.549562*

Gilles, L. & Ellerbroek, B. (2006). Shack-hartmann wavefront sensing with elongated sodium laser beacons: centroiding versus matched filtering, *Appl. Opt.* 45: 6568–6576.

Harding, C. M., Johnston, R. A. & Lane, R. G. (1999). Fast simulation of a kolmogorov phase screen, *Appl. Opt.* 38(11): 2161–2170.
URL: *http://ao.osa.org/abstract.cfm?URI=ao-38-11-2161*

Hardy, J. W. (1998). *Adaptive Optics for Astronomical Telescopes*, Oxford University Press.

Hart, M. L., Jacobsen, A., Angel, J. R. P., Wittman, D., Dekany, R., Mccarthy, D. & Beletic, J. (1995). Adaptive optics experiments using sodium laser guide stars, *Astrophys J* 439: 455–473.

Hartmann, J. (1900). Bemerkungen uber den bau und die justirung von spektrographen, *Z. Instrumentenkd.* 20: 47–58.

Hudgin, R. H. (1977). Wave-front reconstruction for compensated imaging, *J. Opt. Soc. Am.* 67(3): 375–378.
URL: *http://www.opticsinfobase.org/abstract.cfm?URI=josa-67-3-375*

Johansson, E. M. & Gavel, D. T. (1994). Simulation of stellar speckle imaging, 2200(1): 372–383.
URL: *http://dx.doi.org/doi/10.1117/12.177254*

Kellerer, A. N. & Kellerer, A. M. (2011). Error propagation: a comparison of shack–hartmann and curvature sensors, *J. Opt. Soc. Am. A* 28(5): 801–807.
URL: *http://josaa.osa.org/abstract.cfm?URI=josaa-28-5-801*

Leroux, C. & Dainty, C. (2010). Estimation of centroid positions with a matched-filter algorithm: relevance for aberrometry of the eye, *Opt. Express* 18(2): 1197–1206.
URL: *http://www.opticsexpress.org/abstract.cfm?URI=oe-18-2-1197*

López-Quesada, C., Andilla, J. & Martín-Badosa, E. (2009). Correction of aberration in holographic optical tweezers using a shack-hartmann sensor, *Appl. Opt.* 48(6): 1084–1090.
URL: *http://ao.osa.org/abstract.cfm?URI=ao-48-6-1084*

Poyneer, L. A. (2003). Scene-based shack-hartmann wave-front sensing: Analysis and simulation, *Appl. Opt.* 42(29): 5807–5815.
URL: *http://ao.osa.org/abstract.cfm?URI=ao-42-29-5807*

Primmerman, C. A., Murphy, D. V., Page, D. A., Zollars, B. G. & Barclay, H. T. (1991). Compensation of atmospheric optical distortion using a synthetic beacon, *Nature* 353: 141–143.

Roopashree, M. B., Vyas, A. & Prasad, B. R. (2010). Multilayered temporally evolving phase screens based on statistical interpolation, Vol. 7736, SPIE, p. 77363Z.
URL: *http://link.aip.org/link/?PSI/7736/77363Z/1*

Roopashree, M. B., Vyas, A. & Prasad, B. R. (2011). Experimental evaluation of centroiding algorithms at different light intensity and noise levels, AIP.

Schreiber, L., Foppiani, I., Robert, C., Diolaiti, E., Conan, J. M. & Lombini, M. (2009). Laser guide stars for extremely large telescopes: efficient shack-hartmann wavefront sensor design using weighted center-of-gravity algorithm, *Mon. Not. R. Astron. Soc.* .

Sedmak, G. (2004). Implementation of fast-fourier-transform-based simulations of extra-large atmospheric phase and scintillation screens, *Appl. Opt.* 43: 4527–4538.

Sergeyev, A. V. & Roggemann, M. C. (2011). Monitoring the statistics of turbulence: Fried parameter estimation from the wavefront sensor measurements, *Appl. Opt.* 50(20): 3519–3528.
URL: *http://ao.osa.org/abstract.cfm?URI=ao-50-20-3519*

Shack, R. V. & Platt, B. C. (1971). Production and use of a lenticular hartmann screen, *J. Opt. Soc. Am.* 20(61): 656–660.

Silbaugh, E. E., Welsh, B. M. & Roggemann, M. C. (1996). Characterization of atmospheric turbulence phase statistics using wave-front slope measurements, *J. Opt. Soc. Am. A* 13(12): 2453–2460.
URL: *http://josaa.osa.org/abstract.cfm?URI=josaa-13-12-2453*

Southwell, W. H. (1980). Wave-front estimation from wave-front slope measurements, *J. Opt. Soc. Am.* 70(8): 998–1006.
URL: *http://www.opticsinfobase.org/abstract.cfm?URI=josa-70-8-998*

Thomas, S. J., Adkins, S., Gavel, D., Fusco, T. & Michau, V. (2008). Study of optimal wavefront sensing with elongated laser guide stars, *Monthly Notices of the Royal Astronomical Society* 387(1): 173–187.
URL: *http://dx.doi.org/10.1111/j.1365-2966.2008.13110.x*

Thompson, L. A. & Castle, R. M. (1992). Experimental demonstration of a rayleigh-scattered laser guide star at 351 nm, *Monthly Notices of the Royal Astronomical Society* 17: 1485–1487.

Vyas, A., Roopashree, M. B. & Prasad, B. R. (2009a). Digital long focal length lenslet array using spatial light modulator, *Proceedings of the international conference on optics and photonics*.

Vyas, A., Roopashree, M. B. & Prasad, B. R. (2009b). Performance of centroiding algorithms at low light level conditions in adaptive optics, *Advances in Recent Technologies in Communication and Computing, 2009. ARTCom '09. International Conference on*, pp. 366–369. DOI:10.1109/ARTCom.2009.30.

Vyas, A., Roopashree, M. B. & Prasad, B. R. (2010a). Centroid detection by gaussian pattern matching in adaptive optics, *International Journal of Computer Applications* 1(25): 30–35. Published By Foundation of Computer Science.

Vyas, A., Roopashree, M. B. & Prasad, B. R. (2010b). Improved iteratively weighted centroiding for accurate spot detection in laser guide star based shack hartmann sensor, *Atmospheric and Oceanic propagation of electromagnetic waves IV*, SPIE, pp. 758806–1–11.

Vyas, A., Roopashree, M. B. & Prasad, B. R. (2010c). Noise reduction in the centroiding of laser guide star spot pattern using thresholded zernike reconstructor, 7736(1): 77364E. URL: *http://dx.doi.org/doi/10.1117/12.856640*

Vyas, A., Roopashree, M. & Prasad, B. R. (2011). Multi-dither shack hartmann sensor for large telescopes: A numerical performance evaluation, *Adaptive Optics: Methods, Analysis and Applications*, Optical Society of America, p. ATuA4. URL: *http://www.opticsinfobase.org/abstract.cfm?URI=AO-2011-ATuA4*

Wei, X. & Thibos, L. (2010). Design and validation of a scanning shack hartmann aberrometer for measurements of the eye over a wide field of view, *Opt. Express* 18(2): 1134–1143. URL: *http://www.opticsexpress.org/abstract.cfm?URI=oe-18-2-1134*

Wu, D., Wu, S.-Z., Niu, L.-G., Chen, Q.-D., Wang, R., Song, J.-F., Fang, H.-H. & Sun, H.-B. (2010). High numerical aperture microlens arrays of close packing, 97(3): 031109. URL: *http://dx.doi.org/doi/10.1063/1.3464979*

Wu, H.-L., Yan, H.-X., Li, X.-Y. & Li, S.-S. (2009). Statistical interpolation method of turbulent phase screen, *Opt. Express* 17(17): 14649–14664. URL: *http://www.opticsexpress.org/abstract.cfm?URI=oe-17-17-14649*

Yin, X., Li, X., Zhao, L. & Fang, Z. (2009). Adaptive thresholding and dynamic windowing method for automatic centroid detection of digital shack-hartmann wavefront sensor, *Appl. Opt.* 48(32): 6088–6098. URL: *http://ao.osa.org/abstract.cfm?URI=ao-48-32-6088*

Zhao, L., Bai, N., Li, X., Ong, L. S., Fang, Z. P. & Asundi, A. K. (2006). Efficient implementation of a spatial light modulator as a diffractive optical microlens array in a digital shack-hartmann wavefront sensor, *Appl. Opt.* 45(1): 90–94. URL: *http://ao.osa.org/abstract.cfm?URI=ao-45-1-90*

Zou, W., Thompson, K. P. & Rolland, J. P. (2008). Differential shack-hartmann curvature sensor: local principal curvature measurements, *J. Opt. Soc. Am. A* 25(9): 2331–2337. URL: *http://josaa.osa.org/abstract.cfm?URI=josaa-25-9-2331*

9

Measurement Error of Shack-Hartmann Wavefront Sensor

Chaohong Li, Hao Xian, Wenhan Jiang and Changhui Rao
[1]*Institute of Optics and Electronics, Chinese Academy of Sciences, Chengdu,*
[2]*School of Ophthalmology and Optometry, Wenzhou Medical School,*
China

1. Introduction

A Shack-Hartmann sensor is one of the most important and popular wavefront sensors used in an adaptive optics system to measure the aberrations caused by either atmospheric turbulence, laser transmission, or the living eye [1-7]. Its design was based on an aperture array that was developed in 1900 by Johannes Franz Hartmann as a means to trace individual rays of light through the optical system of a large telescope, thereby testing the quality of the image.[8] In the late 1960s Roland Shack and Platt modified the Hartmann screen by replacing the apertures in an opaque screen by an array of lenslets [9-10]. The terminology as proposed by Shack and Platt was "Hartmann-screen". The fundamental principle seems to be documented even before Huygens by the Jesuit philosopher, Christopher Scheiner [11].

The schematic of a Shack-Hartman wavefront sensor is shown in Figure 1. It consists of an array of lenses (called lenslets, see Figure 1) of the same focal length. Each is focused onto a photon sensor (typically a CCD array or quad-cell). The local tilt of the wavefront across each lens can then be calculated from the position of the focal spot on the sensor. Any phase aberration can be approximated to a set of discrete tilts. By sampling an array of lenslets, all of these tilts can be measured and the whole wavefront can be approximated. Since only tilts are measured, the Shack-Hartmann can not measure the discontinuous steps of wavefront.

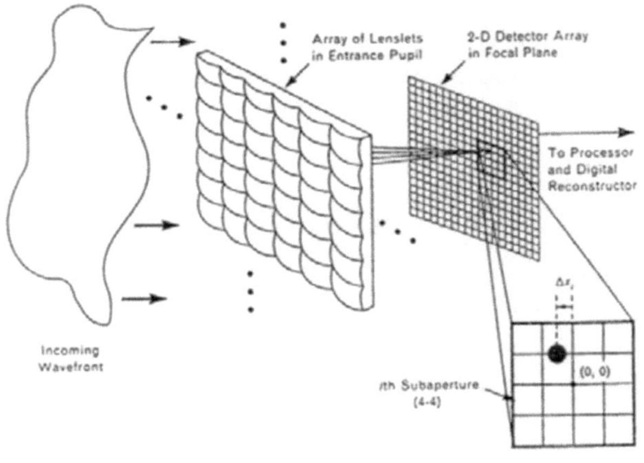

Fig. 1. Schematic of a Shack-Hartmann wavefront sensor

Tyler and Fried have obtained the theory expression, which evaluates the angular position error when a quadrant detector is used in the SHWFS [12]. The formula they obtained, based on circular aperture diffraction, is shown in Eq. (1)

$$\sigma = \frac{3\pi}{16}\frac{1}{SNR}\frac{\lambda}{D} \tag{1}$$

where *SNR* is defined as the ratio of the signal's photoelectron counts to the noise's fluctuation intensity within the detection area, λ is the wavelength, and *D* is the diameter of the aperture. Their analysis did not discuss the size of the incoming light spot on the detection area in detail. The formula was obtained based on a quadrant detector alone. Generally, the theory expression obtained by Tyler and Fried is not suitable for describing the angular position error when the scale of the discrete detector arrays is greater than 2×2 pixels.

Hardy has described formulas that can be used to evaluate the angular position error [13], under the conditions that the photon shot noise of signal is dominant. Although his formulas discussed the size of the diffraction-limited spot on the discrete detector arrays, it is reliable only under the approximation condition that $l/f >> \lambda/D$ or $l/f << \lambda/D$ is satisfied, where l is the length of a pixel, f is the focal length, and V_s is the count of signal photoelectrons. Eq. (2) shows the formulas based on square aperture diffraction.

$$\sigma = \begin{cases} 0.277\frac{\lambda}{D}/V_s^{1/2} & (when\ \ l/f << \lambda/D) \\ 0.500\frac{\lambda}{D}/V_s^{1/2} & (when\ \ l/f >> \lambda/D) \end{cases} \tag{2}$$

Cao *et al.* have also analyzed the measurement error of a SHWFS. Their work emphasized the discrete sampling error of a CCD and obtained a formula which is used to describe the centroid position error induced by the readout noise of the CCD and the photon shot noise of the signal [14]. Their research results are only the approximation of some cases discussed in this article. Jiang *et al.* have partially analyzed the measurement error of SHWFS and performed their method by setting a fixed threshold to suppress the impact of the random noise [15].

In this chapter, the wavefront error of a Shack-Hartmann wavefront sensor was analyzed in detail based on the research results of angular position error and wavefront error [16-17]. and the formula used to evaluate the wavefront error was derived, it concerns with the signal to noise ratio, number of photons and reconstruction matrix also.

2. The angular position error caused by random noise

The wavefront to be measured is segmented into many subwavefronts by lenslet arrays, and the light spots at the focal plane of the subapertures are detected by the CCD. Particularly, the analysis is based on the notion that the wavefront is essentially flat over each subaperture and $r_0 > D$ (r_0 is the coherent length of incoming wavefront). The centroid position can be calculated by Eq. (3) [18]. The detection area of the subaperture is $L_1 \times L_2$ pixels, and x_{nm} and y_{nm} are the (n,m)th pixel's *X* coordinate and *Y* coordinates, respectively. I_{nm} is the total intensity in the (n,m)th pixel, including the signal photons and all other noises.

$$x_i = \frac{\sum_{m=1}^{L_1}\sum_{n=1}^{L_2} x_{nm} I_{nm}}{\sum_{m=1}^{L_1}\sum_{n=1}^{L_2} I_{nm}}, y_i = \frac{\sum_{m=1}^{L_1}\sum_{n=1}^{L_2} y_{nm} I_{nm}}{\sum_{m=1}^{L_1}\sum_{n=1}^{L_2} I_{nm}} \tag{3}$$

The formulas which evaluate the centroid error associated with the signal's photon shot noise and the readout noise of the detector, respectively, have been derived by Cao *et al.* When the photon shot noise of the signal is considered alone, the centroid fluctuation error was obtained by introducing the Gauss width of the signal. When the readout noise of the detector is considered alone, the centroid fluctuation error was also obtained, and its results are shown in Eq. (4) and Eq. (5), respectively [14]. N_r is the rms error induced by the fluctuation of the readout noise in each pixel (with units of photoelectron counts). V_r is the sum of the readout noise's photoelectron count among all of the pixels within the corresponding subaperture.

$$\phi_{cs}^2 = \frac{G_s^2}{V_s} \tag{4}$$

$$\phi_{cr}^2 = \frac{N_r^2}{V_r^2} L_1 L_2 \frac{L_1^2 - 1}{12} \tag{5}$$

ϕ_{cs}^2 is the variance of the centroid fluctuation in one direction (X or Y), induced by the photon shot noise of signal itself, and ϕ_{cr}^2 is the variance of the centroid fluctuation in one direction (X or Y), induced by the readout noise of the detector. ϕ_{cr}^2 and ϕ_{cs}^2 are both defined in pixel^2 units. G_s is the equivalent Gauss width of the signal spot and is defined in pixel units by the expression:

$$G_s = \eta \frac{f}{l} \frac{\lambda}{D} \tag{6}$$

where η is the positive constant. η is 0.353 when the diffraction aperture is square and 0.431 when the diffraction aperture is circular.

Based on Eq. (3), the centroid position in the X direction can be expressed by Eq. (7). The detailed derivation process of this expression is shown in appendix 1.1.

$$x_c = \frac{sbr}{1+sbr} x_{cs} + \frac{1}{1+sbr} x_{cb} \tag{7}$$

where x_c is the calculated centroid position of the signal in the X direction, x_{cs} is the real centroid position of the signal in the X direction, and x_{cb} is the centroid position induced by the total noise, except for the signal in the X direction, where the total noise largely includes the readout noise of the detector and the heterogeneous light noise. $sbr = \sum < S_{n,m} > / \sum < B_{n,m} >$, $<S_{nm}>$ is the collective average of the signal intensity in the (n,m)th pixel, and $<B_{nm}>$ is the collective average of the total noise intensity in the (n,m)th pixel (with units of ADU).

Based on the error transition principles, the rms error of centroid measurement induced by random noise in the X direction can be written in Eq. (8) as:

$$\phi_c = [(\frac{sbr}{1+sbr})^2 \phi_{cs}{}^2 + (\frac{1}{1+sbr})^2 \phi_{cb}{}^2]^{1/2} \tag{8}$$

where ϕ_{cs} is the rms error of centroid measurement in the X direction induced by the signal's photon shot noise and ϕ_{cb} is the rms error of centroid measurement in the X direction induced by all the other noise. The signal is mostly comprised of heterogeneous light and readout noise.

If there were no heterogeneous light and readout noise in the detection area, the signal's photon shot noise should be the unique noise resource which affects centroid measurement. Based on Eq. (4) and Eq. (6), when the discrete sampling error of the detector is ignored, the rms error of angular position in the X direction caused by the photon shot noise of the signal can be written as:

$$\sigma_1 = (\frac{G_s{}^2}{V_s})^{1/2} \frac{l}{f} = \eta \frac{\lambda}{D} V_s{}^{-1/2} \tag{9}$$

When the photon shot noise of the signal is small compared with the readout noise and the heterogeneous light noise, the heterogeneous light noise and the readout noise become the primary noise, which affects centroid calculation. When the heterogeneous light noise can be considered as a uniform noise, like the readout noise of the detector, it exists in each pixel and it has the same fluctuation characteristics among the pixels in the detection area. So, the noise in one pixel (including the heterogeneous light noise and the readout noise of the CCD) can be summed and described by N_b. N_b is defined as the rms error of the heterogeneous light noise and the readout noise photoelectron count in one pixel, and it has the same fluctuation characteristics as the readout noise of the detector. N_b has units of ADU. Subsequently, the rms error of centroid measurement in the X direction caused by the heterogeneous light noise and the readout noise of the CCD can be written as:

$$\begin{aligned} \phi_c &= \frac{1}{1+sbr}\phi_{cb} \\ &= (1+\sum S_{nm} / \sum B_{nm})^{-1}[L_1 L_2 (L_1{}^2 - 1)/12]^{1/2} N_b / V_b \\ &= (\sum B_{nm} / N_b + C_{(\lambda,D,l,f)} \cdot snr)^{-1}[L_1 L_2 (L_1{}^2 - 1)/12]^{1/2} \end{aligned} \tag{10}$$

where $snr = \max(S_{nm}) / N_b$, $V_b = \sum B_{nm}$, and $\max(S_{nm})$ is the signal's peak intensity. snr is defined as the ratio of the signal's peak intensity to the rms error induced by the background noise, B_{nm} is the average intensity of noise in the $(n, m)^{th}$ pixel, which includes the heterogeneous light noise and the readout noise of the CCD. $C_{(\lambda,D,l,f)}$ is the light spot constant which is defined as the ratio of the total signal intensity to the signal's peak intensity in the subaperture, and its value can be measured or calculated exactly by $C_{(\lambda,D,l,f)} = \sum S_{nm} / \max(S_{nm})$.

The intensity distribution of the signal's light spot at the focal plane of the subaperture can be calculated by circular or square aperture diffraction approximations. On the other hand, the Gauss distribution can also be used to approximately describe the intensity distribution of the light spot. The analytic expressions of $C_{(\lambda,D,l,f)}$ are described in Eq. (11) with different approximation conditions. The detailed derivation process is shown in appendix 2.1~2.3. The value of $C_{(\lambda,D,l,f)}$ can be calculated by Eq. (11) .

$$C_{(\lambda,D,l,f)} = \begin{cases} 1/(1-J_0^2(r_1)-J_1^2(r_1)) \\ (circular\ \ aperture\ \ diffraction\ \ approximation) \\ [\frac{\pi}{2}/(-\frac{\sin^2 x_1}{x_1}+\sum_{k=0}^{+\infty}\frac{(-1)^k\times(2x_1)^{2k+1}}{(2k+1)\times(2k+1)!})]^2 \\ (square\ \ aperture\ \ diffraction\ \ approximation) \\ 1/[1-\exp(-\frac{n^2l^2D^2}{2\pi f^2\lambda^2})] \\ (Gauss\ \ distribution\ \ approximation) \end{cases} \tag{11}$$

where $x_1=\frac{\pi D}{\lambda}\sin(\theta)=\frac{\pi D}{\lambda}\sin(\frac{l/2}{f})=\frac{\pi Dl}{2\lambda f}$, $r_1\approx\sqrt{\frac{4}{\pi}x_1}=\sqrt{\frac{4}{\pi}\frac{\pi Dl}{2\lambda f}}$, n is a positive constant, and θ is the diffraction angle.

When the direct current part of the noise (including the heterogeneous light noise and the readout noise of the CCD) is subtracted, it can be considered as white noise, and $\sum B_{nm}=0$. Then, the standard deviation of the angular position error in the X direction caused by noise can be described by Eq. (12):

$$\sigma_2=\frac{[L_1L_2(L_1^2-1)/12]^{1/2}}{C_{(\lambda,D,l,f)}}\frac{1}{snr}\frac{l}{f} \tag{12}$$

When $L_1=L_2=L,$ the Eq. (12) can also be expressed by Eq. (13)

$$\begin{aligned}\sigma_2 &= [L^2(L^2-1)/12]^{1/2}\frac{N_b/(L^2N_b+V_s^{1/2})}{C_{(\lambda,D,l,f)}\max(S_{nm})/(L^2N_b+V_s^{1/2})}\frac{l}{f} \\ &= [(L^2-1)/(12L^2)]^{1/2}\frac{L^2N_b}{(L^2N_b+V_s^{1/2})}\frac{l/f}{\lambda/D}\frac{1}{SNR}\frac{\lambda}{D} \\ &= \omega\frac{1}{SNR}\frac{\lambda}{D}\end{aligned} \tag{13}$$

where SNR has the same definition as in Eq. (1). L^2N_b is the sum of the rms error of total noise among all of the pixels within the detection area, and expresses the total intensity of noise fluctuation. $V_s^{1/2}$ expresses the photon shot noise induced only by the incoming signal.

ω is the position error constant , and it is weighted by the intensity of background noise and the intensity of the signal's photon shot noise (defined in Eq. (14)):

$$\omega = [(L^2-1)/(12L^2)]^{1/2} \frac{l/f}{\lambda/D} \frac{L^2 N_b}{(L^2 N_b + V_s^{1/2})} \tag{14}$$

Substituting Eq. (13) and Eq. (9) into Eq. (8), with the assumed condition that there are no correlations among the photon shot noise of the signal, the heterogeneous light noise, and the readout noise of CCD, then the total rms error of angular position in the X direction caused by random noise can be obtained:

$$\begin{aligned} \sigma &= (\sigma_1{}^2 + \sigma_2{}^2)^{1/2} \\ &\approx \omega \frac{1}{SNR} \frac{\lambda}{D} + \eta \frac{\lambda}{D} V_s{}^{-1/2} \end{aligned} \tag{15}$$

Eq. (15) is the desired result which can be used to precisely describe the angular position error of a Shack-Hartmann wavefront sensor caused by random noise, and therefore, the centroid algorithm is used to calculate the spot position of the incoming light. Generally, when the ideal detector with very small readout noise is used and there is no background light noise ($\omega \to 0$), the photon shot noise of the signal becomes the theoretical limits imposed on the angular position measurement. Eq. (9) showed this expression. In practice, the theoretical limits may not be achieved for the hardware and environment limitations. When the photon shot noise is small enough compared with the heterogeneous light noise and readout noise, it could be ignored in Eq. (15), and Eq. (13) could be used to describe the angular position error caused by the random noise approximately. Commonly, it has enough accuracy. The position-error constant ω described in Eq. (14) is concerned with the scale of the discrete detector arrays in the detection area, the noise characteristics of the detector, and the system parameters. Clearly, the formula based on a quadrant detector obtained by Tyler and Fried is only a special case in this article. On the other hand, the formula obtained in Eq. (13) is suitable to evaluate the angular position error for both a circular and square aperture.

3. Wavefront measurement error caused by centroid position random error

In this chapter, Zernike modes are used as the basis for wavefront reconstruction. The wavefront measurement error can be written as [13, 19]

$$\begin{aligned} \Delta\phi &= \phi - \phi' \\ &= \sum_{j=1}^{P} (a_j - a_j{}')Z_j \\ &= \sum_{j=1}^{P} \Delta a_j Z_j \end{aligned} \tag{16}$$

where $\Delta\phi$ is the wavefront measurement error induced by centroid position random error, ϕ is the wavefront to be measured, ϕ' is the wavefront detected, P is the total number of Zernike modals, aj is the jth Zernike coefficient, and Z expresses the Zernike polynomial.

Then, the mean-square of wavefront measurement error can be written as shown in Eq. (17) [20]. The angle brackets denote a collective average.

$$\begin{aligned}\sigma_\phi^2 &=<\phi^2>-<\phi'^2> \\ &=\sum_{j=1}^{P}<|\Delta a_j|^2>\end{aligned} \tag{17}$$

Based on the principles of the Zernike modal wavefront reconstruction algorithm [20], the Zernike-coefficients vector of a wavefront can be obtained:

$$A = E \cdot H \tag{18}$$

where E is the modal reconstruction matrix and H is the wavefront slope vector.

Therefore, the variance of the modal Zernike coefficients that describe the wavefront measurement error can be written as:

$$\begin{aligned}<|\Delta a_j|^2> &=<\left|\sum_{k=1}^{2Q} e_{j,k}\cdot\Delta h_k\right|^2> \\ &=\sum_{k=1}^{2Q}\sum_{l=1}^{2Q} e_{j,k}e_{j,l}<\Delta h_k \cdot \Delta h_l>\end{aligned} \tag{19}$$

where Q is the total number of subapertures, $e_{j,k}$ is the element of E, and Δh_k is the error of the kth slope element.

In order to simplify analysis, we assume that there are no correlations among the different slope vectors in the corresponding subapertures and the intensity of the signals are uniform and isotropic among the different subapertures Subsequently, the following expression can be obtained:

$$<\Delta h_k \cdot \Delta h_l>=\frac{\sigma_c^2}{f^2}\delta(j_k-j_l) \tag{20}$$

where σ_c^2 is the variance of centroid position random error induced by random noise, f is the focal length of lenslets, $\delta(x,y)$ is the Kronecker delta function [21], and j and k are the subapertures which are connected with the slope h_k and h_l. Substituting Eq. (20) and Eq. (19) into Eq. (17), the mean-square of wavefront measurement error can be written as:

$$\begin{aligned}\sigma_\phi^2 &=\sum_{j=1}^{P}<|\Delta a_j|^2> \\ &=\sum_{j=1}^{P}\sigma_g^2\cdot K(j,Q) \\ &=\sum_{j=1}^{P}(\frac{\sigma_c}{f}\cdot f_0)^2\cdot K(j,Q)\end{aligned} \tag{21}$$

where σ_g is the wavefront average slope of the corresponding subaperture in the unit circle, $K(j,Q)=\sum_{k=1}^{Q}(e_{j,2k-1}+e_{j,2k})^2$. It is concerned with the subaperture segmentation number and the distribution of subapertures. f_0 describes the normalized relationship between the real wavefront slope vector and the normalized wavefront slope vector in the unit circle, and is defined by the expression:

$$f_0=\frac{D}{2\cdot\lambda} \tag{22}$$

where D is the diameter of the aperture and λ is the measuring wavelength.

Then, the root mean square value of wavefront measurement error caused by centroid position random error is obtained:

$$\sigma_\phi=\sigma_c\frac{D}{2\lambda f}[\sum_{j=1}^{P}\sum_{k=1}^{Q}(e_{j,2k-1}+e_{j,2k})^2]^{1/2} \tag{23}$$

Eq. (23) is the desired expression used to evaluate the wavefront measurement error associated with the centroid position random error. σ_c is the standard deviation in pixels of centroid position random error caused by random noise. The formula described in Eq. (23) can help us to decide what the wavefront measurement error will be when the centroid position randomly fluctuates due to random noise, and it may be a factor that must be considered during the design of the SHWFS.

4. Wavefront measurement error analysis based on Zernike modal reconstruction

In a Shack-Hartmann wavefront sensor, the angular position can be calculated from the centroid position in each subaperture and is proportional to the centroid position. The relationship between centroid and angular position can be described by

$$\sigma=\frac{\sigma_c}{f} \tag{24}$$

In Eq. (15), the angular position error caused by random noise was obtained. In Eq. (23), the wavefront error caused by random centroid error was obtained. Therefore, the total wavefront measurement error can be described by Eq. (25):

$$\begin{aligned}\sigma_\phi&=\sigma\cdot f\frac{D}{2\lambda f}[\sum_{j=1}^{P}\sum_{k=1}^{Q}(e_{j,2k-1}+e_{j,2k})^2]^{1/2}\\&=\frac{1}{2}(\frac{1}{SNR}+\eta/\sqrt{V_s})\sqrt{\sum_{j=1}^{P}\sum_{k=1}^{Q}(e_{j,2k-1}+e_{j,2k})^2}\end{aligned} \tag{25}$$

In this formula, we can determine the wavefront measurement error concerned with SNR (see the definition in Eq. (1)), aperture of lenslets (see the definition in Eq. (6)), counts of effective signal, and the reconstruction matrix parameters (see the definition in Eq. (19)).

5. Conclusions

In this chapter, the exact formula (Eq. (25)), which evaluates the Shack-Hartmann wavefront sensor's measurement error associated with the signal to noise ratio of effective signal, was derived in detail. This study was performed based on a modal wavefront reconstruction with Zernike polynomials, and provided an exact and universal formula to describe the wavefront measurement error of a Shack-Hartmann wavefront sensor with discrete detector arrays. It is critical to an adaptive optics system when the Shack-Hartmann sensor is used as the wavefront sensor, and it provides a reference when designing a Shack-Hartmann wavefront sensor and calculating its reconstruction matrix.

6. Acknowledgments

We would like to give our thanks to Shanqiu Chen, Li Shao, Daoai Dong, and Xuejun Zhang for their great discussion and assistance. We will also give our special thanks to Kevin M. Ivers for his great help in writing this chapter.

7. References

[1] J. Liang, B. Grimm, S. Goelz, and J. F. Bille, "Objective measurement of the wave aberrations of the human eye with the use of a Hartmann-Shack wave-front sensor," J. Opt. Soc. Am. A 11, pp. 1949–1957 (1994).

[2] S. Panagopoulou, "Correction of high order aberrations using WASCA in LASIK for myopia," Fall World Refractive Surgery Symposium, Dallas, TX (October 19–21, 2000).

[3] Liang, J., Williams, D. R. & Miller, D. T. Supernormal vision and high-resolution retinal imaging through adaptive optics. J. Opt. Soc. Am. A 14, 2884-2892 (1997).

[4] Roorda, A. & Williams D. R. The arrangement of the three cone classes in the living human eye. Nature 397, 520-522 (1999).

[5] Andre Fleck and Vasudevan Lakshminarayanan, "Statistical error of a compact high-resolution Shack–Hartmann wavefront sensor with a discrete detector array," Appl. Opt. 49, G136-G139 (2010).

[6] D. R. Neal, D. J. Armstrong and W. T. Turner, "Wavefront sensors for control and process monitoring in optics manufacture," SPIE 2993, pp. 211–220 (1997).

[7] R. R. Rammage, D. R. Neal, R. J. Copland, "Application of Shack-Hartmann wavefront sensing technology to transmissive optic metrology," SPIE 4779-27 (2002).

[8] Hartmann "Bemerkungen über den Bau und die Justirung von Spektrographen." Z. Instrumentenkd 20:47 (1900).

[9] Platt, Ben C.; Ronald Shack, "History and Principles of Shack-Hartmann Wavefront Sensing". Journal of Refractive Surgery 17 (2001).

[10] Shack and Platt "Production and use of a lenticular Hartmann screen," JOSA 61:656, (1971).

[11] Scheiner, "Oculus, sive fundamentum opticum," Innspruk (1619).

[12] Glenn A. Tyler and David L. Fried, "Image-position error associated with a quadrant detector". J. Opt. Soc. Am. 72(6), 804-808 (1982).

[13] John W. Hardy, *Adaptive Optics for Astronomical Telescopes* (New York. Oxford University Press, 1998).

[14] Genrui Cao, and Xin Yu, "Accuracy analysis of a Hartmann-Shack wave-front sensor operated with a faint object", Optical Engineering, 33, 2331-2335 (1994).

[15] Wenhan Jiang, Hao Xian, and Feng Sheng, "Detecting error of Shack-Hartmann wave-front sensor". Proc. SPIE, 3126, 534-544 (1997).

[16] C. Li, H. Xian, C. Rao and W. Jiang "Measuring statistical error of Shack–Hartmann wavefront sensor with discrete detector arrays" J. Mod. Opt., 55:14, 2243-2255 (2008).

[17] C. Li, H. Xian, C. Rao and W. Jiang "wavefront error caused by random noise" J. Mod. Opt., 55:1, 127-133 (2007).

[18] Renzhong Zhou, Adaptive Optics (National Defense Industrial Press Beijing, 1999).

[19] Robert J. Noll, J. Opt. Soc. Am. A. 66(3), 207 (1976).

[20] Xinyang Li, and Wenhan Jiang, Proc. SPIE. 4825 121 (2002).

[21] Roddier N, Optical Engineering. 29(10) 1174 (1990).

10

Acceleration of Computation Speed for Wavefront Phase Recovery Using Programmable Logic

Eduardo Magdaleno and Manuel Rodríguez
University of La Laguna
Spain

1. Introduction

Atmospheric turbulence introduces optical aberration into wavefronts arriving at ground-based telescopes. Current adaptive optics (AO) systems use vector-matrix-multiply (VMM) reconstructors to convert gradient measurements to wavefront phase estimates. Until recently, the problem of an efficient phase recoverer design has been implemented over PC or GPU platforms. As the number of actuators n increases, the time to compute the reconstruction by means of the VMM method scales as $O(n^2)$. The number of actuators involved in AO systems is expected to increase dramatically in the future. For instance, the increase in the field of astronomy is due to increasing telescope diameters and new higher-resolution applications on existing systems. The size increase ranges from hundreds up to tens of thousands of actuators and requires faster methods to complete the AO correction within the specified atmospheric characteristic time. The next generation of extremely large telescopes (with diameters measuring from 50 up to 100 meters) will demand important technological advances to maintain telescope segment alignment (phasing of segmented mirrors) and posterior atmospheric aberrations corrections. Furthermore, an increase in telescope size requires significant computational power. Adaptive optics includes several steps: detection, wavefront phase recovery, information transmission to the actuators and their mechanical movements. A quicker wavefront phase reconstruction appears to be an extremely relevant step in its improvement. For this reason other hardware technologies must be taken into account during the development of a specific processor.

In recent years, programmable logic devices called FPGA (Field Programmable Gate Array) are seriously taken into account like a technological alternative in those fields where fast computations are required. The FPGA technology makes the sensor applications small-sized (portable), flexible, customizable, reconfigurable and reprogrammable with the advantages of good customization, cost-effectiveness, integration, accessibility and expandability. Moreover, an FPGA can accelerate the sensor calculations due to the architecture of this device. In this way, FPGA technology offers extremely high-performance signal processing and conditioning capabilities through parallelism based on slices and arithmetic circuits and highly flexible interconnection possibilities (Meyer-Baese, 2001 and Craven & Athanas 2007). Furthermore, FPGA technology is an alternative to custom ICs (integrated circuits) for implementing logic. Custom integrated circuits (ASICS) are expensive to develop, while generating time-to-market delays because of the prohibitive design time (Deschamps 2006).

Thanks to computer-aided design tools, FPGA circuits can be implemented in a relatively short space of time. For these reasons, FPGA technology features are an important consideration in sensor applications nowadays.

In particular, the algorithm of the phase recoverer is based on the estimation of Fast Fourier Transforms (FFT) (Roddier & Roddier, 1991). For this reason, we have to do a careful choice of the architecture of the FFT. An efficient design of this block can result in substantial benefits in speed in comparison with the GPU, DSP or CPU solution.

This chapter presents the design of a fast enough wavefront phase reconstruction algorithm that is based on FPGA technology, paving the way to accomplish the extremely large telescope's (ELT) number of actuators computational requirements within a 6 ms limit, which is the atmospheric response time. The design was programmed using the VHDL hardware description language and XST was used to synthesize into Virtex-6 and Virtex-7 devices. The FPGA implementation results almost 30 times faster than using GPU technology for a 64x64 phase recoverer. The chapter presents a comparative analysis of used resources for several FPGAs and time analysis.

2. A descriptive approach to the wavefront phase recovery

The Shack-Hartmann wavefront sensor samples the signal $\Psi_{telescope}(u,v)$ (complex amplitude of the electromagnetic field) to obtain the wavefront phase map: $\Phi(u,v)$. This is only possible if the sampling is done by microlenses or subpupils with r_0 dimension (that is, inside the phase coherence domain). An array of microlenses is used to sample the wavefront. This array is a rigid piece that fixes the sampling rate. Each *(i,j)* microlense produces a spot (figure 1):

$$I^{ij}(x,y) = \left|FFT[\psi^{ij}_{telescope}(u,v)]\right|^2 \quad (1)$$

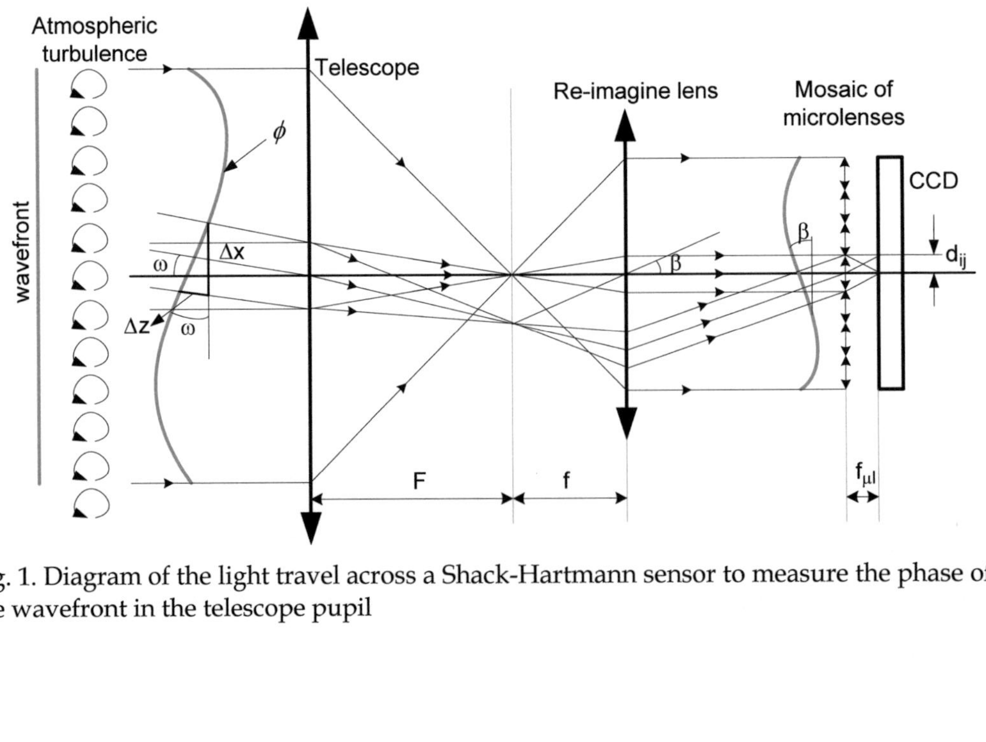

Fig. 1. Diagram of the light travel across a Shack-Hartmann sensor to measure the phase of the wavefront in the telescope pupil

The displacements d_{ij} of the spots centroid with regard to the references centroid (associated with a plane wavefront phase), are a proportional estimation of the average subpupil phase gradient:

$$d_{ij} = K \cdot \frac{\partial \langle \varphi \rangle_{subp_{ij}}}{\partial \vec{r}} \tag{2}$$

Where $\vec{r}$ is the position with components (u,v), and K is a constant. K depends on the wavelength and the focal distances of the telescope, reimaging lens and microlenses. From these gradients estimation, the wavefront phase $\Phi(u,v)$ can be recovered using an expansion over complex exponential polynomials (Poyneer et al., 2002):

$$\varphi(u,v) = \sum_{p,q=0}^{N-1} a_{pq} Z_{pq}(u,v) = \sum_{p,q=0}^{N-1} a_{pq} \frac{1}{N} e^{\frac{2\pi i}{N}(pu+qv)} = IFFT(a_{pq}) \tag{3}$$

The gradient is then written:

$$\vec{S}(u,v) = \vec{\nabla}\varphi(u,v) = \frac{\partial \varphi}{\partial u}\vec{i} + \frac{\partial \varphi}{\partial v}\vec{j} = \sum_{p,q} a_{pq} \vec{\nabla} Z_{pq} \tag{4}$$

Making a least squares fit over the F function:

$$F = \sum_{u,v=1}^{N} [\vec{S}(u,v) - \sum_{p,q} a_{pq}(\frac{\partial Z_{pq}}{\partial u}\vec{i} + \frac{\partial Z_{pq}}{\partial v}\vec{j})]^2 \tag{5}$$

where $\vec{S}$ are experimental data, the coefficients, a_{pq}, of the complex exponential expansion in a modal Fourier wavefront phase reconstructor (spatial filter), can be written as:

$$a_{pq} = \frac{ipFFT\{S^x(u,v)\} + iqFFT\{S^y(u,v)\}}{p^2 + q^2} \tag{6}$$

The phase can then be recovered from the gradient data by transforming backward those coefficients:

$$\varphi(u,v) = FFT^{-1}[a_{pq}] \tag{7}$$

A filter composed of three Fourier transforms therefore must be calculated to recover the phase. In order to accelerate the process, an exhaustive study of the crucial FFT algorithm was carried out which allowed the FFT to be specifically adapted to the modal wavefront recovery pipeline and the FPGA architecture.

3. A descriptive approach to the FPGA architecture

A FPGA device is essentially a matrix of logic cells (called slices). These slices are connected among themselves and with input/output blocks (IOB) through routing channels. These channels are distributed in the FPGA in horizontal and vertical form and its connexions are fixed using a programmable switch matrix (figure 2).

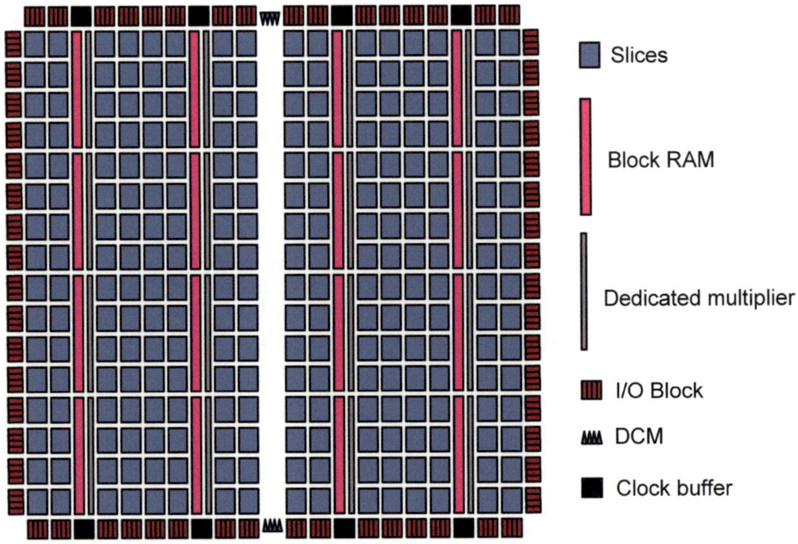

Fig. 2. FPGA generic architecture

Each slice is formed by look-up tables (LUT) and several flip-flops and programmable multiplexers (figure 3). The number and features of these elements depend on the FPGA family. Slices can implement memory (called distributed memory) or simple logic functions. Complex functions and parallel circuits are implemented when they are associated among themselves. Furthermore, FPGAs contain specific RAM blocks (BRAM) and arithmetic circuits.

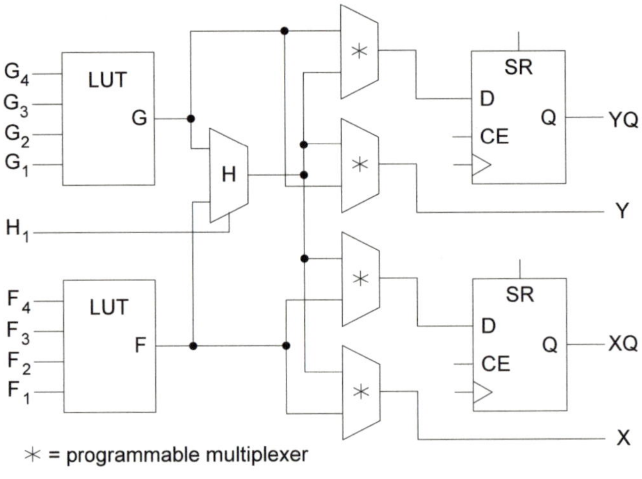

Fig. 3. Block diagram of a slice

In particular, FPGAs allow a wide variety of computer arithmetic implementations for the desired digital signal processing algorithms like FFT, because of the physical bit-level programming architecture. FFT algorithm can be parallelized and we can implement a full pipeline architecture using FPGA. This feature contrasts with DSPs and GPUs, with the fixed multiply accumulator core.

In addition, Xilinx Virtex FPGA devices incorporate DSP48 arithmetic modules. Each DSP48 slice has a two-input multiplier followed by multiplexers and a three-input adder/subtracter. The multiplier accepts two 18-bit, two's complement operands producing a 36-bit, two's complement result. The result is sign extended to 48 bits and can optionally be fed to the adder/subtracter. The adder/subtracter accepts three 48-bit, two's complement operands, and produces a 48-bit two's complement result. Two DSP48 slices, a shared 48-bit C bus, and dedicated interconnect form a DSP48 tile (figure 4).

The DSP48 slices support many independent functions, including multiplier, multiplier-accumulator (MAC), multiplier followed by an adder, three-input adder, barrel shifter, wide bus multiplexers, magnitude comparator, or wide counter. The architecture also supports connecting multiple DSP48 slices to form wide math functions, DSP filters, and complex arithmetic without the use of general FPGA fabric (Hawkes, 2005).

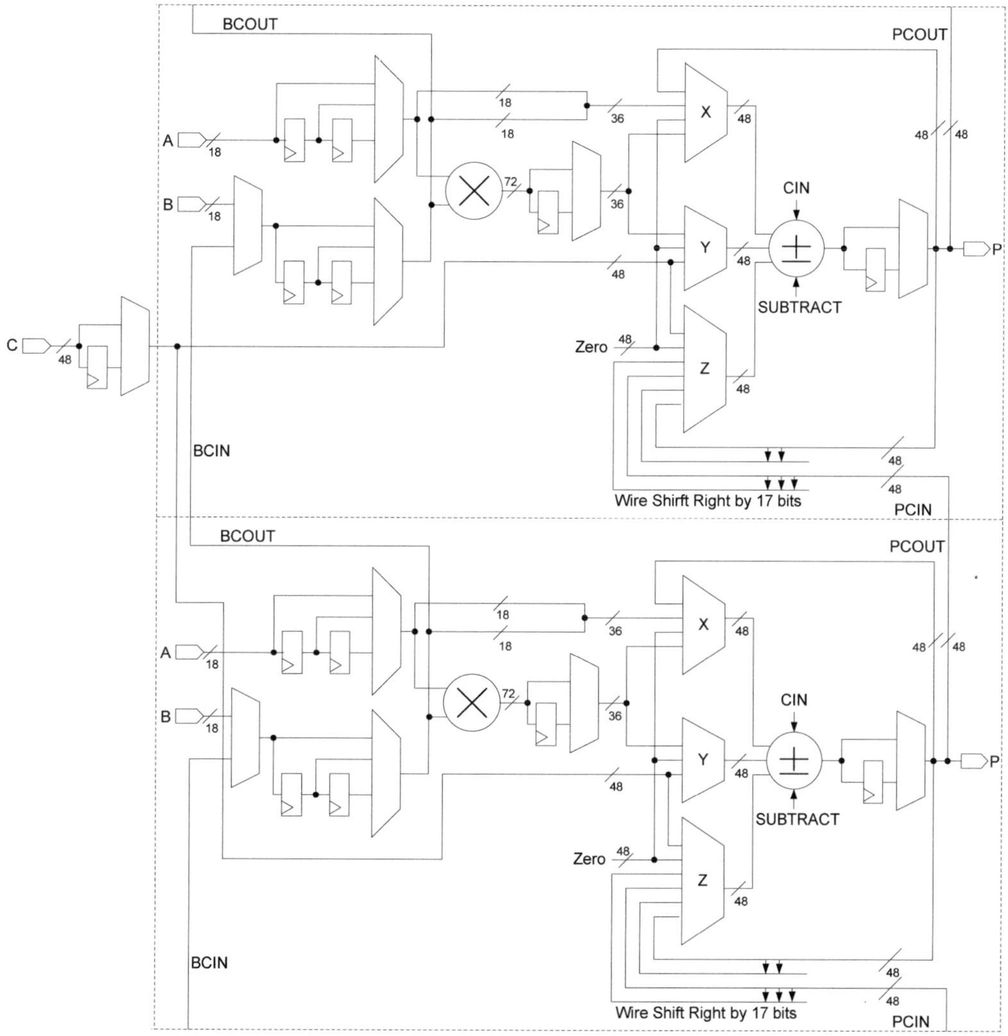

Fig. 4. Two DSP48 arithmetic modules with a shared 48-bit C bus

4. Design of an efficient one-dimensional FFT

The discrete Fourier transform (DFT) of an N-point discrete-time complex sequence x(n), indexed by n=0, 1, N-1, is defined by

$$X(k) = \sum_{n=0}^{N-1} x(n) \cdot W_N^{kn}, \; k = 0,1,\ldots N-1 \tag{8}$$

where $W_N = e^{-j2\pi/N}$ and is referred to as the twiddle factor. The number of complex multiply and add operations for the computation of an N-point DFT is of order N^2 but the problem is alleviated with the development of special fast algorithms, collectively known as fast Fourier transforms (Cooley & Tukey, 1965). These algorithms reduce the number of calculations to $N\log_2 N$.

In the decimation-in frequency (DIF), the FFT algorithm starts with splitting the input data set X(k) into odd- and even-numbered points,

$$\begin{aligned} X(k) &= \sum_{n\ even} x(n) \cdot W_N^{kn} + \sum_{n\ odd} x(n) \cdot W_N^{kn} = \\ &= \sum_{m=0}^{(N/2)-1} x(2m) \cdot W_N^{2mk} + \sum_{m=0}^{(N/2)-1} x(2m+1) \cdot W_N^{k(2m+1)} \end{aligned} \tag{9}$$

So, the problem may be viewed as the DFT of N/2 point sequences, each of which may again be computed through two N/4 point DFTs and so on. This is illustrated in the form of a signal flow graph as an example for N=8 in figure 5.

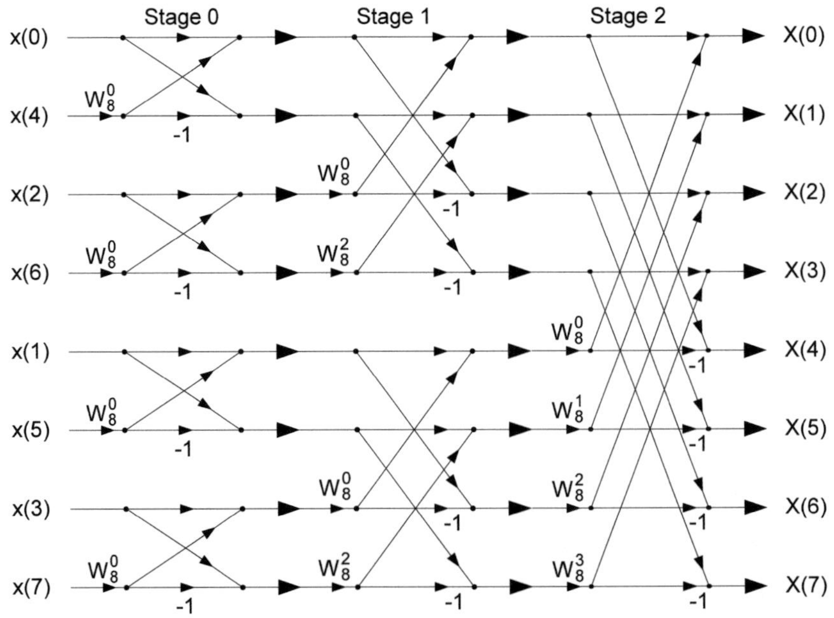

Fig. 5. Computation of FFT by decimation in frequency for N=8

Finally, for the radix-2 FFT algorithm, the smallest transform or butterfly (basic computational unit) used is the 2-point DFT.

Generally, each butterfly implies one complex multiplier and two complex adders. In particular, multipliers consume much silicon area of FPGA because they are implemented with adder trees. Various implementation proposals have been made to save area removing these multipliers (Zhou et al. 2007, Chang & Jen, 1998, Guo, 2000 and Chien et al., 2005). Instead, DSP48 circuits allow some internal calculations of the Fourier transform algorithm or the filter of the phase reconstruction to be parallel, such as in complex multiplications. In this way, the use of these components accelerates the FFT calculation in comparison with a sequential algorithm or adder trees solutions. A complex pipeline multiplier is implemented using only four DSP48 (figure 6) to calculate:

$$P_{real} = A_{real} \cdot B_{real} - A_{imag} \cdot B_{imag}$$
$$P_{imag} = A_{imag} \cdot B_{real} + A_{real} \cdot B_{imag} \quad (10)$$

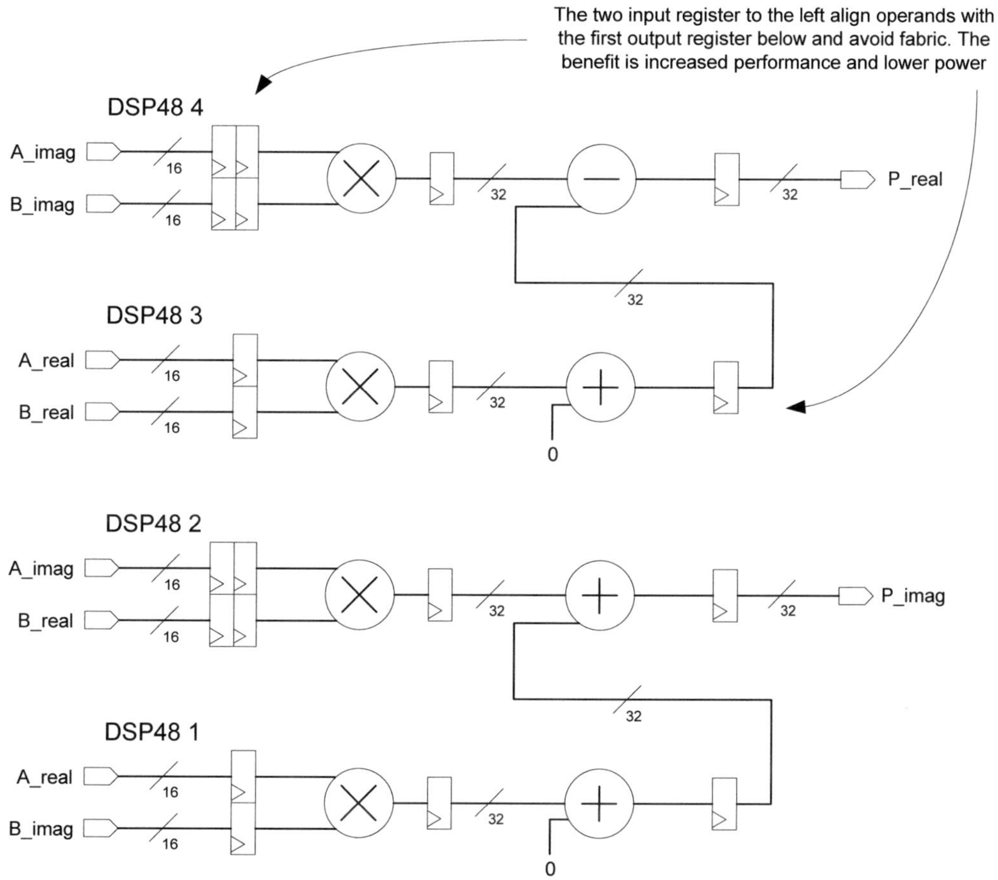

Fig. 6. Pipeline, complex multiplier with four parallel real multipliers

The real and imaginary results use the same DSP48 slice configuration with the exception of the adder/subtracter. The adder/subtracter performs subtraction for the real result and addition for the imaginary one. This implementation only needs four clock cycles to calculate the complex multiplication with up to 550 MHz in XC4VSX35 Virtex-4 (Hawkes, 2005).

The complete pipeline radix-2 butterfly can be easily implemented with this specialized multiplier. It is necessary to use a FPGA Look-Up Table (LUT) (configured as SRL16 3-bits shift register) to preserve the synchronism. The butterfly implemented is depicted in Figure 7 and it needs only seven clock cycles to carry out the computation.

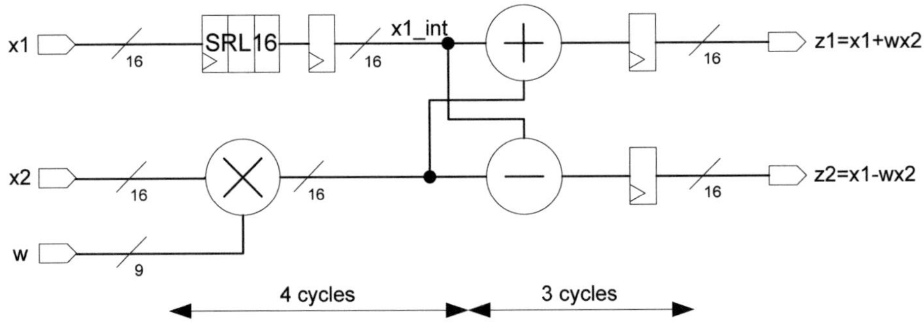

Fig. 7. Pipeline radix-2 butterfly in FPGA

A pipeline radix-2 FFT can be implemented using one butterfly in each stage. The twiddle coefficients used in each stage are stored in twiddle LUT ROMs in the FPGA. The logic resources and the clock cycles of the FFT module is reduced in our implementation using specific butterflies modules at the first and second stages. The first stage utilizes the feature of the twiddle factors related to the first stages of the pipeline.

$$W_N^{N/2} = 1 \tag{11}$$

So, the first stage can be implemented in a very simple way with an adder/subtracter. In the second stage, the next twiddle factors are

$$W_N^{N/4} = j \tag{12}$$

This twiddle suggests a similar splitting structure in the second pipeline stage as in the first one; however, the imaginary unit imposes a special consideration: two additional multiplexers change real and imaginary data, and the pipeline adder/subtracter works according equation 13.

$$(a+bj)j = aj - b = -b + aj \tag{13}$$

Taking into account these features, the 1D-FFT architecture implementation is depicted in figure 8. The swap-blocks arrange the data flow (according figure 5) and preserve the pipeline feature. It consists of two multiplexers and two shift registers. These shift registers are implemented using look-up tables (LUT) in mode shift register (SRL16) for synchronization (figure 9).

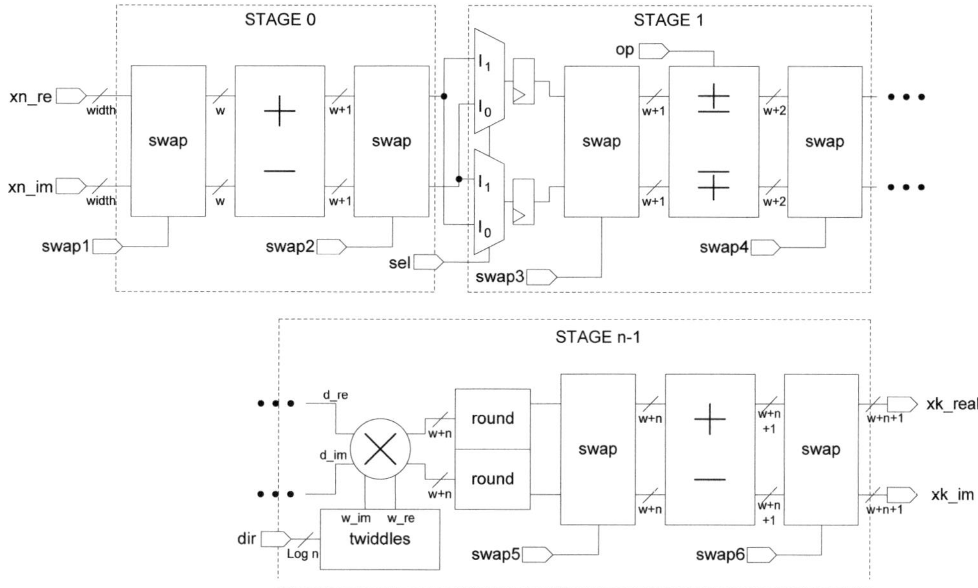

Fig. 8. Architectural block diagram of a radix-2 pipeline FFT

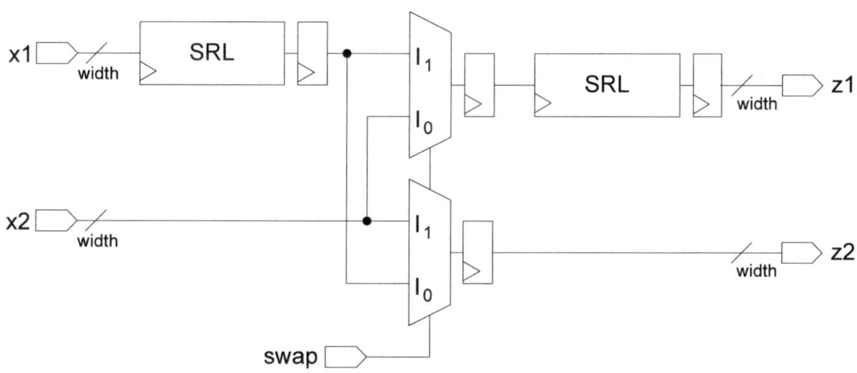

Fig. 9. Swap unit

The system performs the calculation of the FFT with no scaling. The unscaled full-precision method was used to avoid error propagations. This option avoids overflow situations because output data have more bits than input data. Data precision at the output is:

$$output\ width = input\ width + \log_2 points + 1 \tag{14}$$

The number of bits on the output of the multipliers is much larger than the input and must be reduced to a manageable width. The output can be truncated by simply selecting the MSBs required from the filter. However, truncation introduces an undesirable DC data shift. Due to the nature of two's complement numbers, negative numbers become more negative and positive numbers also become more negative. The DC shift can be improved with the use of symmetric rounding stages (figure 8). The periodic signals of the swap units, op signal, and the address generation for the twiddle memories are obtained through a counter module that acts like a control unit.

The pipeline 1D-FFT architecture design is completely parametrisable and portable. The VHDL module has three generics in order to obtain a standard module: *fwd_inv*, *data_width* and *lognpoints* (the logarithm in base 2 of the number of elements that fit with the number of stages in the 1-D FFT calculation). These generics then select direct or inverse transform, data value precision and the transform length.

4.1 Temporal analysis for the radix-2 pipeline FFT module and superior radix

In Table 1 is depicted the latency for each stage of an 8-points FFT (figure 8).

Stage	Module	Cycles
0	Swap 1	5
	Adder 1	2
	Swap 2	5
1	Multiplexers	1
	Swap 3	3
	Add/Sub 2	2
	Swap 4	3
2	Multiplier	4
	Round	1
	Swap 5	2
	Adder 3	2
	Swap 6	2
	Total	**32**

Table 1. Latency of each module in a pipeline 8-points FFT

Taking into account the clock cycles of each block in table 1, the latency of the N-points FFT module ($\log_2 N$ stages) can be written as

$$latency = 2\left(\frac{N}{2}+1\right)+2+1+2\left(\frac{N}{4}+1\right)+2+ \\ +\sum_{n=2}^{(\log_2 N)-1}\left[7+2\left(\frac{N}{2^{n+1}}+1\right)\right], \quad n=2,3,\ldots(\log_2 N)-1 \tag{15}$$

where the two first stages are considered separately, and N and n are the number of points of the transform and the number of stages of the module respectively. Operating,

$$\begin{aligned} latency &= 9+\frac{3N}{2}+\sum_{n=2}^{(\log_2 N)-1}\left[9+\frac{N}{2^n}\right]= \\ &= 9+\frac{3N}{2}+9(\log_2 N-2)+\sum_{n=2}^{(\log_2 N)-1}\frac{N}{2^n}= \\ &= \frac{3N}{2}+9\log_2 N-9+N\sum_{n=2}^{(\log_2 N)-1}\frac{1}{2^n}= \\ &= \frac{3N}{2}-\frac{N}{2}+9\log_2 N-9+N\sum_{n=1}^{(\log_2 N)-1}\frac{1}{2^n}= \\ &= N+9\log_2 N-9+N\sum_{n=1}^{(\log_2 N)-1}\frac{1}{2^n} \end{aligned} \tag{16}$$

The last summand of this equation is a geometric series with common ratio equal to ½. This is a convergent series and the partial sum to r of the series is

$$S_r=\sum_{n=1}^{r}\frac{1}{2^n}=\frac{2^r-1}{2^r} \tag{17}$$

Where in this case $r=log_2N-1$, so

$$S_{(\log_2 N)-1}=\frac{2^{(\log_2 N)-1}-1}{2^{(\log_2 N)-1}}=\frac{2^{\log_2 N}2^{-1}-1}{2^{\log_2 N}2^{-1}}=\frac{\frac{N}{2}-1}{\frac{N}{2}}=\frac{N-2}{N} \tag{18}$$

Adding the geometrical series in equation 17 and grouping, finally

$$latency=2N+9\log_2 N-11\,,\quad N=8,16,32,\ldots \tag{19}$$

When the number of points of the FFT is a power of 4 is computationally more efficient to use a radix 4 algorithm instead of used radix 2. The reasoning is the same than radix 2 but subdividing iteratively a sequence of N data into four subsequences, and so on. The radix-4 FFT algorithm consists of $\log_4 N$ stages, each one containing N/4 butterflies. As the first weight is $W_N^0=1$, each butterfly involves three complex multiplications and 12 complex sums. Performing the sum in two steps, according to Proakis & Manolakis (1996), it is possible to reduce the number of sums (12 to 8). Therefore, the number of complex sums to perform is the same ($N\log_2 N$) than the algorithm in base 2, but the multipliers are reduced by 25% (of $(N/2)\log_2 N$ to $(3N/8)\log_2 N$). Consequently, the number of circuits for use DSP48 is reduced proportionately.

For number of point power of 4, the pipeline radix-4 FFT module has half of arithmetic stages, but the swap modules need twice clock cycles to arrange the data. Then, the latency is expressed as

$$\begin{aligned} latency(radix4) &= 2\left(\frac{3}{4}N+1\right)+2+ \\ &+\sum_{n=1}^{\log_4 N-1}\left[7+2\left(\frac{3}{4^{n+1}}N+1\right)\right]= \\ &=9\log_4 N-5+6N\sum_{n=1}^{\log_4 N-1}\frac{1}{4^n} \end{aligned} \tag{20}$$

Again, the series can be estimated for *log₄N-1* terms, and finally

$$\begin{aligned} latency(radix4) &= 2N+9\log_4 N-13, \\ N &= 16, 64, 256, 1024, \ldots \end{aligned} \tag{21}$$

This time estimation has been realized for other radix, as shown in the following equations:

$$\begin{aligned} latency(radix\,8) &= 2N+9\log_8 N-21, \\ N &= 64, 512, \ldots \\ latency(radix\,16) &= 2N+9\log_{16} N-37, \\ N &= 2^8, 2^{12}, \ldots \end{aligned} \tag{22}$$

Generalizing

$$\begin{aligned} latency(radix\,i) &= 2N+9\log_i N-5-2i, \\ N &= i^{n+1}, n=1,2,3,\ldots \\ i &= 4,8,16,\ldots \end{aligned} \tag{23}$$

In Figure 10 is depicted the clock cycles of each algorithm and a proposed resolution with an orange line. All implementations are close to 2 when the number of points grows (figure 10a). The improvement in terms of computing speed of the algorithm using other radix is relevant when the number of samples is small. For example, the improvement factor for a 1024-point FFT is less than 7% using a radix-32 algorithm and less than 3% using a radix-4 (Figure 10b). However, in our astronomical case the proposed size is relativity small and the improvement using superior radix is relevant. Examining figure 10b, we can observe that the improvement factor is about 20% using radix-8 and 30% using radix-16. Thus, we are considering implementing these algorithms in the future.

4.2 Resources analysis

Several FFTs were implemented over a XC6VLX240T Virtex-6 device and numerical results were satisfactorily compared with Matlab simulations. Resources in this FPGA include 301440 flip-flops, 150720 6-LUTs, 416 BRAM and 768 DSP48s. The syntheses were achieved by changing the size of the FFT and the data precision.

Figure 11 shows the resource utilization to implement a pipeline 1024-FFT when the input data precision is between 8 and 24 bits. The resource utilization is greater when the FFT has more precision due to the increase of the intrinsic complexity found when the precision is increased. From 17 bits of precision, DSP48 circuits are increased and the maximum operating frequency is drastically smaller. It is due to high-precision complex multipliers need eight DSP48 circuits instead four in Figure 6. These high-precision multipliers are increased in each stage of the FFT from 17 bits data precision.

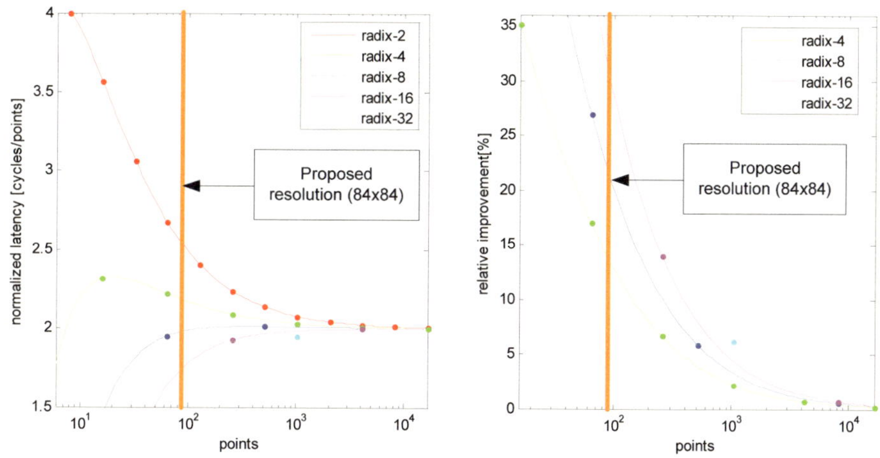

Fig. 10. (a) Normalized latency. (b) Relative improvement respect of radix-2 algorithm

In figure 12 is depicted the resource utilization to implement one-dimension FFT when de number of points of the transform is increased in a power of 2.

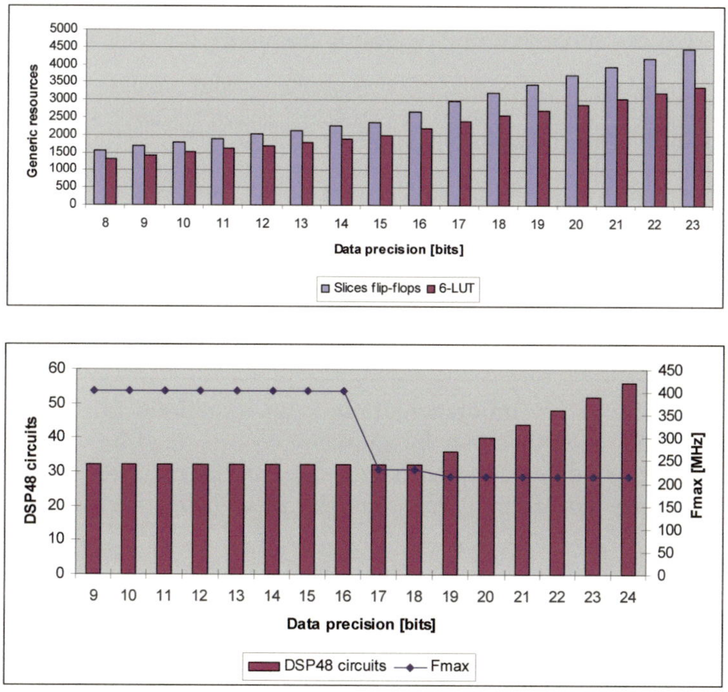

Fig. 11. FPGA Resources and maximum operating frequency for 1024-FFT

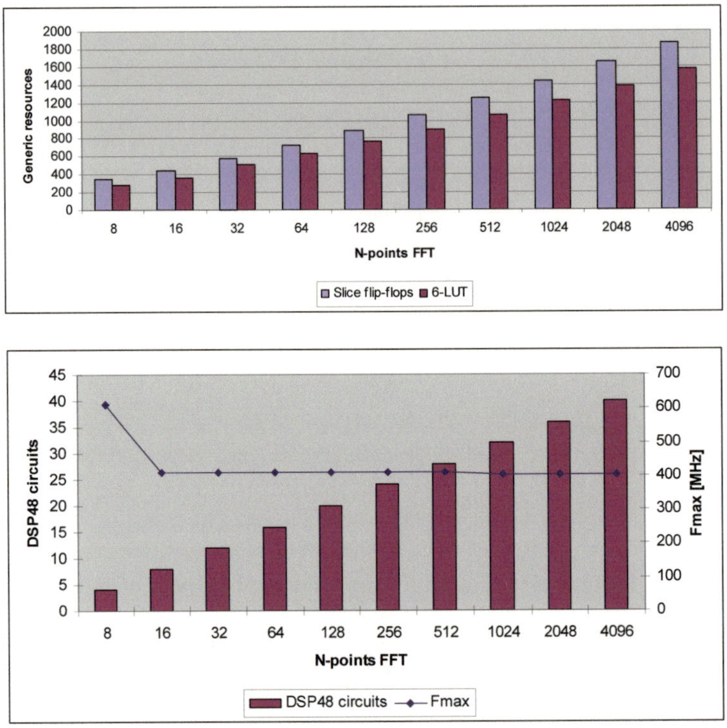

Fig. 12. FPGA Resources and maximum operating frequency for N-FFT with N from 8 to 4096

4.3 Comparison with others implementations

A comparison has been carried out between our design and others implementations. The combined use of the FPGA technology and the developed architecture achieves an improved performance compared to other alternatives. This is showed in figure 13 where our implementation executes an 1024-point FFT operation in 10.64 µs at 200 MHz.

5. Wavefront phase recovery FPGA-implementation

We will focus on the FPGA implementation from equations (6) and (7) to improve processing time. These equations can be implemented using different architectures. We could choose a sequential architecture with a unique 2D-FFT module where data values use this module three times in order to calculate the phase. This architecture represents an example of an implementation using minimal resources of the FPGA. However, we are looking for a fast implementation of the equations in order to stay within the 6 ms limit of the atmospheric turbulence. Given these considerations we chose a parallel and completely pipeline architecture to implement the algorithm. Although the resources of the device increase considerably, we can maintain time restrictions by using extremely high-performance signal processing capability through parallelism. We therefore synthesize three 2D-FFTs instead of one 2D-FFT.

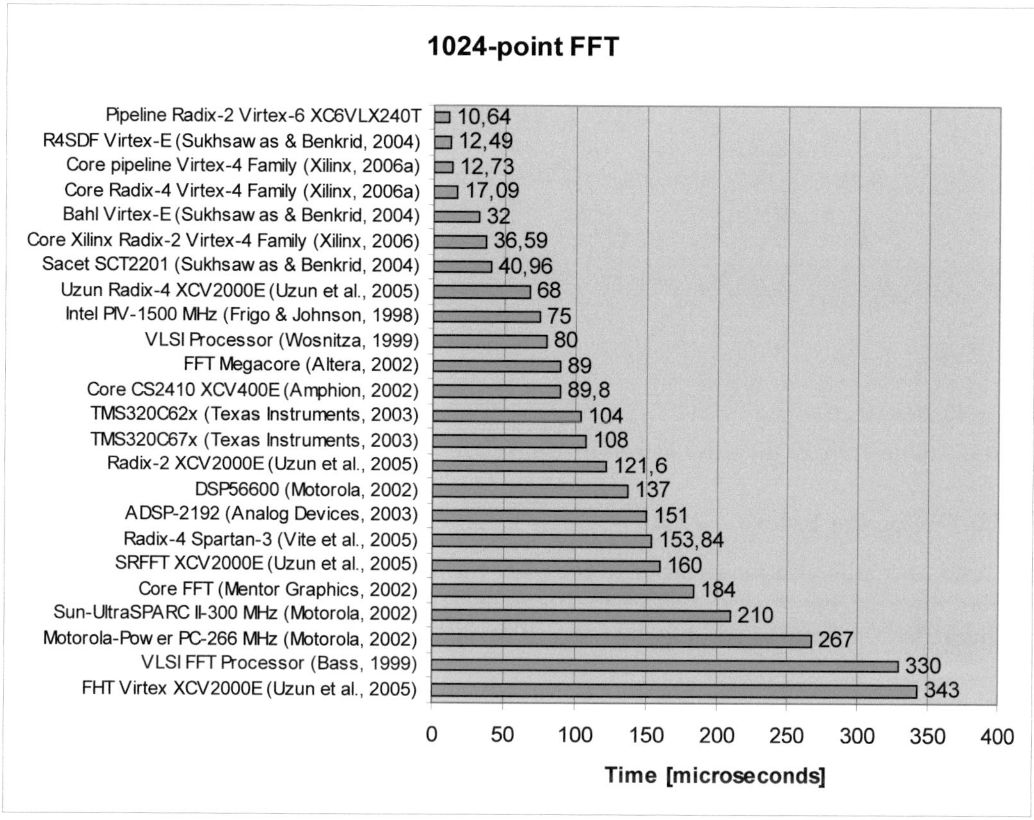

Fig. 13. FPGA Resources and maximum operating frequency for N-FFT with N from 8 to 4096

The block diagram of the designed recoverer is depicted in figure 14 where S_x and S_y represent the image displacement into each subpupil. The bidimensional transforms of S_x and S_y have to be multiplied by ip/p^2+q^2 and iq/p^2+q^2 respectively according to equation 6. These two matrices are identical if we change rows by columns. We can therefore store a unique ROM. The results of the adders (a_{pq} coefficients) are rounded appropriately to obtain 16 bits data precision according with the data input width of the inversed bidimensional transform that is executed at the next stage.

An analysis of the equations and a parallel architecture of its implementation are taken into account. We then break down the design into the following steps or stages:

1. Compute two real forward 2D FFT that compute FFT(S_x) and FFT(S_y).
2. Compute the complex coefficients
3. Carry out a complex inverse 2D FFT on a_{pq}.
4. Flip data results.

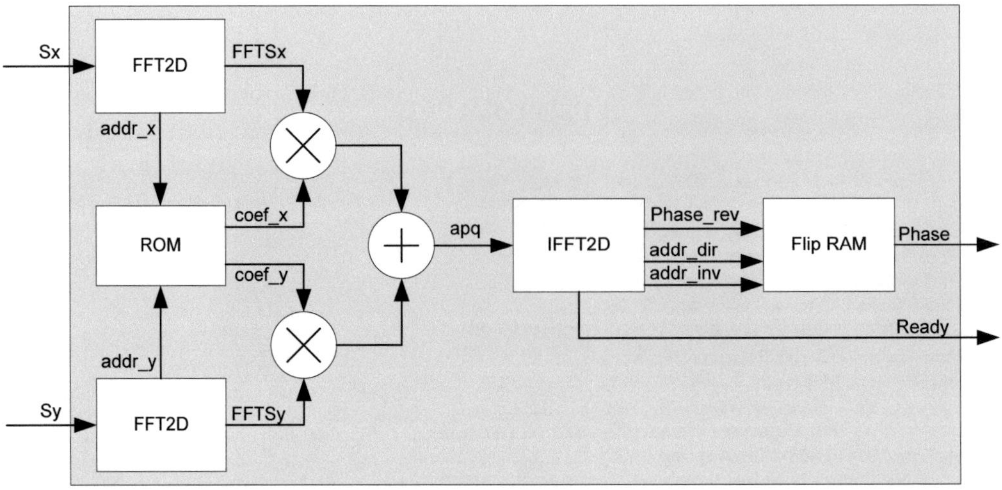

Fig. 14. Architecture of the synthesized phase recovery

5.1 2D-FFT on the Virtex-6 FPGA

The fundamental operation in order to calculate the 2DFFT is equivalent to doing a 1D-FFT on the rows of the matrix and then doing a 1D-FFT on the columns of the result. Traditionally, the parallel and pipeline algorithm is then implemented in the following four steps.

1. Compute the 1D-FFT for each row
2. Transpose the matrix
3. Compute the 1D-FFT for each row
4. Transpose the matrix

Figure 15 depicts the diagram of the implemented transform module. The operation of the developed system takes place when image data is received in serial form by rows. These data are introduced in a block that carries out a one dimensional FFT. As this module obtains the transformed data, the information is stored in two double-port memories (real and imaginary data). To complete the bidimensional FFT, the stored data is introduced in a second 1D-FFT in column format. The 2D-FFT is then obtained from the output of this block.

Continuous data processing using a single dual-port memory (real and imaginary) is not possible. In that case, the new transformed data must wait for the old data to be introduced in the second FFT block. Otherwise data are overwritten. As a result, the pipeline property of the FFT architecture can not be used. This problem can be averted by using two memories instead of one, where memories are continuously commuting between write and read modes. When the odd memory is reading and introducing data values in the second FFT module, the even memory is writing data which arrives from the first FFT. So, data flow is continuous during all of the calculations in the bidimensional transform. The memory modes are always alternating and the function is selected by the counter. The same signal is used to commute the multiplexer that selects the data that enter the column transform unit.

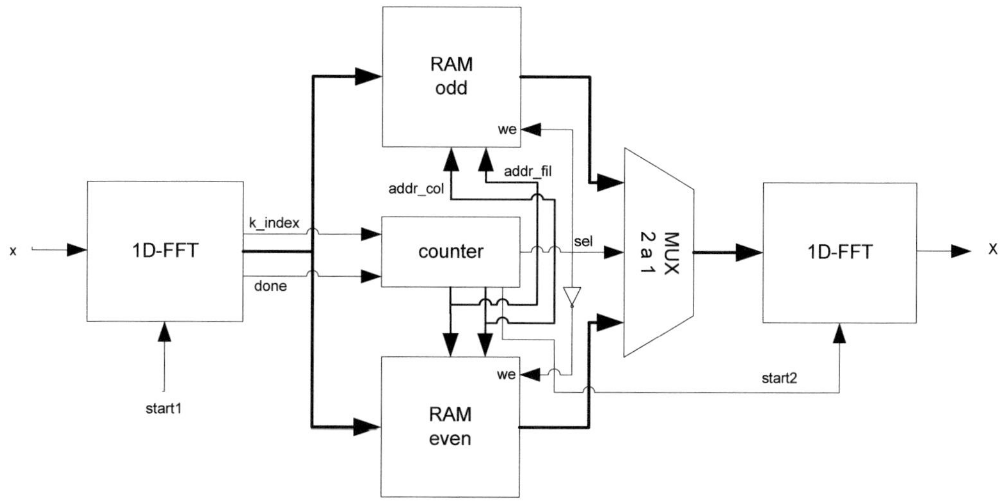

Fig. 15. Block diagram of the implemented 2D-FFT

It is worth mentioning that the transposition step defined in the above (step 2) is implemented simultaneously with the transfer of column data vector to the memory with no delay penalty. In this way, the counter acts as an address generation unit. The last transposition step (step 4) is not implemented in order to save resources and obtain a fast global system. So, the last transposition step is taken into account only at last of the algorithm described in eq. (4)-(5) and showed in figure 13 (Flip-RAM module).

Row 1D-FFT block and column 1D-FFT block are not identical due to the unscaled data precision. For example, a 64x64 2D-FFT for the phase recoverer must meet certain requirements. If the precision of data input is 8 bits, the output data of 1D-FFT of the rows has to be 15 bits according equation 13. 1D-FFT of the columns accepts a 15 bits data format and 22 bits at the output.

Taking into account the latency of the FFT (equation 18) and the pipeline operation of the memory modules, the latency of the 2D-FFT module can be written as

$$latency = N^2 + 4N + 18\log_2 N - 22\,, \quad N = 8, 16, 32, \ldots \tag{24}$$

In Figure 16 is depicted the latency for different sizes of 2D-FFT according to equation 18.

Table 2 shows a performance comparison of existing 2D-FFTs implementations using FPGA and other technologies for matrix sizes 64x64 and 128x128. Rodríguez-Ramos et al. 2007 implemented 2D-FFT on a CPU AMD XP 3500+, 2211 GHz, with 512 KB L2 cache and on a GPU nVidia GeForce 7800 GTX graphical engine with 256 MB RAM. Uzun et al. 2007 implemented several algorithms on a Virtex-2000E FPGA chip where the fastest design is depicted in the table. It can be seen that our design shows improvements when compared to [3] and [15] in terms of frame rate performance. Magdaleno et al. 2010 implemented 2D-FFT in a XC4VSX35 Virtex-4 operating at 100 MHz. Virtex-6 family can operate twice faster and the time operation is decreased in this way.

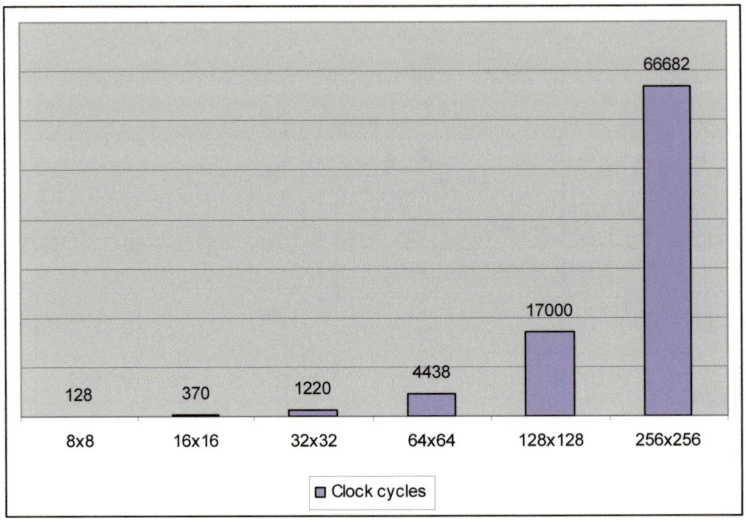

Fig. 16. Estimation of latency for several 2D-FFT

2D-FFT	CPU Rodríguez-Ramos et al. [2007]	GPU Rodríguez-Ramos et al. [2007]	FPGA Uzun et al. [2007]	FPGA Magdaleno et al. [2010]	FPGA Proposed
64x64	114.5 μs	1.58 ms	-	44.4 μs	22.2 μs
128x128	811.0 μs	1.68 ms	2.38 ms	170.8 μs	85.0 μs
256x256	-	-	-	-	333.4 μs

Table 2. 2D-FFT performance comparison with other designs

In figure 17 is depicted the resource utilization to implement 2D-FFT when de number of points of the transform is increased in a power of 2. The relevant resource is the Block-RAM. The 256x256 FFT occupies the 34% of BRAM for a XC6VLX240T Virtex-6 and the maximum operating frequency is only 297 MHz.

5.2 An 64x64 phase recoverer implementation

For a first prototype of the phase recoverer, we have selected a plenoptic sensor with 64x64 pixels sampling each microlens.

The first step is a direct use of the proposed 2D FFT. To accelerate the process, the two real transforms are executed simultaneously through a parallel implementation. Observe how the two 2D-FFT of phase gradients S_x and S_y are multiplied by some constant factors according to the formula of the phase recoverer (Figure 14). An adder is necessary in the following stage to calculate the frequency coefficients and achieve an inversed 2D-FFT. Phase values are then transposed, which require an intermediate memory to properly show output data (flip-ram module). The direct 2D-FFTs are real, so the imaginary components are zero. The inversed transform allows complex input data.

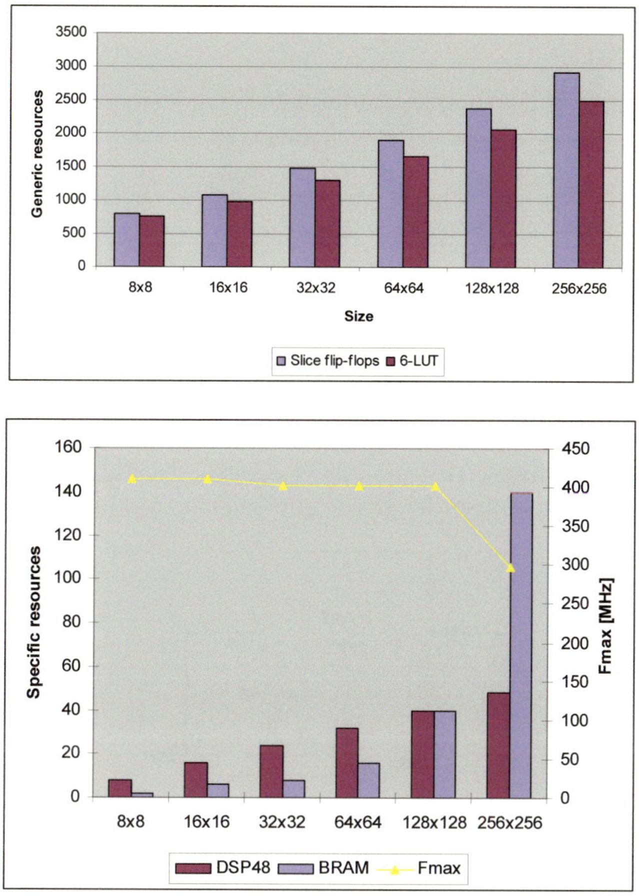

Fig. 17. Estimation of latency for several 2D-FFT

The importance of the parallel execution of the S_x and S_y transforms in the algorithm needs mentioning. This feature allows both inputs to be simultaneously received and to obtain the calculation of both data just as in the previous case at the output of the 2-D FFT units.

The bidimensional transforms of S_x and S_y have to be multiplied by ip/p^2+q^2 and iq/p^2+q^2 respectively according to Equation 6. These two 64x64 points matrix are identical if we change rows by columns. We can therefore store a unique ROM (two 64x16 bits ROM, one ROM for the real component of the factor and the other for the imaginary part). The addresses of these ROMs are supplied by the 2D-FFT module through *cnt_fil* and *cnt_col* signals. These signals are obtained from a built-in counter in this module. The ROM module has two generics (*data_width* and *addr_width*) in order to synthesize a standard single-port memory of any size. However, every time we change these generics, we have to use mathematical software (Matlab in this case) in order to calculate the elements and initiate ROM memory.

There are two complex multipliers. Each one performs the complex multiplication of the constant stored in the ROM by the 2-D FFT result (previously rounding to 18 bits). The complex multiplication needs four DSP48 circuits. Inside of this circuit, the complex multiplication uses multipliers, internal registers and adders/subtractors. These modules are completely standard using the *a_width* and *b_width* generics that select the precisions of the signals to multiply.

The sum of the outputs of the complex multipliers is implemented using slices exclusively. We implemented two adders, one for the real components and other for the imaginary ones. This module is also configured with a generic parameter that supplies data precision. The results of the adders (*apq* coefficients) are rounded appropriately to obtain 8 bits data precision according with the data input width of the inversed bidimensional transform that is executed at the next stage.

The FLIP-RAM module synthesizes a double dual-port memory similar to the memories that were described in the 2D-FFT section. While a memory is in read mode, the other one is in write mode. With this consideration, the total implemented system continues being pipeline. The addressing of the memories is obtained in a similar form, through a counter that is included in the inversed 2D-FFT. In this case, the memories only store real data (data values of the recovered phase). This module is necessary because the phase data that the inversed 2D-FFT provides are disorderly. The implemented module is depicted in figure 18.

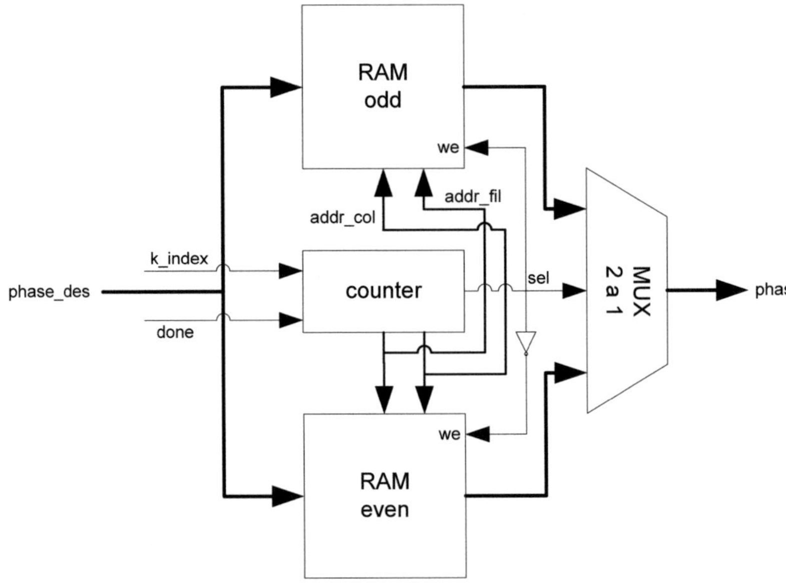

Fig. 18. Block diagram of the output stage of the wavefront recoverer (Flip-RAM)

The design of a 64x64 phase recoverer was programmed using the VHDL hardware description language (IEEE, 2000) and XST (Xilinx, 2006b) was used to synthesize these modules into a XC6VLX240T Virtex-6 FPGA.

The complete system was successfully tested in circuit using ChipScope Pro software (using phase gradients obtained in simulations) that directly inserts a logic analyzer and bus analyzer into the design, allowing any internal signal to be viewed. Signals are captured at operating system speed and brought out through the programming interface. Captured signals can then be analyzed with the PC that acts as a logic analyzer. The numeric results were also successfully compared with those obtained in Matlab. Figure 19 shows an example of a wavefront reconstruction using a 64 × 64 subpupil recoverer. The two first images show the phase gradients (S_x, S_y) given to the module. The last picture is the recovered phase using the implemented module.

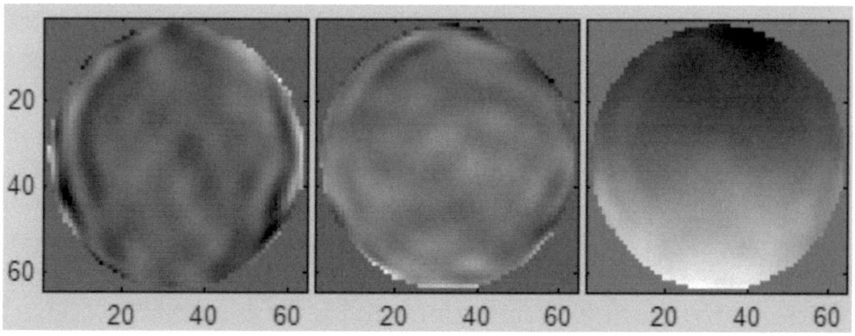

Fig. 19. Phase gradients (S_x and S_y) and the recovered phase for a Shack-Hartmann sensor with 64x64 subpupils

Table 3 shows the total time broken down into the stages of the total system (depicted in figure 13). 12980 clock cycles are necessary for phase recovery, starting with data reception to the activation of the *ready* signal. This value is the latency time for the phase recoverer. At a 200 MHz frequency clock, the system needs less than 65 μs to recover the phase.

Module	Cycles	Duration (@ 200 MHz)
2D-FFT (S_x and S_y)	4438	22.19 μs
Multipliers	4	0.02 μs
Adder	1	0.005 μs
IFFT2D	4438	22.19 μs
Flip-RAM	4096	20.48 μs
Rounding (3)	3	0.015 μs
Total	**12980**	**64.90 μs**

Table 3. Execution time (latency) for the different stages of the 64x64 phase recoverer

Generalizing, the latency for a generic NxN phase recoverer is determined as follows:

$$\begin{aligned} latency &= 2\left(N^2 + 4N + 18\log_2 N - 22\right) + 4 + 1 + 3 + N^2 = \\ &= 3N^2 + 8N + 36\log_2 N - 36\,, \quad N = 8, 16, 32, \ldots \end{aligned} \tag{25}$$

Table 4 shows XC6VLX240T Virtex-6 resource utilization and the maximum operating frequency (pre-synthesis).

Slices FFs	6-LUT	DSP48	BRAM	Fmax [MHz]
5819 (1%)	5000 (3%)	104 (13%)	58 (13%)	402.188

Table 4. XC6VLX240T Virtex-6 resource utilization to implement the phase recoverer prototype.

The implemented architecture is pipeline. This architecture allows phase data to be obtained for each 4,096 clock cycles (this number coincides with the number of points of the transforms, that is, the number of subpupils, 64 × 64, of the Shack-Hartmann sensor). Using the 200 MHz clock, the prototype provides new phase data each 20.48 μs.

These results can be compared with other works. Rodriguez-Ramos *et al.* (2007) implemented a 64 × 64 phase recoverer using GPU. In this technology, the wavefront reconstruction needs 3.531 ms. The FPGA implementation results almost 30 times faster. Baik *et al.* 2007 implemented a wave reconstruction with a 14 × 14 Shack-Hartmann array, an IBM PC and a Matrox Meteor-2 MC image processing board. The wavefront correction speed of the total system was 0.2 s. Although the system includes the gradient estimation, it can be seen that the execution times are slower than in the proposed implementation. Seifer *et al.* 2005 used a sensor with 16 × 12 subpupils and a Pentium M, 1.5 GHz. The wavefront reconstruction in this case was 50 ms using Zernike polynomials to adjust to the coefficients of the aperture function. Again, our implementation using FPGA technology is comparatively faster.

6. Conclusion

A wavefront phase can be recovered from a Shack-Hartmann sensor using FPGA as exclusive computational resource. Wavefront phase recovery in an FPGA is an even more satisfying computational technique because recovery times result faster than GPU or CPU implementations.

A 64 × 64 wavefront recoverer prototype was synthesized with a Xilinx XC6VLX240T Virtex-6 as sole computational resource. This FPGA is provided in a ML605 evaluation platform. Our prototype was designed using ISE Foundation 13.1. The system has been successfully validated in the FPGA chip using simulated data.

A two-dimensional FFT is implemented as nuclei algorithm of the recoverer: processing times are really short. The system can process data in much lower times than the atmospheric response. This feature allows more phases to be introduced in the adaptive optical process. Then, the viability of the FPGAs for AO in the ELTs is assured.

Future work is expected to be focused on the optimization of the 2D-FFT using others algorithms (radix-8, radix-16). Finally, next-generation Virtex-7 devices provide enough DSP48 resources in order to implement all the butterflies in the 64-FFT algorithm. Using these devices, phases in the adaptive optic process could be estimated in much lower times.

7. References

Analog Devices (2003). *Analog Devices DSP Selection Guide 2003 Edition*, Analog Devices, 2003, Available from www.analog.com/processors.

Amphion (2002). CS248 FFT/IFFT Core Datasheet, Amphion Ltd., 2002. Available from www. datasheetarchive.com/500—Amphion-datasheet.html

Altera (2002). *FFT Megacore Function User Guide*, 2002, Available from www.altera.com

Bass B. (1999). A low-power, high-performance 1024-point FFT processor, *IEEE J. Solid-State Circuits*, 34, 380-387, ISSN 0018-9200.

Baik, S. H.; Park, S. K.; Kim, C. J. & Cha, B. (2007). A center detection algorithm for Shack–Hartmann wavefront sensor, *Optics & Laser Technology*, 39, 262-267, 2007, ISSN 0030-3992.

Chang T.S. & Jen C.W. (1998). Hardware efficient transform designs with cyclic formulation and subexpression sharing, *Proceedings of th 1998 IEEE ISCAS 1998*, 2, 398-401, ISBN 0-7803-4455-3, Monterey, California, USA, May 31-Jun 3, 1998.

Chien C. D; Lin C. C.; Yang C. H. & Guo J. I. (2005). Design and realization of a new hardware efficient IP core for the 1-D discrete Fourier transform *IEE Proc. Circuits, Devices and Systems*, 152, 247-258, ISSN 1350-2409.

Cooley, J.W. & Tukey, J.W. (1965). An algorithm for the machine calculation of complex Fourier series, *Mathematics of Computation*, 1965, 19, 297-301, ISSN 0025-5718.

Craven, S. & Athanas (2007), P. Examinig the Viability of FPGA Supercomputing, *EURASIP Journal on Embedded Systems*, 2007, 8, ISSN 1687-3963.

Deschamps, J.; Bioul, G. & Sutter, G. (2006). *Synthesis of Arithmetic Circuits. FPGA, ASIC and Embedded Systems*, Wiley-Interscience, 2006, ISBN 0-471-68783-9.

Frigo M. & Johnson S. (1998). FFTW: An adaptive software of the FFT, *Proceedings of IEEE International Conference On Acoustics, Speech, and Signal Processing 1998*, 1381-1384, ISBN 0-7803-4428-6, Seattle, Washington, USA, May 12-15.

Guo, J. I. (2000). An efficient parallel adder based design for one dimensional discrete Fourier transform, *Proc. Natl. Sci. Counc. ROC*, 2000, 24, 195-204.

Hawkes G. C. (2005). *DSP: Designing for Optimal Results. High-Performance DSP Using Virtex-4 FPGAs*, Xilinx, Available from www.xilinx.com.

IEEE (2000). *Standard VHDL Language Reference Manual, IEEE-1076- 2000*, IEEE, 2000, ISBN 0-7381-3326-4.

Magdaleno, E., Rodríguez, M., Rodríguez-Ramos, J.M. (2010). An efficient pipeline wavefront phase recovery for the CAFADIS camera for extremely large telescopes, *Sensors*, ISSN 1424-8220.

Mentor Graphics (2002). *FFT/WinFFT/Convolver Transforms Core Datasheet*, 2002, Available from www.mentor.com.

Meyer-Baese, U. (2001). *Digital Signal Processing with Field Programmable Gate Arrays*, Springer-Berlag, 2001, ISBN 3-540-72612-8.

Motorola (2002). *Motorota DSP 56600 16-bit DSP Family Datasheet*, Motorola, 2002, Available from www.digchip.com.

Rodríguez-Ramos, J. M.; Marichal-Hernández, J. G. & Rosa F. (2007). Modal Fourier wavefront reconstruction using graphics processing units, *Journal of Electronic Imaging*, 16, 123-134, ISSN 1017-9909.

Poyneer, L. A.; Gave, D. T. & Brase, J. M. (2002). Fast wave-front reconstruction in large adaptive optics systems with use of the Fourier transforms, *Journal of the Optical Society of America A*, 2002, 19, 2100-2111, ISSN 1084-7529.

Roddier, F. & Roddier, C (1991). Wavefront reconstruction using iterative Fourier transforms, *Applied Optics*, 30, 1325-1327, ISSN 2155-3165.

Proakis, J.G. & Manolakis, D.K. (1996). *Digital Signal Proccesing. Principles, Algorithms and Applications*, Prentice Hall, 1996, ISBN 0-13-187374-1.

Seifert, L.; Tiziani, H. J. & Osten W. (2005). Wavefront reconstruction with the adaptive Shack–Hartmann sensor, *Optics Communications*, 245, 255-269, 2005, North Holland, ISSN 0030-4018.

Sukhsawas S. & Benkrid K. (2004). A high-level implementation of a high performance pipeline FFT on Virtex-E FPGAs, *Proceedings of the IEEE Computer Annual Symposium on VLSI Emerging Trends in VLSI Systems Design 2004*, ISBN 0-7695-2097-9, Lafayette, Louisiana, USA, February 19-20.

Texas Instruments (2003). Texas Instruments C62x and C67x DSP Benchmarks, Texas Instruments, 2003, Available from www.ti.com.

Uzun I. S.; Amira, A & Bouridane, A. (2005). FPGA implementations of fast Fourier transforms for real-time signal and image processing, *IEE Proceedings - Vision Image and Signal Processing*, 152, 283-296, ISSN 1350-245X.

Vite J. A.; Romero R. & Ordaz A. (2005). VHDL Core for 1024-Point Radix-4 FFT Computation, *Proceedings of the 2005 International Conference on Reconfigurable Computing and FPGAs*, Puebla, Mexico, September 28-30, ISBN 0-7695-2456-7.

Wosnitza, M. (1999). *High precision 1024-point FFT processor for 2-D object detection*, PhD thesis, Harung-Gorre Verlag, Germany, 1999.

Xilinx (2006a). Fast Fourier Transform v3.2, 2006, Available from www.xilinx.com.

Xilinx (2006b). XST User Guide, Xilinx, 2006, pp. 118-217, Available from www.xilinx.com.

Zhou Y.; Noras, J.M. & Shepherd S. J. (2007). Novel design of multiplier-less FFT processors, *Signal Processing*, 87, 1402-1407, ISSN 0165-0684.

11

Innovative Membrane Deformable Mirrors

S. Bonora[1], U. Bortolozzo[2], G. Naletto[1,3] and S.Residori[2]
[1]CNR-IFN, Laboratory for UV and X-Ray and Optical Research, Padova,
[2]INLN, Université de Nice-Sophia Antipolis, CNRS, Valbonne,
[3]Department of Information Engineering, University of Padova, Padova,
[1,3]Italy
[2]France

1. Introduction

Nowadays adaptive optics (AO) is a very powerful technique for several scientific and technological applications, and with important economical perspectives. In the last decade, AO technology has fostered new solutions to improve adaptive components performance. While the scientific line has grown up quickly, AO has not experienced yet a wide spread commercial diffusion because a reasonable trade-off between performance and costs of adaptive components has not been found yet. This aspect makes AO a technology very attractive but still under development. In fact, many different AO technologies have been studied in the last years, as for example electrostatic membranes, electromagnetic fields, piezoelectricity, electro-striction, liquid crystals, MEMS etc. All these new technologies have peculiar properties which make them very useful and interesting for specific applications; however, the "universal" component, the one which could simultaneously satisfy all the requirements, has not been found yet.

To demonstrate its large and still undiscovered potentialities, we can remember that AO is presently a source of a wealth of experimental activities in scientific applications, as witnessed by the recent rich literature. In some cases AO is a totally new experimental tool which brings unexpected results; in others, its addition to experimental setup is the key for increasing the system performance. The main limitation of AO in many research activities is that AO systems still often require specialized operators, space for allocating devices and dedicated control PC/drivers.

In the last decade the benefit of AO in scientific experiments has been largely demonstrated. The main topics which have received a strong impulse from the introduction of deformable mirrors (DM's) are astronomy (Hardy, 1998), optical communications (Tyson, 1999), microscopy (Rueckel et al., 2006), quantum engineering (Bonato et al, 2008), coherent control (Bartels et al., 2000, Bonora et al., 2010), and femtosecond lasers for pulse compression (Brida et al., 2010) and focalisation (Villoresi et al., 2004). However, the use of DM's in these experiments differs case by case: in fact, each experiment has its own specific requirements depending on correction speed, system stability, amount and resolution of the correction and beam size. For this, recent researches report on the realization of many different DM technologies (Dalimier & Dainty, 2005) in order to address either higher stroke (Bonora & Poletto, 2006), or higher resolution (Bifano, 2011; Bortolozzo et al., 2010), or different shapes.

Among the several AO recent technological developments, we can highlight the capability of deforming a correction mirror in a suitable way, but without having the information about the actual necessary deformation to optimally correct the wavefront. On this respect, let us remind as an example that the most classical AO scientific application, that is astronomy, uses DM's to reduce the impact of atmospheric turbulence on image quality; since this is a rapidly variable and a-priori unknown phenomenon with frequencies up to a few kHz, usually the DM deformation is commanded by means of the information provided by a fast wavefront sensor (WS) which monitors in real time the difference between the actual and the ideal wavefront. Such AO systems need the real-time knowledge of the deformation to be induced to the DM, and could not work without a WS. On the other hand, there are many other AO applications where the necessary mirror deformations are nearly static, and for which the use of WS can just be an additional experimental complexity (Rueckel et al., 2006). In these cases, it can be more convenient to monitor just an experimental parameter, and to optimize a suitable merit function depending on it. For this and other reasons, sensorless AO techniques (that is AO setups which do not use a WS) have been developed, obtaining extremely good results, usually at the expenses of only a reduced correction speed. Among the others, the sensorless AO techniques which makes use of optimization algorithms have been very successful: for example, since their first use (Judson & Rabitz 1992) random search algorithms (genetic, simplex or simulated annealing algorithms) demonstrated to be very effective in femtosecond lasers for both parametric amplifiers pulse compression (Bartels et al., 2000), focalization (Villoresi et al., 2004), and microscopy (Wright et al., 2007). Also the capability of realizing a fast analysis of a reference point image has been successfully used in some applications (Naletto et al., 2007; Grisan et al., 2007).

Among the various techniques developed during the last years to realize deformable optics, membrane DM's is probably becoming the most popular one, because of its extremely valuable performance and potentialities, its relative easiness of use, its handiness, and its low price. Presently, membrane DM's have been successfully used in many scientific applications, such as visual optics and laser focalisation or femtosecond laser compression experiments, and can now be found either in round or rectangular shape (see Fig. 1). Thanks to the rapid increase of fields of applications, and to the continuous technological development, it is reasonable to expect a rapid expansion of their use also in more diffused and commercial applications.

The technology of electrostatic membrane DM's consists in realizing a thin metal (silver, gold or aluminum) coating over a nitrocellulose, polyamide or silicon membrane. The membrane is then usually pre-tensioned and glued over a metallic frame (Bonora et al., 2006), and the frame is finally faced to the electrode pads by means of calibrated spacers to maintain the nominal distance between membrane and electrodes. The number of electrodes usually ranges between 19 and 64, to form a round or rectangular footprint as illustrated in Fig. 1. The electrodes are connected to a voltage amplifier which generates independent voltage signals between 0 V and 300 V. The membrane DM controller has to independently manage variable voltages for all the electrodes, on the basis of the information provided either by a suitable WS or by the adopted sensorless technique.

1.1 Applications of deformable membrane mirrors

Many are the scientific applications in which membrane DM's are presently successfully used. As an example, (Villoresi et al., 2004, Bonora et al., 2011b) reports on the optimization of laser high order harmonics generation (HHG) by the interaction of a femtosecond laser beam and a gas jet. In these experiments, the HHG signal is generated by the multi-photon ionization of gas atoms driven by the strong electric field of the laser. This extreme non-linear effect is very sensitive to laser intensity, and both WS and sensorless AO techniques have been used to improve laser focalisation. Fig. 2 shows the differences between the two experimental setups used for the optimization of the laser focusing, with and without WS. These experiments demonstrated that the sensorless technique can be very performing, even if the convergence of the algorithm requires a very high signal stability and a high laser repetition rate. To be effective in this kind of setup, fast DM systems based on embedded electronics architecture have been developed (Bonora et al., 2006a).

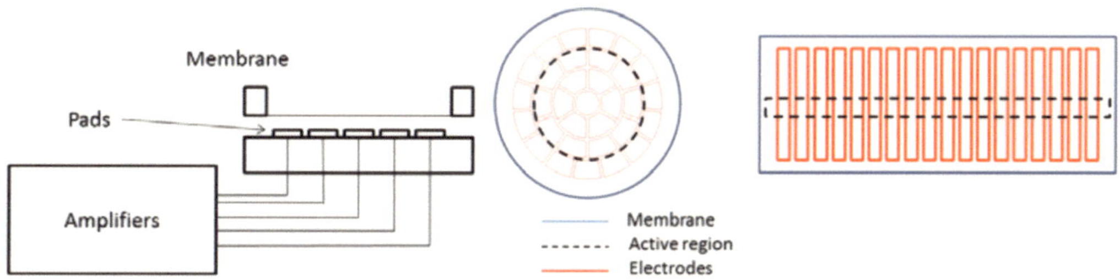

(a) (b)

Fig. 1. Left panel: electrostatic membrane mirror section view. Right panel: membrane active region (delimited by the dashed lines) and actuator footprint; the shown shapes of the actuators are only indicative examples.

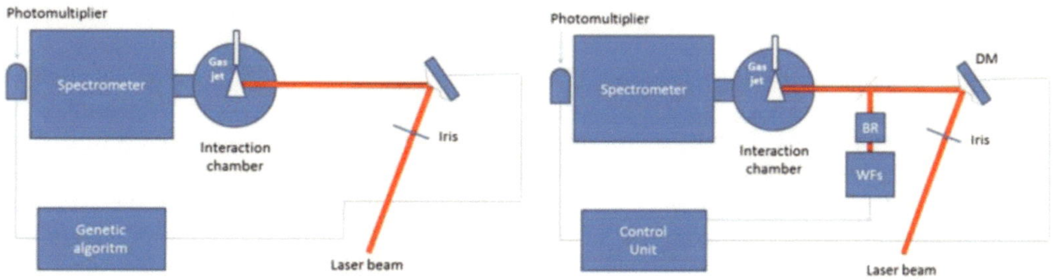

(a) (b)

Fig. 2. (a): schematic of the high order harmonics generation (HHG) experimental setup with sensorless AO system. (b): schematic of the HHG experimental setup for wavefront sensor (WS) AO closed loop system. DM, Deformable Mirror; BR, Beam Reducer; WFs, Wavefront Sensor.

Another interesting application of membrane DM's has been realized in ultrafast laser pulse compression experiments, through a suitable preliminary characterization of the experimental setup by means of interferometric measurements. The review of Brida et al. (2010) reports many examples of pulse compression of parametric amplifiers based on spectral phase measurement, using the frequency resolved optical gating technique in different spectral regions. In these experiments, a rectangular membrane DM is used to change the Optical Path Length (OPL) of each spectral component of the parametric source in order to cancel the dispersion and to achieve the shortest laser pulse duration (the so-called transform limited pulse). Fig. 3 illustrates the experimental layout of the adaptive compressor: it is composed of a dispersing element, either a prism or a diffraction grating, followed by a spherical collimating mirror which reflects each colour component in a different position on a rectangular membrane DM. The spectral phase added by the DM to each colour component is given by the formula

$$\Delta\varphi(x) = 2\frac{2\pi}{\lambda(x)}\delta z(x), \tag{1}$$

where $\delta z(x)$ is the membrane displacement in the position x relative to the wavelength $\lambda(x)$.

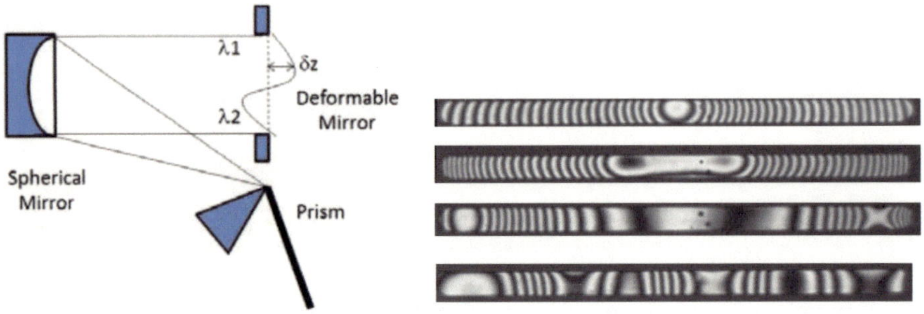

(a) (b)

Fig. 3. (a): 4-*f* arrangement of a prism compressor for femtosecond optical parametric amplifiers. (b): examples of interferograms of the rectangular deformable mirror during the experiment characterization: these interferograms allowed to determine the mirror deformations necessary to optimize the pulse compression.

Another successful application of membrane DM's, where the shape to be given to the DM was known a-priori, and the deformation set to be given to the DM was pre-determined by interferometric measurements, was carried out in quantum optics experiments (Bonato et al., 2006; Bonato et al., 2010). For example, in Ref. (Bonato et al., 2010), a membrane rectangular DM was used for the stabilization of the phase of a two-photon entangled state. The experimental setup (see Fig. 4) consisted in an interferometer in which random fluctuations of the OPL resulted in detrimental fluctuations of the relative phase of the quantum superposition. The membrane DM was used as the optical element able to modulate the relative OPL of the two arms of the interferometer. In order to achieve that goal, two small areas on the linear DM were used as flat plane mirrors whose relative distance could be adjusted. The corresponding DM deformation was obtained thanks to the preliminary determination of the influence functions by means of an interferometer and then by finding the electrode voltages providing the deformations, as shown in Fig. 5. The thick blue lines in this figure represent the two flat mirror areas where the beams are reflected.

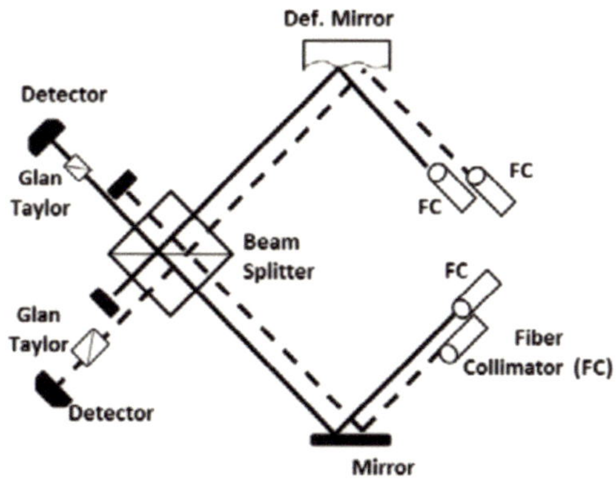

Fig. 4. Experimental apparatus for the quantum interferometer stabilisation.

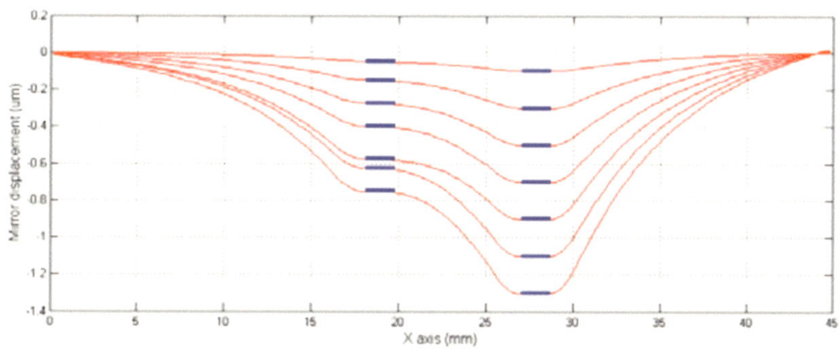

Fig. 5. Measurement of the set of deformations which create a displacements between the blue reference planes.

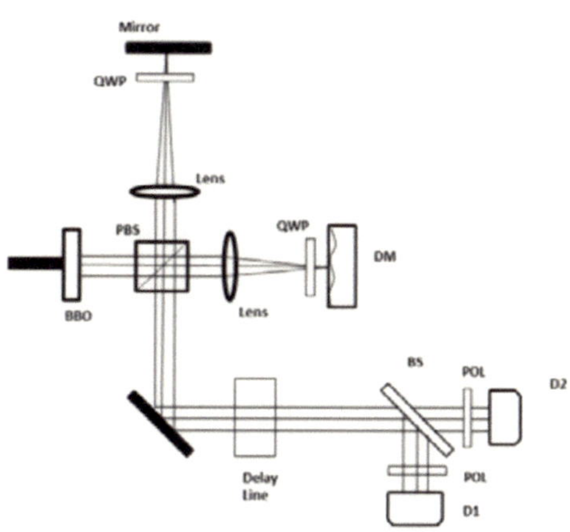

Fig. 6. Hong-Ou_Mandel quantum interferometer which was used to demonstrate the even order aberrations cancellation. The membrane DM is positioned on one arm of a 4-*f* quantum interferometer (BS: beam splitter, QWP: quarter waveplate, BBO: beta barium borate crystal, Pol: polarizer, D: detector).

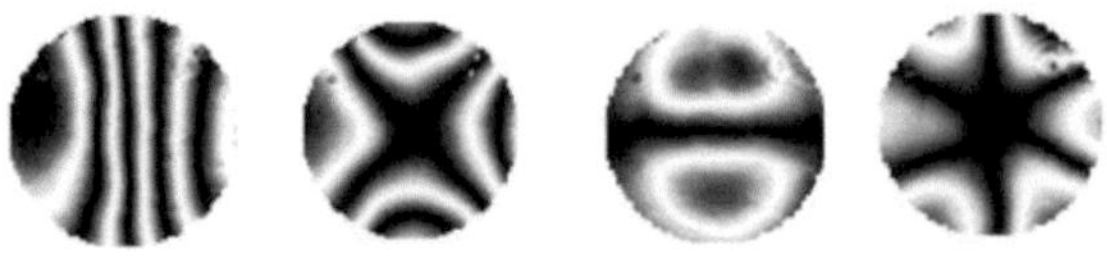

Fig. 7. Measurements of the membrane deformations given by the round membrane DM to demonstrate the even order aberration cancellation.

Another experiment which was carried out by preliminary membrane DM calibration is shown in Fig. 6. This experiment reports on the the first experimental demonstration of the even-order aberration cancellation effect in quantum interference applications. A round membrane DM was used in a 4−*f* AO system to modulate the distribution of wavefronts emitted by a spontaneous parametric down conversion process, so effectively introducing a controllable degree of wavefront aberration. This was used to show that even-order aberrations, such as astigmatism and polynomial (4,4), do not affect the interferograms, and that the quality of the quantum interference only depends on odd-order aberrations (Bonato et al., 2006). Fig. 7 shows the interferometric results of the measurements of the membrane deformations for some of the considered aberrations.

1.2 Membrane deformable mirror control

To control a membrane DM, with either round or rectangular shape, it is necessary to characterize it through the acquisition of the influence function matrix (IFM). Since the electrostatic membrane mirror can just pull the membrane, for allowing the aberration compensation on both directions it is necessary to work biasing the membrane to its middle: with respect to this bias ("starting") position, it is possible to both pull and push the membrane. The IFM can be considered as a basis of linear independent vectors to represent all the possible deformations of a DM. More in detail, given N actuators and a bi-dimensional wavefront domain of rank $M = m_x \times m_y$, the $M \times N$ IFM A is obtained by placing in each column the wavefronts w_1, w_2, ..., w_N, corresponding to the measured wavefronts w_i, a vector of rank M, obtained by activating the ith actuator with a unitary voltage, and called "influence function". By using this definition, and remembering that the electrostatic pressure depends on the square of the applied voltage, it is easy to write any mirror deformation as a linear combination $w = Av$, where v is an M rank vector of the square applied voltages. Fig. 8 shows some examples of influence function interferograms obtained by applying a unitary voltage to the membrane DM's shown in Fig. 1.

In order to know which voltages have to be applied to the actuators to obtain a specific DM deformation, the problem of the IFM inversion has to be solved. This is realized, for example, with the so-called pseudo-inversion standard technique, by means of the Singular Value Decomposition (SVD), as follows. Given the matrix A, it can be factorized as

$$A = U\Sigma V^* , \tag{2}$$

where U is an $m_x \times m_x$ unitary matrix, the $m_x \times m_y$ matrix Σ has non-negative numbers on the diagonal and zeros off the diagonal, and V^* denotes the conjugate transpose of V, an $m_y \times m_y$

unitary matrix. At this point, we can use the rule for which the pseudo-inverse of the matrix A with singular value decomposition $A = U\Sigma V^*$ is:

$$A^+ = V\Sigma^+ U^*, \tag{3}$$

where Σ^+ is the transpose of Σ with every nonzero entry replaced by its reciprocal.

The application of this technique to the specific problem is that given the IFM A of the influence functions, the matrix V represents an orthogonal base composed by the modes of the mirror. The values in the diagonal matrix Σ are the gains of each mode. The gain values are in decreasing order, and the modes are ordered from the lowest spatial frequency to the highest one. Therefore, the larger gain corresponds to the lower spatial frequency mode. Fig. 9 shows the interferograms of the modes for a round membrane DM.

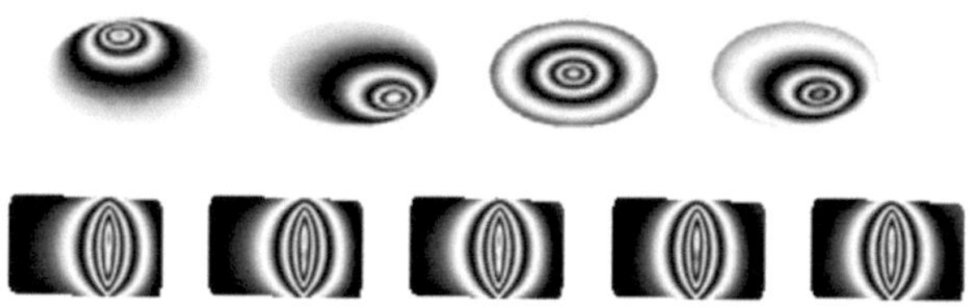

Fig. 8. Examples of interferograms of the influence functions generated by the actuators for a round membrane DM (top) and for a rectangular membrane DM (bottom).

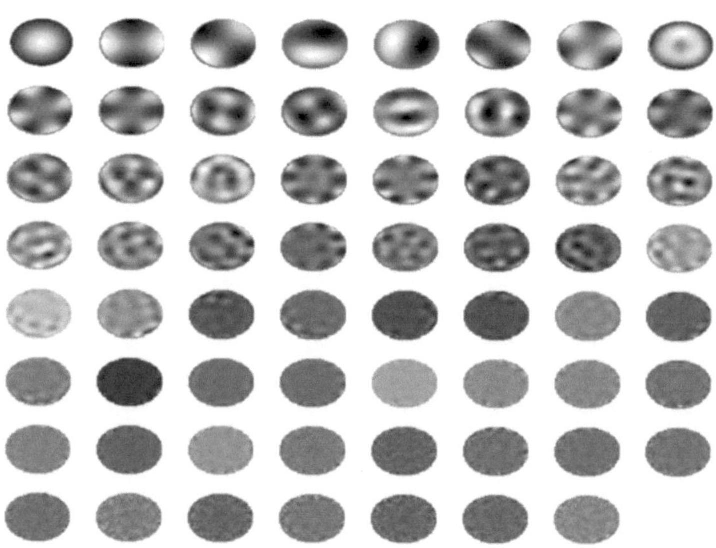

Fig. 9. Modes relative to a 61 electrode membrane DM.

At this point, the fit of the wavefront passes through the determination of the ideal number of necessary modes. Ideally the higher the number of modes the better the reconstruction, because using a high number of modes means to be able to reproduce with better accuracy the high spatial frequency mode. However, in practise there are two main limitations. The first is that the voltage range is usually limited: so, since the high order modes have a small gain, using a large amount of modes usually makes the algorithm saturating the voltages. The second is the presence of noise: usually noise measurements or artefacts get coupled with the highest orders, driving the algorithm to wrong solutions and to voltage saturation.

Fig. 10 shows the root mean square (RMS) deviation of the generated wavefront from the ideal correction against the number of used modes. It is evident that using just a few modes is not sufficient to fit the solution with enough accuracy, while on the contrary the reconstruction error can be very low if a number of modes between 10 and 22 is used. However, as explained before, it is also clear from this plot that the noise tends to reduce the system performance when a higher number of modes is used. The standard deviation of the voltages with respect to the bias level against the number of used modes is illustrated in Fig. 11 (the maximum voltage with respect to bias is 64 V). Comparing Figures 10 and 11, it is easy to note that after about 20 modes the voltage level increases with a deterioration of the results.

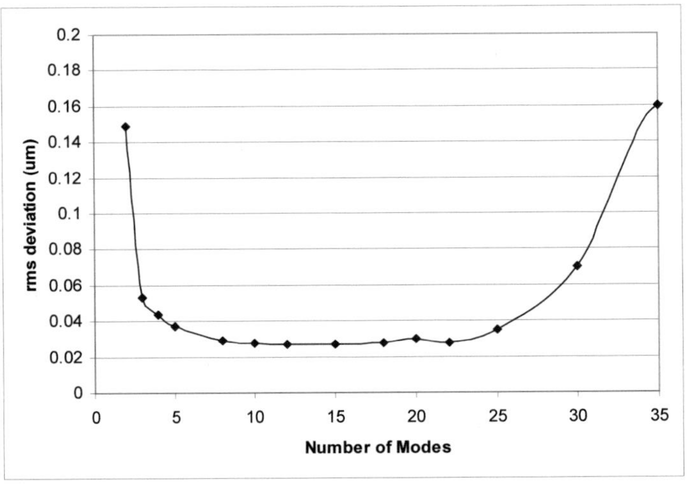

Fig. 10. RMS error in the fit of the solution against the number of modes used.

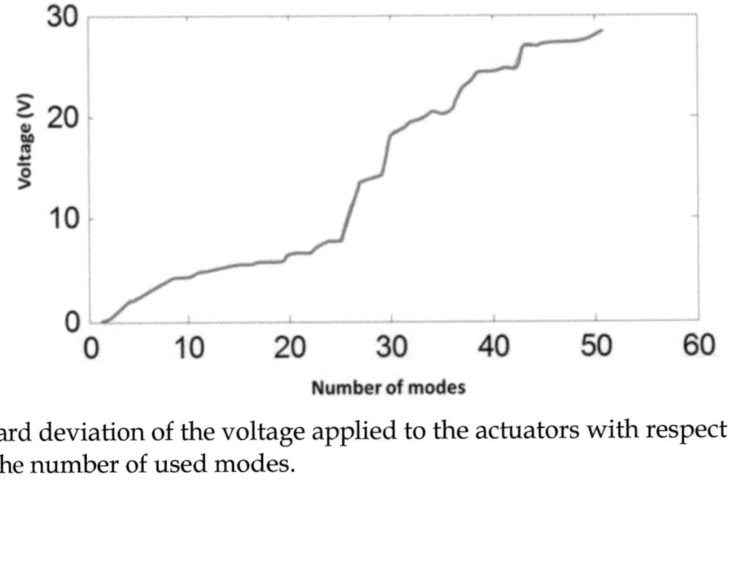

Fig. 11. Standard deviation of the voltage applied to the actuators with respect to the bias level, versus the number of used modes.

2. Resistive actuators deformable mirror for sensorless applications

Sensorless applications play an important role in the AO development thanks to the reduced hardware complexity. With this technique it is possible to improve the image quality or, for example, to increase the efficiency of a laser process affected by aberrations, through the optimization of a specific merit function. The resistive Modal Deformable Mirror (MDM) is an electrostatic membrane DM where the actuators are composed by a resistive layer which continuously distributes the electrostatic pressure on the membrane. An example of actuators layout optimized for the direct generation of aberrations with the minimum number of actuators is illustrated in Fig. 12 (a). Its convenience and flexibility of use have been recently demonstrated in two completely different fields such as visual optics for the optimization of image quality (Bonora, 2011a) and non-linear ultrafast optics for the optimization of the harmonics generation in gases.

The MDM addresses two problems: 1) the use of the smallest possible number of actuators, 2) the introduction of the DM modal control in a continuous actuators arrangement. The main feature of the resistive MDM is that the actuators response is directly related to the optical aberrations allowing for a more versatile and straightforward use than conventional discrete actuators deformable mirrors. In the MDM, the voltages necessary to drive the mirror can be directly computed from the Zernike decomposition terms of the target aberrations, leading to an ideal device for modal control in sensorless applications.

As an example of resistive MDM, we can consider a membrane electrostatic DM composed by a silvered 5 μm thick nitrocellulose membrane suspended 70 μm above the actuators by some spacers. The prototype described in ref. (Bonora, 2011a) mounts a membrane of 19 mm diameter designed for an optimal deformable active region of 10 mm diameter. This prototype is driven by a DM multichannel electronic driver (Adaptica IO32) which can supply up to 260 V over 32 channels; in order to generate both positive and negative deformations, the membrane is connected to a voltage reference of 130 V.

The DM is actuated at the position (x,y) by the electrostatic pressure $p(x,y)$ between the actuators and the metalized membrane which deforms the mirror surface $M(x,y)$ according to the Poisson equation

$$\Delta M(X,y)=\frac{1}{T}p(x,y)\,, \tag{4}$$

where T is the mechanical tension of the membrane.

The device is composed by 3 actuators placed on three concentric rings (see Fig. 12). The actuators are composed by a 35 μm thick graphite layer which presents a sheet resistance of 1 MΩ/inch^2 which continuously distributes the voltage. The estimated current for each channel when applying the maximum voltage, is about 60 μA with a power consumption for each actuator of about 10 mW.

The voltage distribution over the resistive layer with resistivity ρ can be computed solving the Laplace equation for the scalar electric potential U:

$$\frac{1}{\rho}\Delta U(x,y)=0\,. \tag{5}$$

By a proper design of the MDM, actuator 3 (see Fig. 12(a)) can be used to generate piston, tilt and astigmatism, actuator (2) can be used for the generation of coma and defocus, and actuator 1 for the generation of spherical aberration. In order to solve both formulas (4) and (5) we applied the recursive finite difference method. Fig. 13 reports some examples of simulated voltage distributions, and of the corresponding electrostatic pressure and mirror deformation for each actuator ring.

Following Bonora (2011a), the resistive MDM response can be directly derived from the Zernike decomposition of the incoming wavefront, realizing a modal control of the DM, rather than from the influence functions (zonal control). Thus, the voltages V_{MDM} which generate the aberrations described by the Zernike coefficients are determined by the algebraic sum of the voltages which generate each single aberration defocus, astigmatism, coma and spherical aberration:

$$V_{MDM} = V_{defocus} + V_{astigmatism} + V_{coma} + V_{spherical}\,. \tag{6}$$

By using (6), it is possible to generate any aberration starting from the knowledge of its Zernike spectrum.

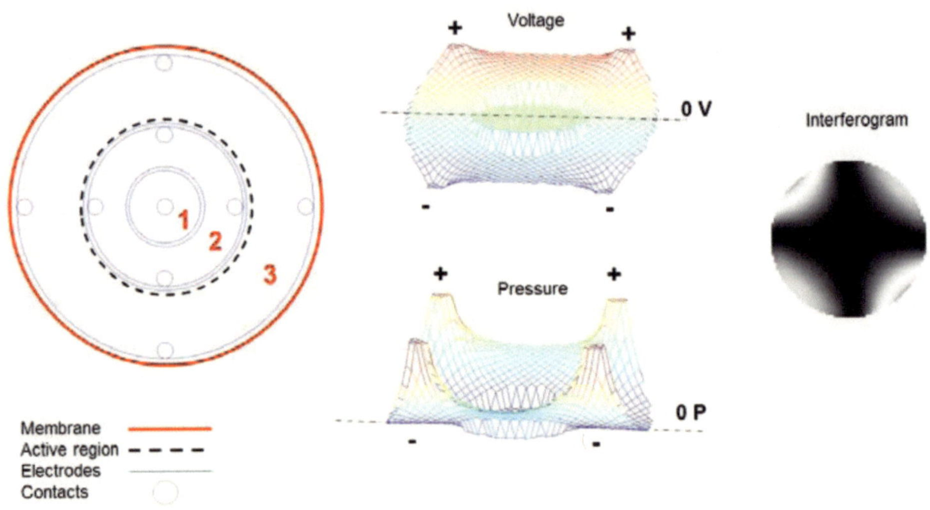

(a) (b) (c)

Fig. 12. (a): Layout of the electrodes of the resistive MDM. The two external ring actuators, number 2 and 3, have four symmetrical electrical contacts each, while the central one, number 1, has a single central electrode. (b): voltage distribution and relative induced electrostatic pressure generating an astigmatic mirror deformation. (c): interferogram of the MDM surface deformed by the electrostatic pressure shown in (b). The application of both positive and negative voltages generates a zero voltage distribution (dotted line 0 V) in the middle of the membrane which corresponds to a zero electrostatic pressure (dotted line 0 P).

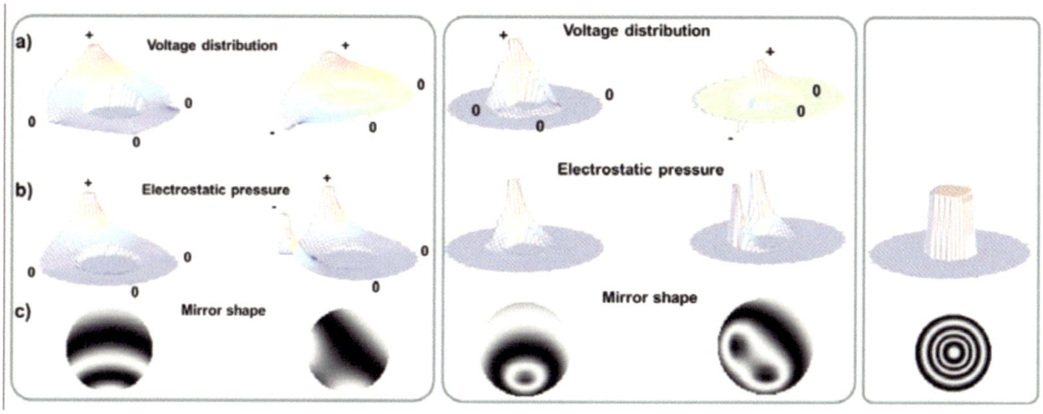

Fig. 13. Simulation results. Left box: examples of voltage distribution (top), electrostatic pressure (middle) and membrane shape (bottom) using the outer ring. Central box: as for the left box, using the central actuator ring only. Right box: as for the previous, applying voltage only to the central actuator.

2.1 Control algorithm

The properties of the MDM allow to use a sequential aberration correction algorithm, that is an algorithm which allows to correct each single aberration in a sequence, as illustrated by the block diagram of Fig. 14. As reported by Debarre & Booth, 2007, image sharpening functions related to low frequencies can be expanded as the quadratic sum of the aberration coefficients. Exploiting this property, it is possible to see that the aberration correction: a) can be operated by the maximisation of the sharpening function of each single aberration; b) is independent of the order of the optimization sequence; c) can be resolved by three single measurements. The optimization strategy starts with the optimization of a merit function applying a defocus to the MDM; then, fixed the best defocus value, the optimization is carried out to astigmatism, then to spherical aberration and finally to coma. This algorithm has been applied successfully in visual optics (see section 2.3.2) and laser focalisation experiments (see section 2.3.3).

2.2 Examples of application of the resistive modal deformable mirror

2.2.1 Application of resistive MDM for eye aberration correction

To evaluate the performance of a resistive MDM for the compensation of ocular aberrations, we used the above established principles, as reported by Thibos et al. (2002). To have a better idea about the actual potentialities of the MDM, a comparison of its correction performance has been realized with a commercially available 32-actuator electrostatic membrane DM (Pan DM, Adaptica srl) (Bonora et al., 2006a). The actuator layout of the two DM's is shown in Fig. 15.

The performance comparison has been realized by analysing the DM's maximum stroke necessary to correct a sample eye population with the same average residual. The results reported in Fig. 16 show the RMS wavefront deviation before and after correction, for a population of 50 eyes. From this comparison, it can be noticed that the resistive MDM is equivalent to the commercial Pan DM with a stroke reduced by 25%.

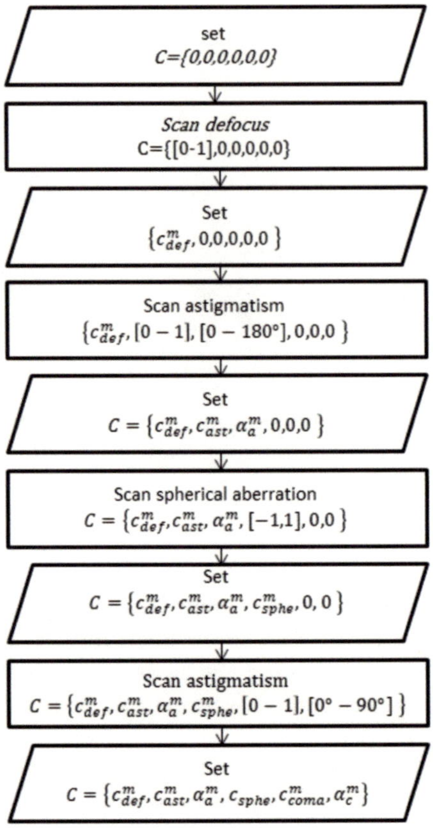

Fig. 14. Block diagram showing the sequence of the aberration correction algorithm implemented on the MDM.

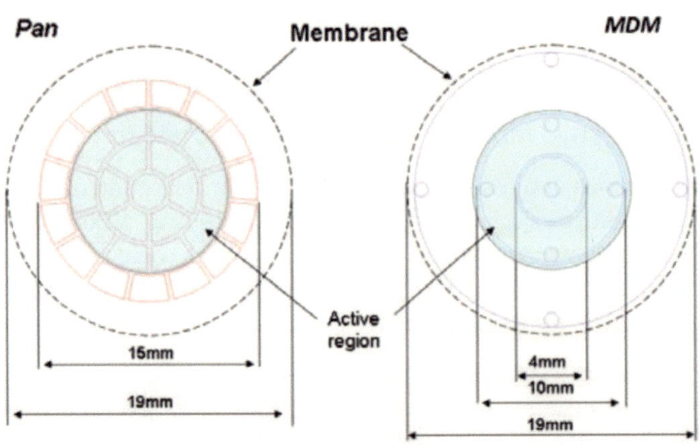

Fig. 15. Actuator layout for both the commercial Pan DM (left) and the resistive MDM (right).

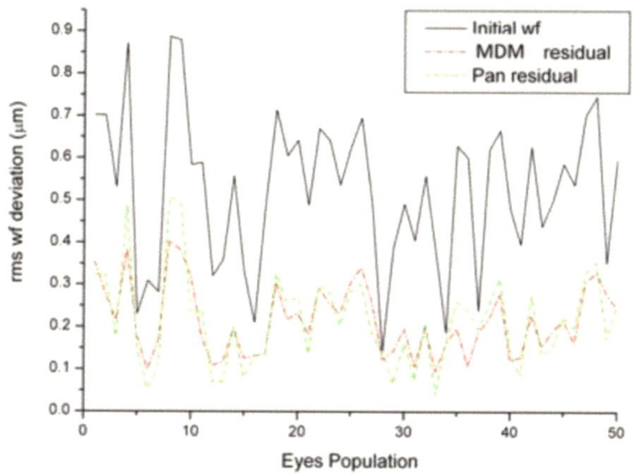

Fig. 16. RMS wavefront deviation for the aberrations of a sample of 50 eyes before and after DM correction. The two DM's under test are a resistive MDM and a commercial discrete electrodes DM.

2.2.2 Visual optics application of resistive MDM

An experiment of sensorless optimization using a resistive MDM is reported by Bonora, 2011. In this experiment, a sample image (a honeybee leg) has been altered by the introduction of an aberrated phase plate. The result is a degradation of the image quality as shown in Fig. 17: the top row shows an interferogram, the image and the Point Spread Function (PSF) after the introduction of the aberrated phase plate and defocus correction; the bottom row shows the same optical parameters at the end of the correction, realized through a 30 acquisition process.

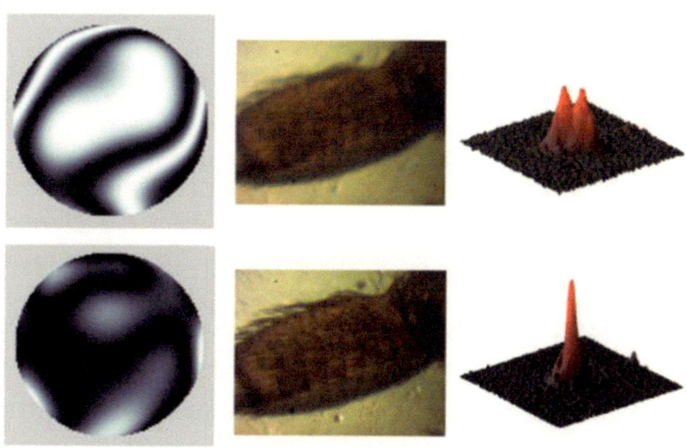

(a) (b) (c)

Fig. 17. Improvement of the quality of an aberrated honeybee leg image by the use of a resistive MDM. The top row corresponds to the situation after the defocus correction, while the bottom one corresponds to the status at the end of the correction process. Column (a) shows the interferograms of a central image point; column (b) shows the images; column (c) shows the PSF at the center of the field.

As can be clearly seen, by the application of the low spatial frequency image sharpening function and of the algorithm described in Fig. 14, it has been possible to largely increase the image quality, with a significant reduction of the image aberrations.

Fig. 18 shows the evolution of the selected merit function during the scan of the defocus, astigmatism, spherical aberration, and coma performed during the correction process. Each scan was carried out with 5 acquisitions and interpolated with a polynomial curve to find the aberration value which maximizes the merit function.

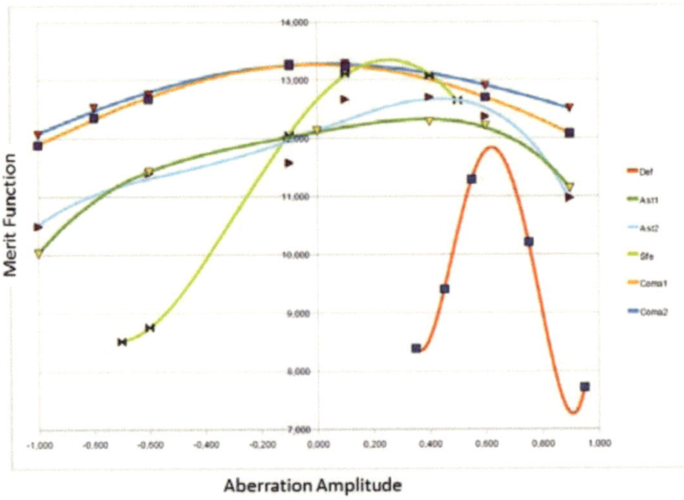

Fig. 18. Evolution of the merit function during the scan of the defocus, astigmatism, spherical aberration, and coma during the correction process.

2.2.3 Resistive MDM for laser focalisation application

Another interesting application of a resistive MDM was carried out to optimize the harmonic generation from the interaction of a tunable high energy mid-IR optical parametric amplifier (OPA) with a krypton gas jet (Vozzi at al., 2007, Bonora et al., 2011b). In this case, a fast sensorless optimisation system is the key for the aberration correction: in fact, the alternative solution with a mid-IR WS is possible but at the price of a higher cost and an increase of the experimental complexity; moreover, the relatively low repetition rate of the amplified systems effectively results in the practical impossibility of using other algorithms, as for example the random search one, which needs hundreds of iterations.

The experimental setup used for this application is illustrated in Fig. 19. To demonstrate the easiness of implementation of this device within the experiment, the optical path before the DM is shown with a dotted line. The additional elements are simply a plane mirror and the resistive MDM, which have been introduced without complex operations. The obtained system optimization consisted in an increase of the harmonic signal detected by the photomultiplier at the output of the monochromator. This result is illustrated in Fig. 20, where it is possible to see that the photon flux on the photomultiplier is doubled with respect to the one obtained after the defocus correction.

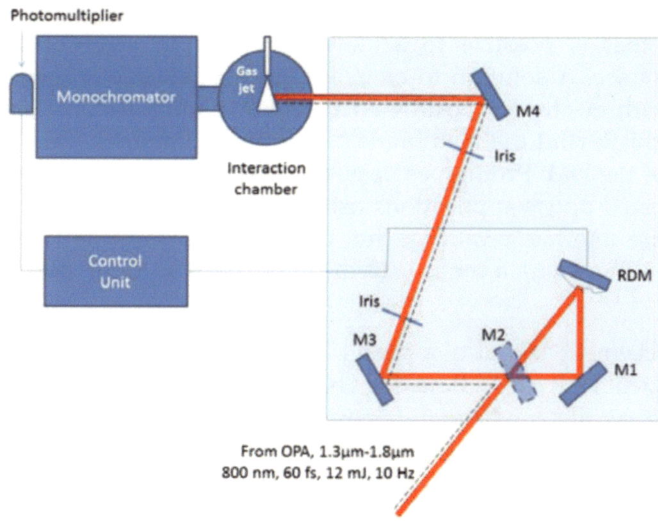

Fig. 19. Experimental setup for the generation of harmonic from a femtosecond tunable high energy mid-IR OPA. Dotted line: optical path before the experiment with the MDM. Red line: optical path realized for the experiment with the deformable mirror.

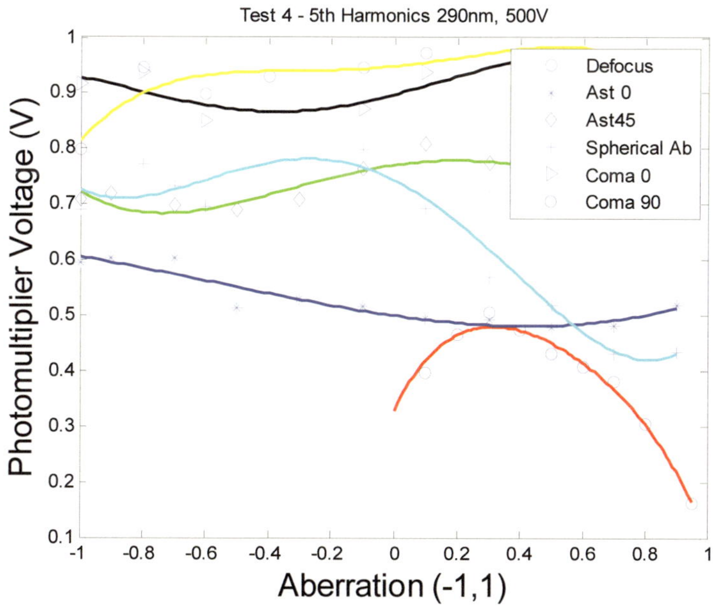

Fig. 20. Optimization of the voltage generated by the photomultiplier over a 50 Ω load for the 5th harmonic at 290 nm, obtained by the use of krypton gas.

3. Push-pull membrane deformable mirror

Membrane DM's are rather diffused thanks to their simplicity, low cost and the bipolarity of the reproduced deformation that allows closed loop operations without biasing. A disadvantage which can be encountered when using membrane DM's is the not so large amount of stroke that is possible to achieve, and that in some cases can limit the instrument performance. A solution to overcome this problem has been found realizing membrane DM's with electrodes positioned both on the bottom and on the top of the membrane. Using this layout the membrane can be pulled on both sides, thus increasing the optical power of the DM. Prototypes of push-pull membrane electrostatic mirror have been realized for visual optics applications using a transparent front electrode (Bonora & Poletto, 2006), while another prototype for pulse compression and shaping has been realized in a square shape with a central slit to get the light onto the membrane (Bonora et al., 2006b).

Fig. 21 shows the schematic layout of a push-pull membrane DM. In this DM, the top-side electrodes are realized with an Indium-Tin-Oxide (ITO) coating, which is transparent to visible light and electrically conductive, deposited on a front disc glass, whereas the back electrodes are printed on a standard electronics Printed Circuit Board (PCB). A sketch of the electrode patterns is depicted in Fig. 22. As demonstrated in Vdovin et al., 2006, a membrane DM is more effective when outside the active area there is at least one ring of actuators. For this reason the active region of this DM corresponds to the glass window in the front side actuators. Some examples of electrodes deformations are given in Fig. 22 for both the front and the back side electrodes.

To generate controlled large mirror deformations with push-pull membrane DM's, Bonora & Poletto, 2006 reports an interesting algorithm which exploits electrodes saturation (see block diagram in Fig. 23). In this algorithm, the electrostatic pressure p, which induces the membrane shape, is initially calculated by the pseudoinversion of the influence functions matrix A starting with the target shape Z; then all the electrode voltages that should nominally be above the saturation threshold are fixed to the saturation value, and all the others are iteratively used to decrease the membrane shape residual. A comparison between the pull-only electrostatic DM and the push-pull electrostatic DM implementing this algorithm has been carried out for the aberrations up to the 4th order: as illustrated in Fig. 24, the push-pull DM demonstrated to have nearly twice the optical power than the pull-only DM.

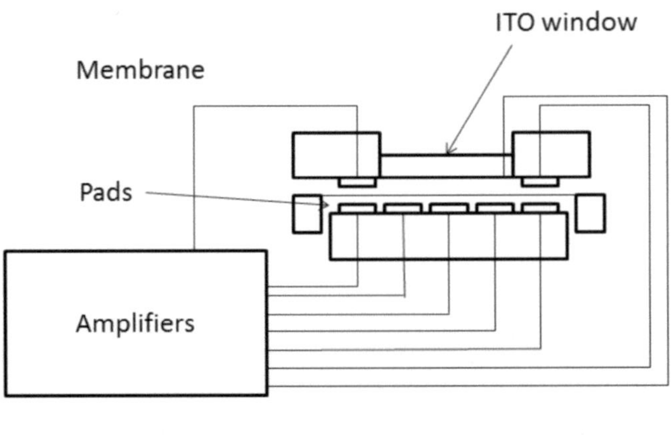

Fig. 21. Schematic layout of a push-pull DM cross section.

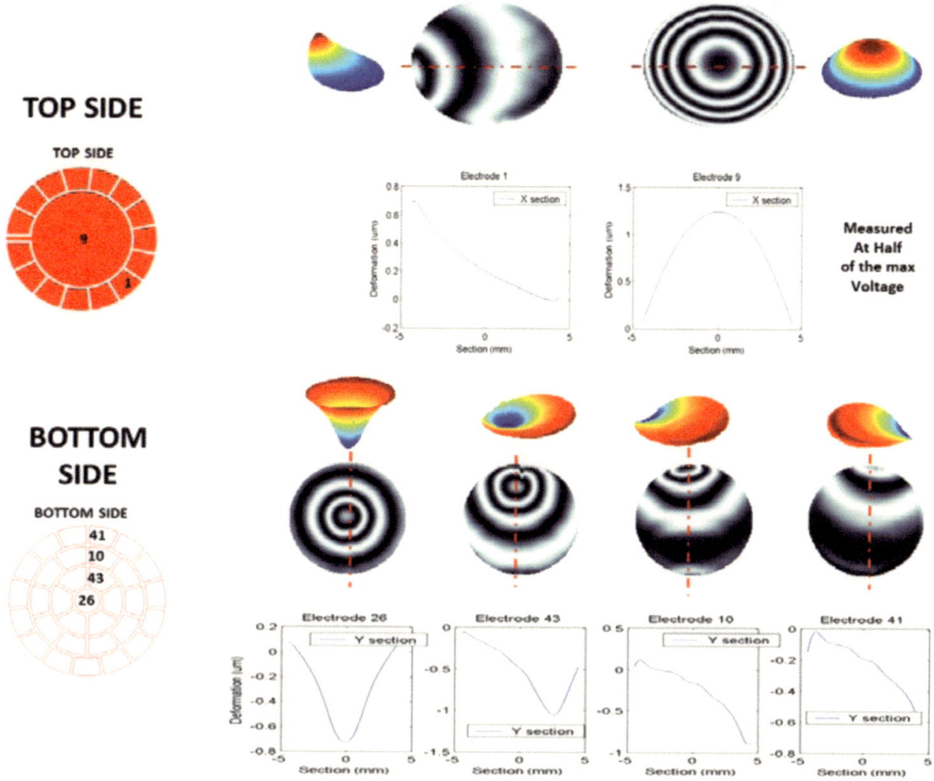

Fig. 22. Front side and back side electrodes footprint (left) and influence functions (right).

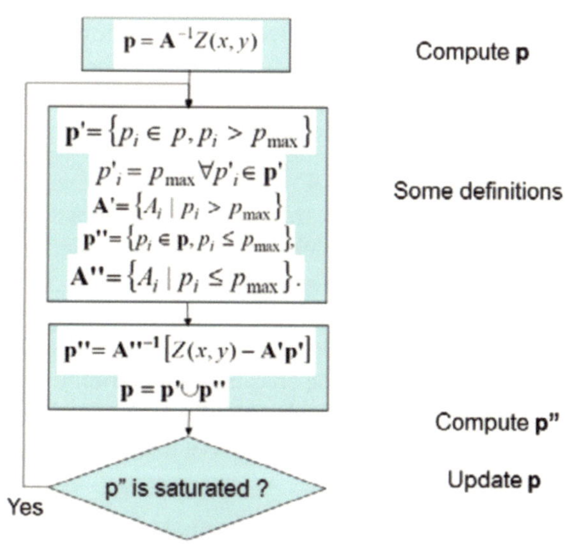

Fig. 23. Block diagram for the exploitation of the saturation in the calculation of the electrostatic pressure "*p*". *A* is the influence functions matrix, and *Z*(*x*,*y*) is the aberration target.

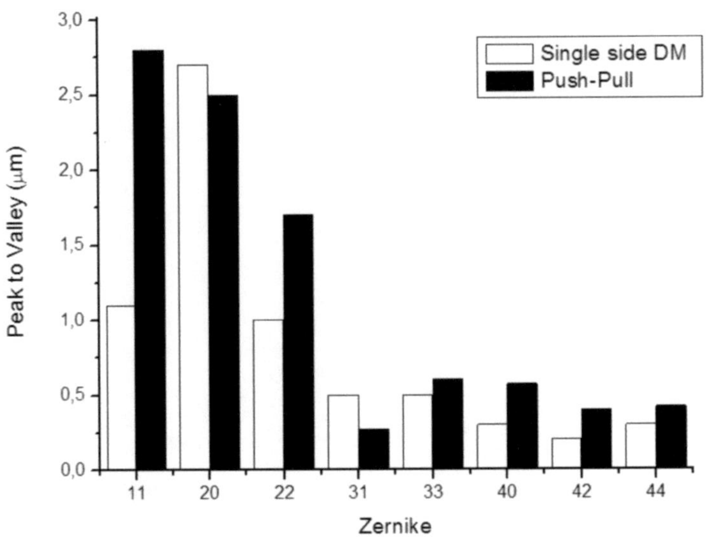

Fig. 24. Comparison between the pull-only and the push-pull membrane DM's in the generation of Zernike polynomials.

3.1 An application of the push-pull membrane DM

Push-pull linear membrane DM's have been found an interesting application in femtosecond parametric amplifiers compression and shaping (Weiner, 2000). In fact, as reported by Brida et al., 2010, using DM's is the optimal way to manipulate the broadband radiation coming from an ultrabroadband OPA: it has been shown (see for example Zeek et al., 1999, Brida et al., 2010) that the use of a DM as spectral phase modulator has significant advantages with respect to acousto-optics modulators and liquid crystals, as for example high efficiency and achromaticity, at the expenses of a slightly smaller resolution.

The pulses produced by ultrabroadband OPA with the linear membrane DM's have found many important scientific applications in material science to study physical phenomena which happen on the femtosecond timescale, such as for example the study of the quantum dynamics of photoelectrons in Mott insulators (Wall et al., 2001). Unfortunately the limited stroke of the used membrane DM's allowed just to realize pulse compression and not also pulse shaping. The recent use of push-pull membrane linear DM's (Brida et al., 2010) sorted this limit out, demonstrating that also pulse shaping can be obtained with a push-pull DM. The device layout is illustrated in Fig. 25: the back side electrodes are rectangular and fill the whole membrane area, while the front electrodes have a small slit in the centre to allow the OPA light to reach the reflective membrane.

Using this push-pull DM, it was demonstrated that it is possible both to compress ultrabroadband pulses and to shape their temporal profile. Couples of 20 fs pulses with temporal separation tunable between 0 to 150 fs have been measured and used to control the dynamics of LD690 dye molecules.

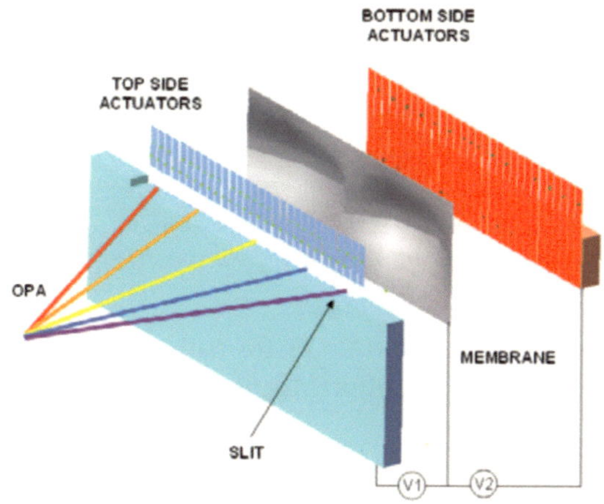

Fig. 25. Layout of the push-pull deformable mirror for pulse shaping with OPA.

4. Photo-controlled deformable membrane mirror

A new type of membrane DM is made by associating a metalized membrane with a monolithic non-pixelated photoconductive substrate (Bortolozzo et al., 2010). The assembly constitutes a continuous photo-controlled deformable mirror (PCDM), which is driven by sending suitable light intensity distributions onto its photoconductive side, opposite to the reflection side. This approach, eliminating the spatial segmentation of the driving elements, provides a continuous photo-addressing of the mirror and thus largely simplifies the driving electronics; in addition, at the same time, it allows to realize more flexible configurations.

As an example, a PCDM made by using a photorefractive $Bi_{12}SiO_{20}$ (BSO) crystal as a photosensitive element is depicted in Fig. 26. The BSO is cut in the form of a thin disk, 1 mm thickness and 35 mm diameter, coated on one side with an Indium-Tin-Oxide (ITO) transparent electrode; the facing membrane is a nitrocellulose layer, 19 mm diameter and 5 μm thickness, metalized by an Ag coating and mounted on a rigid aluminum ring. Mylar spacers are introduced between the BSO and the ring supporting the membrane, in order to provide a gap of a few tens of microns. When an ac voltage is applied to the PCDM, the electrostatic pressure across the gap attracts the membrane towards the BSO substrate. Since the impedance of the BSO decreases with the light intensity, when the BSO is uniformly illuminated a large deformation in the form of a paraboloid is induced on the membrane. Once the membrane has reached an equilibrium position, further deformations can be superimposed by local point-like illuminations of the BSO. PCDM with gaps of 50 μm were realized as a good compromise between the optimization of the capacitive effect and the maximum allowable deformation before the membrane snaps down on the photoconductive substrate. The membrane deformation was measured by using a visual interferometric profilometer (Zygo GP-LC). When the voltage is applied across the mirror, the membrane deformation is directly seen as a radial displacement of the fringe pattern.

From the measured phase change $\Delta\varphi$, the maximum membrane deformation Δx can be derived as:

$$\Delta\varphi = \frac{2\pi}{\lambda} 2\Delta x \,, \tag{7}$$

where λ is the optical wavelength. Uniform illumination of the BSO side with an expanded laser beam at 474 nm wavelength, which is inside the range of maximum response of the BSO (Gunter & Huignard, 2006), leads to membrane deformation of the order of a few microns. The maximum deformation Δx is plotted in Fig. 27 both as a function of the light intensity I_{PC} on the BSO side at different amplitudes of the applied voltage V_o, and as a function of the operating frequency f.

The model accounting for the equilibrium deformation of the membrane is derived by considering that the membrane deformation $M(\rho,\theta)$ obeys a Laplace equation

$$\nabla^2 M(\rho,\theta) = \frac{\varepsilon_o}{2T} \frac{V_{GAP}^2}{d^2} \,, \tag{8}$$

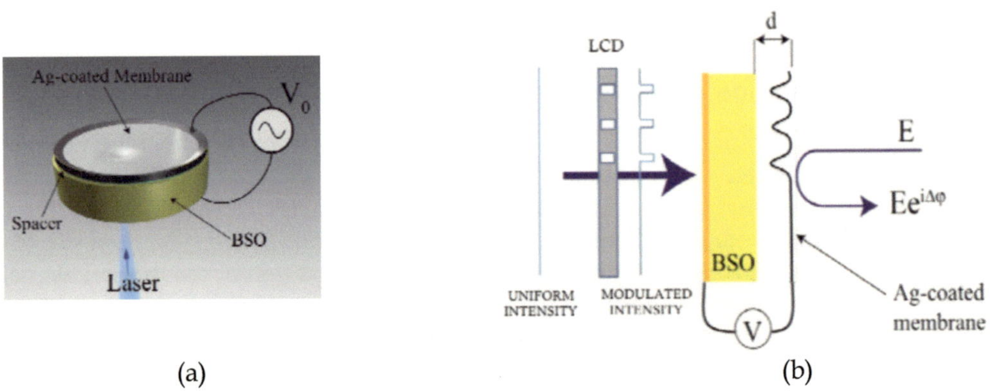

Fig. 26. (a): Schematic representation of a photocontrolled deformable membrane mirror. (b): selective optical addressing by the use of a liquid crystal display (LCD); the membrane deformation follows the intensity distribution on the BSO and the reflected beam acquires a correspondent phase shift $\Delta\varphi$.

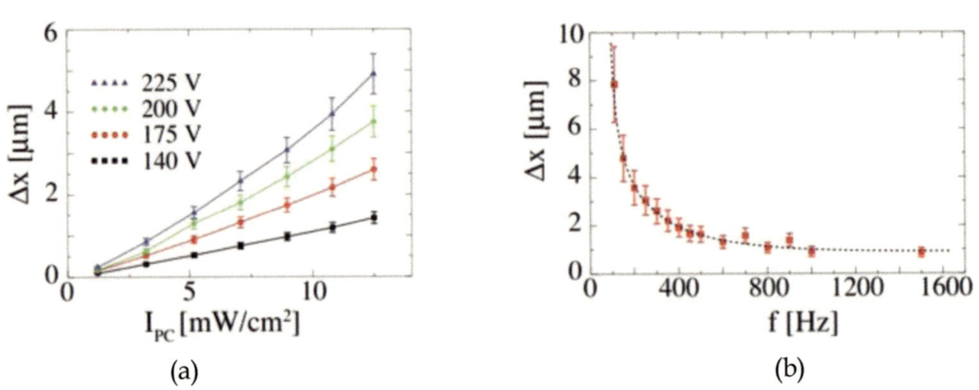

Fig. 27. (a): Membrane maximum deformation Δx as a function of the light intensity I_{PC} on the photoconductor side of the mirror; $f = 1$ kHz. (b): membrane maximum deformation Δx as a function of the frequency f of the applied voltage; $V_o = 140$ V peak-to-peak.

where ρ and θ are the radial and angular directions, respectively, T is the membrane tension factor, V_{GAP} is the effective voltage that drops across the empty gap of the mirror and d is the thickness of the gap. By approximating the membrane deformation with a parabolic profile, and by taking the appropriate boundary conditions, we obtain that the maximum membrane deflection $M(0,\theta) \equiv \Delta x$ occurs at the center, $\rho = 0$, and is given by

$$\Delta x = \frac{\varepsilon_o}{32T} \frac{a^2 V_{GAP}^2}{d^2}, \tag{9}$$

where a is the diameter of the membrane (Efron & Dekker, 1995). By approximating the BSO response with a linear function (Bortolozzo & Residori, 2006), we have that $V_{GAP} = \Gamma V_o + \alpha I_{PC}$, where V_o is the voltage externally applied to the mirror, Γ the dark impedance of the BSO, I_{PC} the intensity of the photo-addressing beam and α a phenomenological parameter that can be deduced from the mirror characteristics. By developing Δx at the first order approximation, we obtain

$$\Delta x = \frac{\varepsilon_o}{32T} \frac{a^2 \Gamma^2 V_o^2}{d^2} \left(1 + 2\frac{\alpha I_{PC}}{\Gamma V_o}\right). \tag{10}$$

Since the phase delay acquired by the probe beam is proportional to Δx, it can be seen from equation (10) that the PCDM provides a phase shift scaling quadratically with V_o and linearly with I_{PC}. For low I_{PC} intensities and for small applied voltages V_o, the linear dependence of Δx from I_{PC} is in agreement with the experimental results (see Fig. 27). When I_{PC} and V_o increase, higher order corrections can take into account the deviations from the linear behavior.

To characterize the spatial resolution, the photoconductive substrate has been addressed by a 474 nm wavelength laser beam with a diameter of 300 μm and intensity $I = 1$ mW/cm². The corresponding local deformation of the membrane is shown in Fig. 28 for three different positions of the photo-addressing beam with respect to the centre of the membrane, $r_1 = 0.05$ mm, $r_2 = 4$ and $r_3 = 6$ mm, respectively. We notice that the membrane rigidity imposes smaller deformations when approaching the boundary.

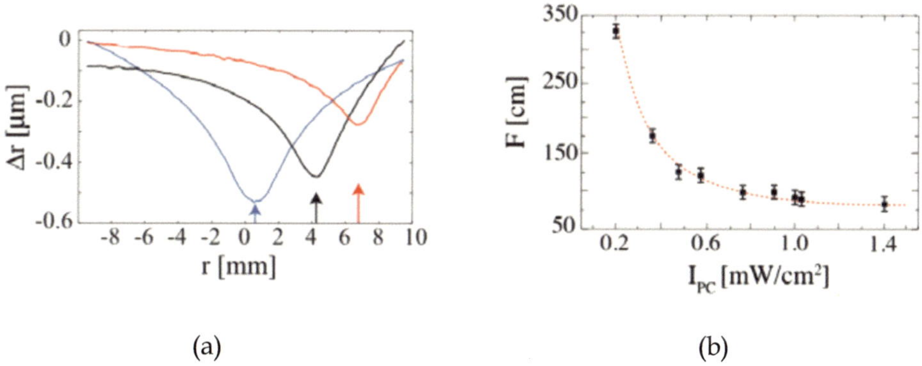

Fig. 28. (a): Membrane deformation Δr as a response to a light spot illuminating the BSO at different distances from the membrane center; $V_o = 165$ V peak-to-peak, and $f = 400$ Hz. (b): focal length F measured for the reflected beam versus I_{PC} light intensity on the BSO side.

When a localized light spot is sent on the BSO, correspondingly the beam reflected by the membrane is focused at a distance changing with the intensity of the addressing beam. In the right panel of Fig. 28, the measured focal distance F is reported as a function of the addressing intensity, for $f = 200$ Hz and $V_o = 210$ V peak-to-peak. A large control of the focal distance is achieved, with F changing from 50 to 350 cm. Other ranges of the focusing distance can be obtained by changing the working point of the PCDM, that is, for other values of f and V_o.

Finally, the performance of the mirror for adaptive optics operations have also been tested. Different target images, corresponding to the Zernike polynomials of the most common mode deformations, have been projected on the photoconductive side of the mirror via a liquid crystal display (LCD). A set of example results is shown in Fig. 29, where the intensity distributions of the beam reflected by the membrane (top row) are shown together with the target patterns (middle), and the corresponding intensity distributions on the BSO (bottom). A very good agreement with the target deformation can be appreciated from the membrane deformation patterns.

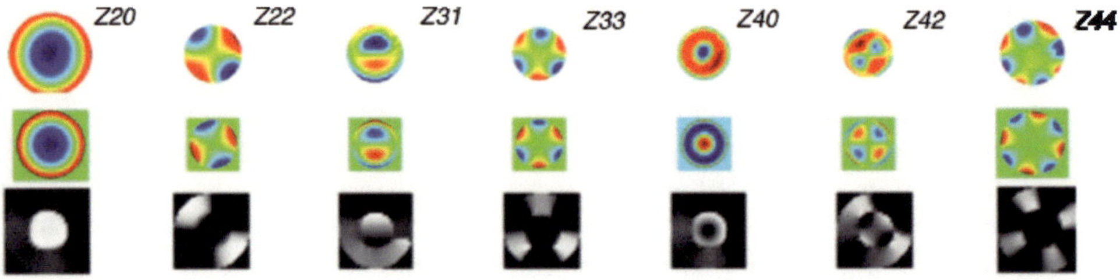

Fig. 29. Example of Zernike polynomial terms generated by the photo-controlled deformable membrane mirror. Top row: membrane deformation; middle row: target deformation; bottom row: corresponding intensity distribution on the photoconductive layer.

5. Conclusions

In this paper we have described some recent technological developments and applications of membrane deformable mirrors. These devices are becoming extremely popular because of their good performance, potentiality, easiness of use, and convenient price.

The described realization technologies and the shown applications of membrane deformable mirrors demonstrate that these adaptive optics devices are presently having a large development activity, with a lot of innovations, with always increasing performance, and with a larger and larger field of applications. All these considerations induce to foresee a rapid extension of the use of membrane deformable mirrors, possibly also in more diffused and commercial applications.

6. References

Bartels R., Backus S., Zeek E., Misoguti L., Vdovin G., Christov I. P., Murnane M. M., Kapteyn H. C., (2000), Shaped-pulse optimization of coherent emission of high-harmonic soft x-rays, Nature, Vol. 406, No. 6792, pp. 164–166, (July 2000), 0028-0836.

Bifano T. (2011), Adaptive Imaging: MEMS deformable mirrors, Nature Photonics, Vol. 5, (2011), pp. 21-23, 1749-4885

Bonato C., Sergienko a., Bonora S., Villoresi P., (2008), Even-order aberration cancellation in quantum interferometry, Physical Review Letters, Vol. 101, No. 23, (December 2008), pp. 233603

Bonato C., Bonora S., Chiuri A.& Mataloni P.& Milani G.& Vallone G.& Villoresi P. (2010), Phase control of a path-entangled photon state by a deformable membrane mirror, JOSA B, Vol. 27, No. 6, (2010), pp. A175-A180, 40-3224

Bonora S., Capraro I., Poletto L., Romanin M.,, Trestino C., Villoresi P. (2006), Fast wavefront active control by a simple DSP-Driven deformable mirror, Review of Scientific Instruments Vol. 77, No. 9, (September 2006), 0034-6748

Bonora S., Poletto L. (2006), Push-pull membrane mirrors for adaptive optics, Optics Express, Vol. 14, No. 25, (December 2006), pp. 11935-11944, 1094-4087

Bonora S., Brida D., Villoresi P., Cerullo G., (2010) Ultrabroadband pulse shaping with a push-pull deformable mirror, Optics Express, Vol. 18, No. 22, pp. 23147-23152, (October 2010), 1094-4087

Bonora S. (2011), Distributed actuators deformable mirror for adaptive optics, Optics Communications, Vol. 284, No. 13, (June 2011), 0030-4018

Bonora S., Frassetto F., Coraggia S., Coreno M., Negro M., Devetta M., Vozzi C., Stagira S., Poletto L. (2011), Optimization of low-order harmonic generation by exploitation of a resistive deformable mirror, Applied Physics B, In press, DOI 10.1007/s00340-011-4820-9

Bortolozzo U., Bonora S., Huignard J.P., Residori S. (2010), Continuous photocontrolled deformable membrane mirror, Applied Physics Letters, Vol. 96, No.25, (January 2010), 0003-6951

Bortolozzo U, Residori S. (2006), Storage of Localized Structure Matrices in Nematic Liquid Crystals, Physical Review Letters, Vol. 96, 037801 (January 2006), 0031-9007

Brida D., Manzoni C., Cirmi G., Marangoni M.,, Bonora S., Villoresi P., De Silvestri S., Cerullo G., (2010), Few-optical-cycle pulses tunable from the visible to the mid-infrared by optical parametric amplifiers, Journal of Optics, Vol. 12, No. 1, (January 2010), 2040-8978

Dalimier E., Dainty C. (2005), Comparative analysis of deformable mirrors for ocular adaptive optics, Optics Express, Vol. 13, No. 11, (May 2005), pp. 4275–4285, 1094-4087

Debarre D., Booth M.J., Wilson T. (2007), Image based adaptive optics through optimisation of low spatial frequencies, Optics Express, Vol. 15, No. 13, (2007), pp. (8176-8190), 1094-4087

Efron U., Dekker M. (1995),Spatial Light Modulators: Materials, Devices, and Systems (Marcel Dekker, New York, 1995)

Grisan E., Frassetto F., Da Deppo V., Naletto G., Ruggeri A. (2007), No wavefront sensor adaptive optics system for compensation of primary aberrations by software analysis of a point source image. Part I: methods, Applied Optics, Vol. 46, No. 25, (2007), pp. 6434-6441, 0003-6935

Günter P., Huignard J.P. (2006), Photorefractive Materials and Their Applications, (Springer, New York, 2006), Vol. 1.

Hardy J.W., 1998, Adaptive Optics for Astronomical Telescopes, Oxford University Press, ISBN-10: 0195090195, USA

Judson R.S., Rabitz H. (1992), Teching lasers to control molecules, Physical Review Letters, Vol. 68, No. 10, (March 1992), pp. (1500-1503), 1079-7114

Naletto G., Frassetto F., Codogno N., Grisan E., Bonora S., Da Deppo V., Ruggeri A. (2007), No wavefront sensor adaptive optics system for compensation of primary aberrations by software analysis of a point source image, Part II: tests, Applied Optics Vol. 46, No. 25, (2007), pp. 6427-643, 0003-6935

Rueckel M., Mack-Bucher J.A., Denk W. (2006), Adaptive wavefront correction in two-photon microscopy using coherence-gated wavefront sensing, PNAS, Vol. 103 , No. 46, (November 2006), pp. (17137-17142), 0027-8424

Thibos L.N., Bradley A., Hong X. (2002), A statistical model of the aberration structure of normal, well-corrected eyes, Ophthalmic, Physiological Optics, Vol. 22, No. 5, (2002), pp. (427-33), 1475-1313

Tyson R., 1999, Adaptive Optics Engineering Handbook , CRC Press, ISBN-10: 0824782755, New York USA

Vdovin G., Soloviev O., Samokhin A., Loktev M. (2008), Correction of low order aberrations using continuous deformable mirrors, Optics Express, Vol. 16, No. 5, (2008), pp. (2859-2866), 1094-4087

Villoresi P., Bonora S., Pascolini M., Poletto L., Tondello G., Vozzi C., Nisoli M., Sansone G., Stagira S., De Silvestri S. (2004), Optimization of high-order-harmonic generation by adaptive control of sub-10 fs pulse wavefront, Optics Letters, Vol. 29, No.2, pp. (207-209), (January 2004), 0146-9592

Vozzi C.& Calegari F., Benedetti E., Gasilov S., Sansone G.& Cerullo, G. Nisoli M., De Silvestri S., Stagira S. (2007), Millijoule-level phase-stabilized few-optical-cycle infrared parametric source, Optics Letters, Vol. 32, No. 20, (2007), pp. 2957-2959, 0146-9592

Wall S., Brida D., Clark S. R., Ehrke, H.P., Jaksch, D., Ardavan, A., Bonora S., Uemura H., Takahashi Y., Hasegawa T., Okamoto H., Cerullo, G., Cavalleri, A. (2011), Nature Materials, Vol. 7, No. 2, (February 2011), pp. (114-118), 1745-2473

Weiner A., Femtosecond pulse shaping using spatial light modulators, Review of Scientific Instruments (2000), Vol. 71, No. 5, (2000) pp.(1929-1962) , 0034-6748

Wright J.& Poland S. P., Girkin J. M., Freudiger C. W., Evans C. L., Xie X. S. (2007), Adaptive optics for enhanced signal in CARS microscopy, Optics Express, Vol. 15, No. 26, (2007), pp. 18209-18219, 1094-4087

Zeek E., Maginnis K., Backus S., Russek U., Murnane M., Mourou G., Kapteyn H., Vdovin G. (1999), Pulse compression by the use of a deformable mirror, Optics Letters, Vol. 24, No. 7, (1999), pp. (493-495), 0034-6748